Seeking Understanding

Seeking Understanding

The Lifelong Pursuit to Build the Scientific Mind

Edited by

Jan Visser and Muriel Visser

Foreword by Walter R. Erdelen

BRILL
SENSE

LEIDEN | BOSTON

All chapters in this book have undergone peer review.

The Library of Congress Cataloging-in-Publication Data is available online at http://catalog.loc.gov

Typeface for the Latin, Greek, and Cyrillic scripts: "Brill". See and download: brill.com/brill-typeface.

ISBN 978-90-04-41678-9 (paperback)
ISBN 978-90-04-41679-6 (hardback)
ISBN 978-90-04-41680-2 (e-book)

This book is printed on acid-free paper and produced in a sustainable manner.

Advance Praise for *Seeking Understanding*

"Samuel Butler once said that 'science, after all, is only an expression of our ignorance of our ignorance', and he was right. Yet science remains a powerful method to know ourselves and to learn about our universe. How are we to continue to kindle the Promethean fire while retaining the modesty so crucial to using it responsibly? In this wonderfully wise and humane collection of essays by scientists, humanists, educators, artists and policy-makers—ranging from discussions of the importance of curiosity and education, to the role of economics in sustainability, to the cultural dimensions of health, and all the way to the function of music and beauty in cultivating our sense of belonging—we encounter an honest, and modest, attempt to make sense of our possible futures, and the ways in which we might all contribute to making those futures more flourishing and meaningful."
– Oren Harman, Bar Ilan University, author of *Evolutions: Fifteen Myths That Explain Our World*

"Accelerating global change has led to a paradoxical situation: the world for which our educational systems were built does no longer exist. The 21st century has led us out of a world in which problems were reduced to ever smaller segments into a world of complexity. The third millennium calls for learning and education systems that cultivate competencies, attitudes and values for our time, the Anthropocene. This is where *Seeking Understanding* comes in, offering a multitude of entry points into this new universe. The diversity of themes discussed reflects the multidimensional and transdisciplinary nature of what is implied in *The Lifelong Pursuit to Build the Scientific Mind*."
– Walter R. Erdelen, author of *Managing Complexity: Earth Systems and Strategies for the Future*

"The contributing authors explore, each in a distinctive voice, with energy and insight a fundamental challenge of our times: how to foster scientific ways of understanding and the spirit of inquiry for all learners."
– David Perkins, Harvard Graduate School of Education, author of *Future Wise*

"Congratulations to Jan Visser and his team of authors for having produced an excellent publication. *Seeking Inderstanding* comes at a crucial moment in which human beings, all equal in dignity, must act as free and responsible

citizens of the world, expressing themselves and seeking understanding, thus forging the transition from the reason of force to the force of reason, from a culture of imposition and domination to a culture of dialogue and conciliation."
– **Federico Mayor, Former Director-General of UNESCO, President, Foundation for a Culture of Peace**

"This book is a meaningful source of inspiration for anyone interested in fostering transformations in the way we, as human beings, relate to the world in which we are living. This collection of contributions—written by scientists, scholars, artists, and educators, from all over the world—provides the reader with a coherent and convincing narrative to envision what is implied by the development of a way-of-being, driven by the advancement of understanding and wisdom. Going beyond a narrow definition of the 'scientific mind,' the work conceives of its development as a lifelong process through which humans cultivate specific habits of thinking, dispositions, values, ethical concerns, aesthetic considerations and attitudes, alongside building mastery of a heterogeneous array of skills and mental capabilities. This volume is the product of transdisciplinary and cross-cultural collaborations nurtured and developed over two decades by a broad community of thinkers. As such, it demonstrates the value and the richness of any scientific, artistic or educational endeavor, with the aim to sustain dialogue through space and time, and link heterogeneous sources of knowledge, in order to critically and creatively engage with the complexity of the world we inhabit."
– **Michel Alhadeff-Jones, Teachers College, Columbia University, author of *Time and the Rhythms of Emancipatory Education***

"Jan and Muriel Visser have put together a remarkable series of essays on the future of the scientific mind. This book, as they point out, is a collaborative creation, and one that radically expands our view of knowledge and what it means to be educated in the 21st century. The essays in this volume challenge the limited and limiting focus on hyper-specialization and disciplinary fragmentation, and urge us to reweave knowledge in a way that acknowledges and integrates complexity, creativity, and wonder. They are radical because they go to the roots of what constitutes knowledge, what is necessary for the 21st century, what is possible, and what is desirable. They are generative because they open up a world of new insights and new directions. This volume makes for compelling reading for anyone interested in the future of science, knowledge, and the development of the transdisciplinary skills and competencies required to thrive in a networked society."
– **Alfonso Montuori, California Institute of Integral Studies, editor of *Journeys in Complexity: Autobiographical Accounts by Leading Systems and Complexity Thinkers***

"It would be hard to find a more diverse and cosmopolitan set of authors than the one assembled by Jan and Muriel Visser here. Listen in as they draw on a variety of disciplines to shed light on the construction of the scientific mind, building on years of group collaboration. Their humanity and optimism shines like a beacon through the political smog of our times."
– **Graham Harman, Southern California Institute of Architecture, author of** ***Object Oriented Ontology: A New Theory of Everything***

Science is built with facts as a house is with stones, but a collection of facts is no more a science than a heap of stones is a house.

JULES HENRI POINCARÉ (1854–1912)

•••

The ideals which have lighted my way, and time after time have given me new courage to face life cheerfully, have been Kindness, Beauty, and Truth.

ALBERT EINSTEIN (1879–1955)

Contents

Foreword

> I make no apology for a fascination with the soft edge of science. It is here, it seems, that we get fleeting glimpses of strange shadows just beneath the surface of current understanding.
>
> LYALL WATSON

∵

The future of Humanity is at a crossroads. As highlighted in the *2030 Agenda for Sustainable Development*, global challenges are manifold, but we are also living in times of immense opportunities. Virtually all of the Agenda's Sustainable Development Goals concern education. "Sustainable development has been the Holy Grail during the last decades when environmental problems, globalization and above all a continuously growing population raised serious questions about the carrying capacity of humans on planet Earth" (Erdelen & Richardson, 2018, p. 210).

One of the key ingredients for humankind's future on planet Earth is to have the 'right' capacities in place: capacities which we need, which anticipate. *Inter alia* these are built and nurtured through education, learning and research, the topics of *Seeking Understanding* by Jan and Muriel Visser. More specifically, the authors aim at "… [inspiring] change in the domains of education and learning … while motivating readers "to *do* things differently and to *think* differently."[1] Isn't this one of the essentials for making sustainable development become reality: through massively not only improving but re-casting of our educational systems? For the simple reason that these systems have not (yet) produced the knowledge societies with citizens literate in matters related to sustainability.

To be even more precise, global change takes place in an evolving, complex and highly dynamic setting, the environment in the Anthropocene. It is up to us citizens of the Earth to secure quality of life and sustain natural resource use. At the same time we are facing the challenges of seeking to live in harmony with nature and preserving Mother Earth's natural and *Homo sapiens'* cultural heritage. This highly dynamic and complex global situation puts additional pressure on making education systems—currently in a massive worldwide crisis—fit for purpose. They need to step up attention to *sapiens'* global concerns in its rapidly changing world, with unprecedented growth in information and knowledge. This requires radical re-conceptualization 'from within,' i.e. of how we learn and educate present and future generations. As Federico Mayor (1999) stated almost twenty years ago already, "Education is the 'force for the future' because it is one of the most powerful instruments of change."

When I met Jan Visser for the first time, in my office at UNESCO headquarters in Paris, in 2004, he introduced me to his thinking about learning and his concept of "Building the Scientific Mind" (BtSM). Over the last twenty years BtSM, an acronym that has since become a household name among members of the community of scientists associated with it, has massively evolved from what it was at the time of its creation in the late 1990s. Thanks to Jan Visser's leadership as founding president, it has become an important focal area of research and development for the *Learning Development Institute.*

At the time we met I was at the helm of UNESCO's Natural Sciences Sector. The Sector's mission has been to further the advancement and sharing of scientific knowledge and to promote application of this knowledge and its understanding to the pursuit of sustainable development.

I felt that Jan's seed may have fallen on fertile ground: we agreed that BtSM could benefit many of the Sector's programs and activities, in modern jargon a 'win-win,' or even 'multiple-win' situation. We had agreed, furthermore, that ideas developed in the course of our discussion might contribute to fostering scientific awareness, identifying new research topics, and for possibly even triggering paradigm changes in education. UNESCO's field offices and their national or regional responsibilities turned out as important 'outreach mechanisms' for translating thought into action.

Indeed, our first exchange showed that the 'chemistry between us' was right. (In fact, this eventually turned into a sustained long-lasting friendship.) We felt we could find a way out of a situation of chronic underfunding and understaffing by directly improving the delivery capacity of the only organization within the UN system that has a mandate in science. Perhaps indirectly, this book may be viewed as a (mental) product of that time, having since 'coevolved' with developments in related fields. These include, *inter alia,* education for sustainable development, life-long learning, ICT-based education, and innovative concepts such as the Anthropocene, planetary boundaries, and tipping points in Earth systems.

The title *Seeking Understanding: The Lifelong Pursuit to Build the Scientific Mind* mirrors what has become ever more important in our highly interconnected and networked world: the need for literacy in various domains to understand our *environment*; the fact that learning has changed into a life-long endeavor; and that we need the 'right mindset,' *sensu* Jan and Muriel Visser's '*the scientific mind*' concept; and a holistic view of *environment,* including our 'internal' as well as the 'external' environment. I cannot resist but reminding us of David Orr's farsightedness in stating, some twenty-five years ago: "All education is environmental education" (1994, p. 12).

BtSM's thought-provoking aspects are that—in Jan Visser's words[2]—it describes a multidimensional and transdisciplinary approach aiming at contributing to a "much needed reimagining and reshaping of the learning landscape in both informal and formal environments" targeted "not about merely serving the scientific community. It is about serving the interests and wellbeing of everyone, whether scientist or not."

This is well reflected in the book's 18 chapters, the 17 *Intermezzi*, and Jan's introductory chapter on the "quest to know." Chapters are truly multidimensional, ranging from music and beauty to education along the entire lifespan, mental models, awareness and understanding of the universe, the importance of curiosity, dialogues and interaction, the popularization of science and mathematics, inquiry and skills development, economics, invention of our future, health and its cultural dimension, psychopathology and learning, complex adaptive systems and wicked problems, and to humanness and the medical sciences.

In conclusion, we may wonder if BtSM does not convey a *broader* message: the quest for developing the 'right' mindsets, mindsets as needed in a world of continued population growth, increasing globalization and complexity, a world in which we risk falling further and further behind, seemingly decoupled from our environment, if we do not keep in permanent harmonious exchange with the world around us. Knowledge is created at unprecedented rates, information about virtually everything is available to us anytime and anywhere. ICTs and AI are increasingly affecting our daily life. At the same time drowning in illiteracy is becoming more and more likely, with unprecedented consequences.

I wish Jan and Muriel's book the best of success: May it become a milestone contribution to building the global society we want, the society for and—eventually—of our future.

Notes

1 Unpublished concept paper guiding the production of the book.
2 Unpublished concept paper guiding the production of the book.

References

Erdelen, W. R., & Richardson, J. G. (2018). *Managing complexity: Earth systems and strategies for the future*. Abingdon & New York, NY: Routledge.

Mayor, F. (1999). Preface. In E. Morin (Ed.), *Seven complex lessons in education for the future*. Paris: UNESCO Publishing.

Orr, D. W. (1994). *Earth in mind: On education, environment, and the human project.* Washington, DC: Island Press.

Walter R. Erdelen
Former Assistant Director-General for Natural Sciences
UNESCO

Acknowledgements

There were hundreds of individuals around the world with whom we had the pleasure to interact, mostly in the framework of the BtSM colloquia. Their creative collaboration was vital to getting us where we are now. But creative collaboration has not remained limited to how single individuals participated in the BtSM effort. Organizations and institutions did so as well. After a successful start of the Building the Scientific Mind colloquium series in 2005, the Learning Development Institute's board of directors decided that BtSM should henceforth be the primary focus of LDI's work. Consequently, the Institute has invested a large proportion of its limited resources into moving the program forward. But we could never have done it alone. The Building the Scientific Mind program would not have met the success it has without the help and support of other institutions and organizations.

First and foremost, we are grateful to UNESCO, the United Nations Educational, Scientific and Cultural Organization, under the patronage of which the program took off in 2005 and continued to advance over the years. Regular contacts with the Organization's Science Sector, in particular with its Assistant Director-General, Dr. Erdelen, were inspiring and were instrumental in mobilizing the cooperation of UNESCO's Field Offices, and occasionally National Commissions for UNESCO, in countries and regions where BtSM was active. The Netherlands, India, Canada, Egypt, and Indonesia are notable cases in point where such cooperation played an important role.

Professional organizations, such as the Association for Educational Communications and Technology and the American Educational Research Association were important fora for putting ideas to the test in symposia, workshops and panel sessions with colleagues. Local entities, such as the Indian National Science Academy, and Ain Shams University in Egypt followed through on colloquia. In addition, we enjoyed the hospitality, cooperation and commitment of institutions where the various colloquia were held: the Institute of Social Studies in The Hague, the Netherlands; the Emily Carr University of Art and Design in Vancouver, Canada; the Friends of Environment and Development Association in Cairo, Egypt; the Sustainability Institute in partnership with Stellenbosch University in Lynedoch Ecovillage, Western Cape, South Africa; The Bosscha Observatory of the Bandung Institute of Technology in Lembang, West Java, Indonesia; and the Fundação Museu do Homem Americano in São Raimundo Nonato, Piauí, Brazil. Local coordinators in all those places contributed greatly to the success of the program. Thanks are due to Jolanda Karada and the typesetters at Brill | Sense for their constructive participation in finalizing this book.

A big thank you to all.

Figures and Tables

Figures

Tables

Notes on Contributors

Matthew Colless

is Director of the Research School of Astronomy and Astrophysics at the Australian National University and formerly Director of the Australian Astronomical Observatory. He was educated at Sydney and Cambridge universities, and held positions at Durham University, Kitt Peak National Observatory, and King's College, Cambridge. His research has involved mapping the large-scale structure of the universe to understand what it is made of, how it formed, and how it has evolved. He has led the construction of new instruments and giant telescopes that allow astronomers to see further into the universe with greater detail. The big questions in science and the extraordinary beauty of the universe have fascinated him since he was a boy with a backyard telescope in rural Australia. He feels blessed that his lifetime has coincided with the greatest-ever expansion of knowledge about the universe in humanity's history.

Martinus Petrus de Wit

holds a doctoral degree in Economics and a Masters of Arts in Theological Studies. He currently serves as Professor in Environmental Governance at Stellenbosch University in South Africa. One of his main challenges is to teach environmental economics in a language that can be understood by students from various disciplines and walks of life. His work on the board of ASSET Research (http://www.assetresearch.org.za/), a not-for-profit organization focused on student development in the fields of economics, ecology and development, is a constant reminder to develop and use workable adult teaching and learning strategies. He finds great inspiration working closely with people passionate about environmental conservation on the ground and who continue to radiate hope for the future in the face of seemingly insurmountable ecological problems. Martin has a great weakness for books, but if not found reading, he will take every opportunity to hike or camp in the outdoors with his family and friends.

Walter R. Erdelen

is a German ecologist and biogeographer. He studied Zoology, Botany, Genetics and Chemistry at the University of Munich. As UNESCO's former Assistant-Director-General he was at the helm of its Natural Sciences Sector. Subsequently, as senior strategic adviser, he provided future-oriented advice, carried out specific analyses, and compiled reports in areas of strategic importance for UNESCO's program sectors and partners. Throughout his almost 50-year professional career Walter has focused on "how and where he could make a difference." Examples include: Support to the Government of Indonesia in analyzing and improving standards at six selected universities, based on a holistic approach. Communicating science in a more attractive and lively manner through linking the arts and sciences. Documenting institutional memory through the book compilation *Sixty Years of Science in UNESCO 1945–2005* with over 60 contributors. Organizing and realizing *Planet Earth: From Space to Place*, an exhibition held at UNESCO's General Conference in 2007. Walter's latest book is titled *Managing Complexity: Earth Systems and Strategies for the Future* (Routledge, 2019).

Carlo Fabricatore
is a computer scientist and human factors scholar, interested in complex adaptive systems and the self-regulatory mechanisms that drive human learning and development in complex scenarios and activities. He believes that the complex problems that dynamize our global society require integrative thinking skills and sensibilities innate in all human beings, but not sufficiently nurtured through formal education. Games can significantly contribute to the development of these skills and sensibilities. In fact, he has seen how people are often better feelers/thinkers/doers when they play or design 'games,' rather than when they play or design 'real life.' This motivated him to engage in human factors and complexity research to study how games and game design can promote engagement in real-world complex problems and foster the development of capabilities and sensibilities required to tackle them.

Robert Greenberg
is a composer, pianist, music historian and teacher who believes that the acts of writing music, performing music, and teaching music are all rooted in storytelling. Born in Brooklyn, New York in 1954, Greenberg has lived in the San Francisco Bay Area since 1978. He received a B.A. in Music, magna cum laude, from Princeton University in 1976 and a Ph.D. in music composition—with distinction—from the University of California, Berkeley in 1984. Across the span of an international career as a composer, pianist and teacher of music, Greenberg has come to understand that music is a mirror of the inner life and experience of its creator as well the outer environment in which that composer lives and works. The metaphors, analogs, and narratives invoked in music (of virtually every sort) can inform, delight, and edify virtually any listener, provided an informed ear and an open mind. Dr. Greenberg is currently Music Historian-in-Residence with San Francisco Performances. He is a bestselling creator of music courses for The Great Courses company and a designated Steinway Artist.

Arvind Gupta

earned a B. Tech from the Indian Institute of Technology (IIT), Kanpur, in 1975. He left a corporate job to pursue his passion—making science fun for children. He has written 30 books on science activities, translated 400 books into Hindi, and presented 125 films on science activities on *Doordarshan*, India's national TV channel. He worked at the Children's Science Center at the Inter-University Center for Astronomy and Astrophysics (IUCAA) in Pune, India, for 11 years. The Center produced 8700 short (2-minute) videos on simple experiments and science toys in 20 different languages. The videos have been viewed by over 73-million children worldwide. He shares his passion for books and toys through his popular website http://arvindguptatoys.com. He has received many honors including the Distinguished Alumnus Award of IIT/Kanpur and the Padam Shri.

Paul E. Heckman

holds a Ph.D. from the University of California, Los Angeles. He has taught at the elementary and secondary levels. He has also held professorships at the University of Southern Maine, the University of Arizona, the University of Washington, and, now, at the University of California, Davis. He spent seven years as Associate Dean in the School of Education at UC Davis. Since 2016, he has returned solely to his teaching and research. His research focuses on school and afterschool programs. In particular, he seeks understandings and actions in educational settings that move schooling away from the standard mental models. He questions standard educational and community change in economically poorest areas of urban centers. His hope is to identify and understand the conditions in these settings that will promote teachers, parents, and students undertaking educational invention and educational success in these settings.

Stephen P. Hinshaw

is Professor of Psychology at UC Berkeley and of Psychiatry UC San Francisco. Focusing on developmental psychopathology and stigma, he has authored over 350 articles/chapters plus 12 books, including *Another Kind of Madness: A Journey through the Stigma and Hope of Mental Illness* (St. Martin's, 2017). He is a 'top 10' most productive scholar in clinical psychology across the past decade. He received a Distinguished Teaching Award from UC Berkeley's Social Sciences Division in 2001; his Teaching Company ('Great Lecture') series, "Origins of the Human Mind," emerged in 2010. Hinshaw's research efforts have been lauded: Distinguished Scientist Award from the Society for a Science of Clinical Psychology, James McKeen Cattell Award from the Association for Psychological Science, and Distinguished Scientific Contributions Award from the

Society for Research in Child Development); he is the only individual ever to have been awarded all three.

Lê Nguyên Hoang
is a very active science communicator. He started blogging in 2012, during his Ph.D. in mathematics. After graduating and a year of postdoc, he started to work full time on online science communication. He launched his YouTube channel Science4All, which has gathered over 140k subscribers and 7M views. He also works at the École polytechnique fédérale de Lausanne (EPFL), where he produces both educational videos for university students and popularization videos featuring EPFL professors. He is also present in other media: he recently published a book on Bayesianism and he co-hosts the podcast Axiome (in French) with a philosopher. Lê is very passionate about science and about sharing his passion for science.

James Lees
is an anthropologist and educator based at the University of the Western Cape in South Africa. After submitting the final draft of his chapter he came across a single quote attributed to several people that captures what he tried to say in so many paragraphs and pages. "The opposite of war is not peace, it is creation." Anisur Rahman, quoted in the chapter, believes that exercising our ability to create is what makes us human. To undermine it is to dehumanize. Following a biomedical model that placed the HIV virus at the center of the epidemic, the world increasingly (and increasingly unsuccessfully) fought AIDS with pharmaceuticals. Our own growth and evolution displaced by big pharma, we the people were not called upon to re-create ourselves in relationship to each other and our sexualities. A chance was lost for us to be better at being human. Given the extraordinary challenges that face our human species today, perhaps the HIV epidemic foreshadowed events to come, highlighting the real issue—our humanness—and the need for all of us to actively create a better version of what it means to be part of the human species.

Ximena López
is a psychologist interested in exploring the role of the human mind in society from different angles and world views. This curiosity emerged in her childhood during precious moments she shared with her father while stargazing the southern skies, feeling our tininess in the immensity of the universe. It sparked her interests in how different people and cultures create realities through languages and social exchange. She has had the fortune of living and working in different countries, which made her aware of the enormous complexity that we face as humans in creating shared worldviews that will help the development

of humane and sustainable societies. She is currently based in the UK, working on projects for people with dementia that may hopefully promote, as long as possible, their inclusion in society.

Federico Mayor
is a Spanish scientist, politician, scholar and social change activist who served as the Director-General of UNESCO (United Nations Educational, Scientific, and Cultural Organization) from 1987 to 1999. Professor of biochemistry and Rector of the University of Granada (1968–1972). He was the promoter of the neonatal diagnosis and treatment of diseases that can produce irreversible neonatal damage. Co-founder in 1973 of the Center of Molecular Biology 'Severo Ochoa' in the Autonomous University of Madrid (AUM). He became Minister of Education and Science in 1981 and was elected to the European Parliament in 1987. Under his leadership, UNESCO developed a Culture of Peace Program that outlined a plan of action addressing education for peace, human rights, and democracy. In 1999 Federico Mayor returned to Spain and started the Foundation for a Culture of Peace. In 2016 he set up in the AUM the Institute of Human Rights, Democracy, Culture of Peace and Non-Violence (DEMOS-PAZ).

George Miley
is an Irish-Dutch astronomer and emeritus professor at Leiden University in the Netherlands. His research focuses on multiwavelength observations of galaxies that are huge luminous sources of radio emission. He was the initiator of the low-frequency radio telescope 'LOFAR,' a unique facility centred in the Netherlands, with stations spread out over Europe. In 2004 he began 'Universe Awareness' (UNAWE), a program aimed at using the fascination of astronomy to inspire disadvantaged young children. UNAWE became a Cornerstone project of the 2009 IAU/UNESCO International Year of Astronomy and UNAWE activities have been carried out in more than 60 countries. As vice-president of the International Astronomical Union from 2006 to 2012, Miley designed the IAU strategic plan 'Astronomy for Development 2010–2020,' to exploit astronomy as a tool for furthering international development at every level and of which UNAWE is an integral part.

Carolina Ödman
is Associate Professor at the University of the Western Cape, South Africa. Trained in physics at EPFL (École polytechnique fédérale de Lausanne) in Switzerland, she holds a Ph.D. in cosmology from Cambridge University, UK. After a Marie Curie research fellowship in Italy, she became the first international project manager of Universe Awareness. She joined the South African Astronomical Observatory as an SKA Research Fellow working at the interface

between astronomy and technology, which she continues to do as 'member of the braintrust' of the .Astronomy conference series (https://www.dotastronomy.com/). She was Director of Academic Development for the Next Einstein Initiative before joining a financial technology start-up as Chief Scientist. In 2018, she joined the Inter-University Institute for Data Intensive Astronomy as Associate Director, Development and Outreach. That same year, she was given a special Award in Astronomy Outreach, Development and Education by the International Astronomical Union in recognition for her pioneering work in those fields.

Premana W. Premadi
thinks that the Universe is too vast and too beautiful to keep to herself. She spends all her adult life trying to share it with practically everyone she meets. Having a Ph.D. in physics in the field of cosmology adds responsibility to passion. You may lose her when she is captivated by her favorite classical pieces. She takes refuge in the extra dimensional spacetime the Chaconne from Bach's Violin Partita No. 2 provides. Otherwise she would be found out and about trying to lend a hand to the world's effort in making it a happier place for all.

Pedro Russo
is the coordinator of the *Astronomy & Society Group* at Leiden University, the Netherlands. Russo was Project Coordinator of the Space Awareness Project and was the International Project Manager of the FP7 EU—Universe Awareness project. He is passionately committed to using astronomy as an educational tool. Until 2010 he was the global coordinator for the largest science outreach network project ever carried out: the International Year of Astronomy 2009 (IYA2009). He worked with data from the *Venus Express* exploration program of the European Space Agency. Since then he has been working for and with various international organizations, such as the European Planetology Network, the International Astronomical Union, the International Astronautical Federation and the International Year of Light 2015 Advisory Board. During his time at the European Southern Observatory in Munich, Russo was the Outreach Scientist of the educational and public outreach department. He is the founder and was the first editor-in-chief of the *Communicating Astronomy with the Public* (CAP) journal.

J. Michael Spector
is Professor of Learning Technologies at the University of North Texas. He was previously Professor of Educational Psychology and Instructional Technology at the University of Georgia. Prior to that, he was Professor of Instructional Systems at Florida State University; he served as Chair of Instructional Design, Development and Evaluation at Syracuse University; and he was Director of the Educational Information Science and Technology Research Program at

the University of Bergen. He earned a Ph.D. from The University of Texas at Austin. His research focuses on intelligent support for instructional design, assessing learning in complex domains, and technology integration in education. Dr. Spector is a Past-President of the Association for Educational and Communications Technology, He edited the third and fourth editions of the *Handbook of Research on Educational Communications and Technology*, as well as the *Encyclopedia of Educational Technology*, and has more than 200 publications to his credit.

Ralf Syring
born in Germany just after the end of World War II, participated in a student exchange with pupils from a Dutch twin city in 1962. The experience made him ask why people from the country where he was born, had occupied this neighboring country and oppressed its citizens. It led him to try to contribute to a situation in which that should not happen again. Several ways to prepare himself for that task and to put it into practice seemed open. They were attempted in sequence. First, he studied theology in Germany and the Netherlands and worked for a short time in the protestant church. Next, he studied social science and education and worked for several years as a teacher in vocational schools. Last, he studied medicine—specializing in paediatrics—and worked to improve public health in different African countries and in Central America. He continues seeking to find out why things happen as they do and how to make them happen in a way such as to produce peace with justice.

Jos van den Broek
was born in Tilburg, the Netherlands, in 1951. He studied biochemistry (M.Sc. in 1977) and pharmacology at Leiden University (Ph.D. in 1981). He has been a science writer and communicator for a large part of his career, with a strong focus on visual communication. In 1989–1990 he earned a Knight Science Journalism Fellowship at MIT (Cambridge, MA). He currently is a professor emeritus in science communication at Leiden University where he is strongly involved in educational programs for disadvantaged kids. He has written a dozen popular-science books, among which several books with children's questions.

Emily Vargas-Barón
is devoted to promoting global, regional and national policies and programs in early childhood development for children of all abilities. Currently, she directs the RISE Institute, conducts research, and consults internationally. From 1994 to 2001, she was Deputy Assistant Administrator of USAID, directing the Center for Human Capacity Development devoted to educational development, training and telecommunications. Previously, she founded and directed a research and development institute for early childhood development and

early childhood intervention in Austin, Texas. She was the Education Advisor of The Ford Foundation's Office in Bogotá, Colombia for the Andean Region and also served as an Educational Planner for UNESCO, Paris. She taught at the University of Washington, University of Texas, and Javeriana University. She holds a Ph.D. in Anthropology from Stanford University, with a secondary focus in educational planning. She is the author of many publications and has worked in all world regions.

Anaïsa Visser
grew up all over the world, namely in Mozambique, the United States, then France, before moving to Vancouver, Canada to attend Emily Carr University of Art and Design. She graduated in 2013 with a BFA in Film, Video and Integrated Media. Anaïsa is currently pursuing her Master's Degree in Film Production and Creative Writing at the University of British Columbia. She dedicates all of her spare time to helping others make their independent films, and to writing, directing, and producing her own independent work. She is primarily interested in work revolving around human rights, gender and racial equality, mental health, and cultural identity. These interests are heavily influenced and informed by her diverse background and multicultural upbringing. Anaïsa also enjoys traveling, photography, reading, spending time with her husband, and spoiling her cats.

Jan Visser
is a theoretical physicist by original vocation and a cognitive scientist by subsequent calling. As a researcher and explorer of the unknown he likes to look at problems and issues of interest as more broadly defined than their apparent boundaries suggest. Science, the arts, and philosophy are, in his view, intimately interwoven expressions of a single human concern—which pervades the long history of Homo sapiens—to come to grips with the enigmas of our origin, our temporality, and the future beyond us. He sees 'constructive interaction with change' as the essential purpose of human learning and the 'quest to deeply understand' as the primary *raison d'être* of the scientific pursuit. The two concepts culminate in his view of 'human life in the Anthropocene,' which he sees as contingent upon, on the one hand, deep understanding of our place and role in the evolution of the universe and life on Earth, and, on the other hand, our ability to restore harmony with the human, social, biological and physical environment of which we are a natural part.

Lya Visser
has spent many years of her professional life concentrating on exploring the role of motivational communication in distance learning environments using not only cognitive but also affective strategies to assist learners in reaching

their learning goals. Her focus was on the beauty of simplicity. Lately she has begun to explore the use of affective communication strategies by expressing her hopes, thoughts and worries through sculpturing in stone. Sculptures do not just happen. They are brought into existence with compassion and determination. She thinks that the same holds for a civilization. In these dark times we have to offer something that is much richer than the present—the opportunity to live in the service not of superficial expectations, but, just as the sculptor and the stone will do, follow the power of life itself with compassion and determination.

Muriel Visser
grew up in Africa. Her early exposure to the continent, led to a life-long passion in development work. Muriel has worked extensively in adult education and teacher training, and used this experience to branch into HIV/AIDS prevention in the late nineties which became the topic of her doctoral thesis. Muriel has written and published extensively in her areas of interest, with a particular focus on research in behavior change communication. For the last 15 years Muriel has worked across Africa, South East Asia, Latin America and the Caribbean as an independent consultant for bilateral and multi-lateral aid agencies. Her main focus has been on conducting evaluations of development interventions and on providing aid agencies with advice on their development policies. In parallel Muriel has collaborated with the Learning Development Institute's projects on human learning. In her leisure time Muriel does pottery. She lives in a nature reserve in the south of France.

Paul Webb
is professor emeritus in Science Education at the Nelson Mandela University, South Africa. His interests are in the promotion of scientific literacy, understandings of the nature of science, alternative worldviews, language, inquiry and the development of scientific thinking skills. Much of the work that has resulted from these interests has been driven by a desire to understand what people believe and why they think in the way that they do. Most people understand that they cannot always be correct, but this does not often translate into recognition that any of their long and dearly held beliefs could possibly be wrong, never mind result in them being open to change their minds. It is the often seemingly inexplicable attachment to different understandings in the face of what, to some, is seen as the incontrovertible evidence that science provides that underpins his chapter on promoting thinking and understanding through dialogue.

INTRODUCTION

The Making of This Book and Its Roots in Creative Collaboration

Jan Visser and Muriel Visser

This book has a long history. It wasn't really planned. It emerged. The need for it to be written was only felt more than ten years after an event at which we now look back as its start: a symposium in Montreal in 1999. The event is mentioned elsewhere in this book, particularly in Chapter 1. There are also some implicit references to that symposium and the colloquia that followed in the 'Intermezzi.'

Vera John-Steiner writes in 2000 in the introduction to her book *Creative Collaboration* that "generative ideas emerge from joint thinking, from significant conversations, and from sustained, shared struggles to achieve new insights by partners in thought" (p. 3). The book appeared at the start of a new century and a new millennium. She argues that this change in the calendar of the western world marks a turning point in how, in that same world, our thoughts have changed about the role played by 'solitary thinkers' on the one hand and 'thought communities' on the other in constructing knowledge and shaping artistic forms that can deepen our understanding. The book you hold in your hands or explore on your screen belongs proudly to the twenty-first century. It wouldn't have existed—and it would certainly not have existed in its current form—had it not, in every respect, been the product of creative collaboration among those who contributed to it in the immediate sense, i.e., its authors and editors. It wouldn't have existed either had there not been hundreds of people around the world who participated in the six colloquia on 'Building the Scientific Mind' held in six different regions of the world, embedded in six different multi-cultural contexts between 2005 and 2015. It would also not have come into being had not eight people come together in 1999 to think collaboratively about what is implied in 'Overcoming the underdevelopment of learning,' the theme of the 1999 symposium held in Montreal.

It was only in 2013 that we, the two editors, sat together for a working lunch in a French restaurant somewhere in Provence. It's the most appropriate region and cuisine to form the setting for a fruitful conversation. We had just concluded, in Lembang, Indonesia, the fifth 'Building the Scientific Mind' colloquium. Together with the prior four colloquia since 2005 it had led to a clear vision of what it meant, for someone, to build their scientific mind and what conditions should be present to facilitate the process of shaping such a mind at both individual and community level. How long should we still go on organizing such colloquia? Was there anything more that commanded our attention? Yes, there was. We decided to do two more things. (1) We would organize one last BtSM colloquium in 2015. It should be of a celebratory nature to mark the conclusion of the BtSM work. It should probably best take place somewhere in Latin America.[1] (2) A book should be written, inspired by the BtSM work over the past ten years, to enthuse future developments. That's this book. In the spirit of widening the circle of the already existing creative collaboration we furthermore thought it would be a good idea not to limit ourselves to selecting authors from among the several hundred people of the then prevailing BtSM network but invite also others whose thoughts we knew to be of relevance. It brought us into contact with some of the finest minds we were aware of. We wish to share the privilege to interact with them, with the readers of this book and the viewers of the video conversations, available on the *Building the Scientific Mind* YouTube channel, www.youtube.com/channel/UCNx_SL357aoLl7i-hqVHXqwQ (Anaïsa Visser, Video Editor), which complement the book.

The book took shape the way a sculptor would work with a rough, still unworked, piece of stone, having initially a vague idea of how to reshape it and, as the work progresses, discovering again and again new ways to express an evolving creative idea. In a similar fashion we kept adding authors and chapters to the book as it gradually took shape. Having Lya Visser, an educational scientist as well as a sculptor in stone, as an editorial advisor added to our team, might have inspired our approach. As you correctly note, that's three people with the same family name. No, wait a moment, it should be four. Anaïsa Visser, independent filmmaker, serves as video editor for the YouTube videos of conversations with the authors.

Now, look back again at Vera John-Steiner's research on 'Creative collaboration.' Many of the cases she analyzes concern collaboration between members of the same family. The four of us who are behind this book represent three generations of one family. We are all established professionals in different, though sometimes related, fields. Questioning our own thoughts and those of others, before either agreeing among ourselves or respecting each other's different views while being aware of their underlying assumptions, has always been part and parcel of how we live, and love each other, together. It stimulates us as much at the dinner table as at scientific conferences where we present.

We hope you enjoy what we did.

Note

1 It ended up bringing us one last time together again at the Serra da Capivara archeological site in Piauí in North-West Brazil in August 2015.

Reference

John-Steiner, V. (2000). *Creative collaboration*. New York, NY: Oxford University Press.

Intermezzi: Thoughts Inspired by the Thoughts of Others

Jan Visser

> In music, intermezzi [singular: intermezzo] are pieces that are complete in themselves and that at the same time fit organically between the more substantive parts of a larger composition.

∵

Creating the book on *Seeking Understanding: The Lifelong Pursuit to Build the Scientific Mind* has been a five-year odyssey for me, which started in 2013 and will end when, in 2019, the book appears in print. I return from the journey enriched and enlightened, thanks to its authors, who accepted to meet the challenge and accompanied me on the journey. My travel log, which begins next, reflects on the journey. It consists of 17 intermezzi, placed between the various chapters. Taken together with this brief introduction, they should be considered a 'distributed chapter.'

We all had to learn, Muriel and I perhaps the most, as she and I started out with preconceptions based on the outcome of five consecutive Building the Scientific Mind (BtSM) colloquia, held every second year since 2005. We would have made our task much easier had we invited only authors from among the key participants in those colloquia. But we didn't. Of the 21 authors who contributed to this book, 11 were entirely new to the issues at hand. The diversity of the author team thus enhanced has been exciting and enriching.

I hope that my personal notes in this travel log help you to make yourself part of the voyage as you read the various chapters. I also hope that it helps you to get a feel for how this book is the product of complex growth and evolution, rather than linear planning. We kept adding authors to the book as we saw it gradually take shape, while opportunities arose to more fully represent the richness of the scientific mind. It's the reason why it took five years for the book to come to fruition. This evolutionary process should not stop with the book. We sincerely hope that appropriation by our readers of the ideas presented in it—and in the accompanying series of video conversations—will lead to improving our work and enhanced recognition of the importance of caring for the scientific mind, its awakening in every newly born human being, and its growth along the lifespan and across the generations.

The collection of 17 intermezzi interspersed between the various chapters is conceived in such a way that, taken together, and read as a separate document, the thoughts expressed in them represent a picture that is more than the simple sum of its parts. As such it should underline the integral nature of the book. The book as a whole presents a vision that transcends the individual chapters.

CHAPTER 1

The Quest to Know: Seeking Understanding and Wisdom

Jan Visser

1 An Irrepressible Urge: *Seeking Understanding*

The desire to know, to understand, distinguishes us, humans, from all other lifeforms in the biosphere. It is what *makes* us human. Among other forms of life that may have emerged elsewhere in the universe, we may or may not be the only intelligent form of life that has evolved. Considering the vastness of the universe and the countless celestial bodies with likely characteristics similar to those that engendered the currently prevailing conditions on planet Earth, it would actually be improbable that we are alone. But we don't know. Therefore, we want to know.

Science, the quest to know, starts with the discovery of ignorance, argues Yuval Harari (2015, chapter 14) in *Sapiens: A brief history of humankind.* We want to know what life really is and how it originated. We want to know who and what we are. Why are we here? Is there a meaning to our being in the world? We want to know our home in the universe, how it is put together, and how we can best live within it, all of us together. Those are questions that have been on the mind of humankind since time immemorial.

FIGURE 1.1 Who are we, among all that lives? Rock art from Serra da Capivara, Piauí, Brazil

 | DOI: 10.1163/9789004416802_001

Since the time, thousands of years ago, when our prehistoric ancestors asked themselves questions about who they were and what their place was among all that lived (Figure 1.1), creating an artful universe of images for their reflection on those questions, we have come a long way in both the arts and the sciences. But those fundamental questions are bound to remain with us. We will never know fully. Therefore, we want to know.

Much later than our ancient forebears, in the year 1897 of the Common Era, the French post-impressionist painter Paul Gauguin was still asking those same questions—"Where do we come from? What are we? Where are we going?"—words he inscribed in French in the top left-hand corner of the painting with that same long name (Figure 1.2).

FIGURE 1.2 Paul Gauguin's 1897 painting "D'où Venons Nous? Que Sommes Nous? Où Allons Nous?" (Source: Wikimedia Commons)

Not only the arts but the sciences as well have contributed to increasing our understanding of who we are and of the world of which we are an inseparable part. Art and science represent different modalities of knowing, allowing us to reflect on our being in the world from different perspectives. We tend to think of them as separate domains, but this has not always been so. Artists and scientists are more alike than different, argues John Maeda (2013) of the Rhode Island School of Design in a *Scientific American* blog post. "In DaVinci's time when expertise in art and science had not yet matured to the polarized state in which they exist today, they coexisted naturally" (n.p.). According to our current perceptions, science is seen as a conglomerate of areas of study and research, neatly divided into disciplines such as physics, chemistry and biology, each of which can be further divided into subdisciplines. Because of the perceived separation of art and science, some people may say that they don't like science or that they don't have a knack for it. Yet they ask themselves questions such as the ones above. Others may say exactly the opposite. They love science and think of themselves as disinclined or unable to appreciate the more artful exploration of those same questions, considering anything outside the realm of mathematics and science as not of their concern. Yet, life confronts them with those same big questions to which science, in the restricted sense we now use the term, has only partial answers. We thus end up having disparate communities, looking at identical questions, seeing the world from different perspectives.

We live in a culturally and intellectually divided world. Snow (1961) referred to this schism more than half a century ago as a "gulf of mutual incomprehension" (p. 4) with representatives of the two disparate cultures—the arts and the sciences—having a "curious distorted image of each other." True science is often seen as that which can be mathematized. No doubt, there is great value in the mathematical expressions that capture the grand visions of how our world (or aspects thereof) works. There may even be strange and exciting beauty in such expressions. The search to reveal such beauty could actually drive the scientist's pursuit, rather than the passion to solve a particular problem in our understanding of nature (see for instance Chandrasekhar, 1987; Colless, Chapter 10, this volume; Farmelo, 2002; Pais, 1998). But the mandatory requirement of mathematization limits our contemplation of and reflection about the world in which we live to what can be described and understood mathematically.

The ubiquitous use of mathematics in the natural sciences, particularly physics, is relatively recent. While it is difficult to imagine that advances in our understanding of nature could have been achieved without it, it is likewise true that great strides were made also in earlier times (see e.g. Figure 1.3).

Though representations of what we see in the sky evolve over time, we share, of course, the same sky with those who came before us, including our most remote ancestors, and those who will populate the earth after us. Not only do we share the sky, it also has inspired us with awe throughout the ages. "Since the earliest times, people everywhere have looked up at the stars and wondered about them," remarks Britain's Astronomer Royal, Martin Rees, in an interview with Michael Proctor (2014, n.p.) for King's Review, the online magazine of King's College, Cambridge, UK. Theoretical physicist Neil Turok expresses a similar thought when he notes in his 2012 Massey lectures that "… throughout history, from Anaximander and Pythagoras to Galileo and Newton, the universe has been an endless source of wonder, inspiring us to rise above our current circumstances and see what lies beyond" (p. 205). Indeed, it inspires reflection about how and where we find ourselves between our distant enigmatic origin and a future that we can only try to imagine. Such reflection gives rise to myths, to stories that provide a certain level of comfort as we contemplate our life and attempt to give it meaning, as well as to serious science.

FIGURE 1.3 A ninth century representation of the positions of the seven planets on 18 March 816. Fascination with what is up in the sky has been a primordial force driving humans throughout history to explore the unknown (Source: Wikimedia Commons)

But what about the *facts* of life? The myths and metaphors with which we live, the stories we tell ourselves, including the beautiful story revealed through disciplined scientific inquiry, help us understand, but do we ever know for sure? Do we *really* understand? "Do facts ever really exist outside of how we make them? Is there such a thing as 'reality,' unambiguously presented to human consciousness and then described by language directly?" asks Oren Harman (2018, p. 4). Or is reality simply 'that what we think,' as the mysterious 'traveler' who occasionally appears in the 'StarTrek—The Next Generation' epic seems to suggests? "Perhaps," says Harman, "our concepts organize our world, before we perceive its so-called certainties. And perhaps those concepts are intimately tied to who we are" (p. 4).

These considerations raise pertinent questions about the issue of objectivity. That is a good thing. One of the most valuable characteristics of science is that it encourages questioning. The scientific mind is, among many other attributes, an inquiring mind. It takes the world seriously and thus questions it. One only asks questions concerning things one considers relevant. The scientific mind also takes *itself* seriously and thus equally asks questions about itself, the regularities according to which science works, and the assumptions underlying it. The notion of objectivity is a clear example. It has the attention of scientists as well as historians and philosophers of science, such as Daston and Galison (2007), who wrote an entire book about the topic.

Robinson (2014) argues, in a podcast lecture on 'Hume and causation' that the meaning of 'objectivity' has dramatically changed over time. In the seventeenth century it included concern with "the full contextual setting in which something is found." In contrast, most scientists today would rather be concerned with how well they have managed to decontextualize the object of their scientific inquiry, so as to study it in the absence of any undue external influences. This keeps the scientist separate from the problem or problems that led to the study in question. The objective of the study then becomes 'mastery' with the purpose of being able to 'intervene' in the world. It leaves the researcher disengaged and uninvolved. Merleau-Ponty summarizes this thought brilliantly in the opening sentence of his essay *L'oeil et l'esprit* which reads: "La science manipule les choses et renonce à les habiter" (1964, p. 9), or, in translation, 'Science manipulates things and refuses to live within them.'

John Shotter (2005) elaborates on the same thought as follows:

> As far as we as investigators are concerned, we remain unchanged as the people we are in the process. We remain 'set over against' or 'outside' the other or otherness we are inquiring into; we are not engaged or involved with it. We acquire extra *knowledge* about it in the form of facts or information with the purpose of gaining *mastery* over it. We do this by searching for regularities in its behavior, by establishing a *single order of connectedness* among what we take to be the stable, identifiable elements making up its nature. But these stable 'elements' are parts of a cognitive framework of our own creation that we impose on the phenomena in question from the outside. As such, they have their being for us within the already well-known realm of measurement, thus to elaborate it further. But also as such it is a form of thought that eliminates the inner dynamics, inner complexities, and internal relations in terms of which both created and self-creating beings come into existence and have their being there. Such reductions of complexities to simplicities can sometimes occur in an instant, in a flash of insight! (p. 153)

It reminds me of past occasions when I was teaching physics at pre-university and college level and frequently found that students' 'vision' remained stuck in the mathematical formulae that describe the relationships in the material world. They were well familiar with how to use those formulae in dealing with problems of the kind that would often appear in exam papers, but they could not see the world behind the mathematical descriptions. It was for them like reading a poem—seeing words, phrases, sentences and stanzas while being aware of the formulaic rules that apply—but being sadly unaware of the deeper meanings and emotions behind the words that appear on paper. There is beauty behind a simple equation like $\mathbf{F} = \mathbf{m} \times \mathbf{a}$. There is beauty also behind the equations of a later date in the history of science that describe nature more comprehensively, taking into account findings that were not yet known in Newton's time. But when I look at Maxwell's equations, or those formulated by Schrödinger, Dirac or Einstein, do I see the same beauty that others perceive?

Perhaps I don't. But does it really matter? The important thing is that we all see beauty. Though our perceptions may be dissimilar—like they are when we collectively listen to the same beautiful musical performance—what matters is that we are able to share the emotion, which helps us to understand the world and, as we understand the world together, to also understand ourselves and each other. The city of Leiden in the Netherlands has understood this very well. It decorates the walls of buildings across town with expressions of beauty, such as those that can be found in poetry, prose and, as Figure 1.4 shows, in mathematical description of how the universe works according to Einstein. By doing so, it turns the public space into an environment that makes one think, at least for those who allow themselves the time and tranquility to walk.

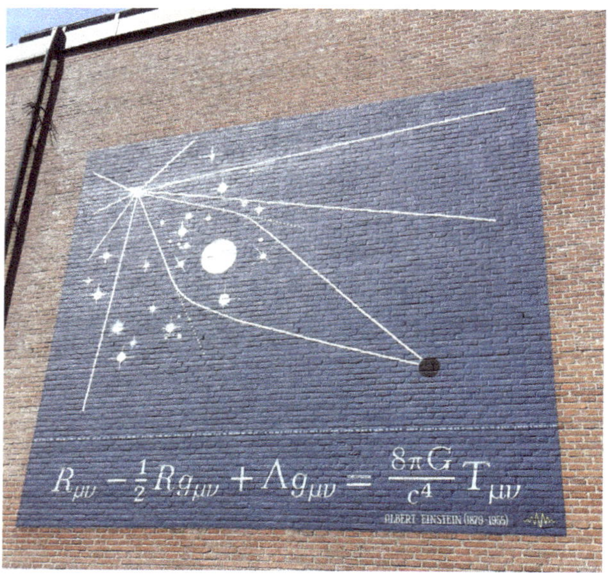

FIGURE 1.4 21st century rock art, featuring Einstein's Field Equations on the wall of the 'Rijksmuseum Boerhave' in the city of Leiden, the Netherlands. (Photograph: Jos van den Broek)

2 There's a Story to Be Told

2.1 *Once upon a Time ...*

... in April 1999, a small group of people from different parts of the world gathered in Montreal, Canada, at a seminar on 'Overcoming the underdevelopment of learning.' The seminar took place in the framework of the Annual Meeting of the American Educational Research Association (AERA) and had been co-organized by UNESCO and the Learning Development Institute (LDI). The seminar brought together a wonderfully diverse community of mind, which included a theoretical physicist with a passion for transdisciplinary approaches towards building knowledge; a communications scholar, president of a higher education institution and social critic; an astrophysicist (program manager of the Mars Exploration Directorate at NASA's Jet Propulsion Laboratory) and ardent activist for educational development in Africa; an educational psychologist with profound interest in the design of learning systems; an experimental physicist (and Nobel Laureate) with a passionate desire to change the order in which the various scientific disciplines are often taught; an innovator in the use of technology to improve the nature and quality of learning; a technologist/educator involved in creating technology-facilitated problem-oriented learning communities; and me. I am both a theoretical physicist and cognitive scientist and have decades-long experience of educational development around the world, particularly in difficult circumstances in Africa. I was the proposer and convener of the seminar.[1]

At the time, I was UNESCO's director for Learning Without Frontiers (LWF).[2] While my work at UNESCO was phasing out, LDI was being phased in as an autonomous not-for-profit entity set up to continue the work started during the 1990s in the context of UNESCO's LWF program. The idea of 'Learning Without Frontiers' had been launched in 1993 by the French philosopher Michel Serres in the context of deliberations by the *Ad Hoc Forum of Reflection*, a collective of thinkers that had been set up to advise UNESCO's Executive Board concerning the Organization's role in the last decade of the twentieth century, with a view to attempting "to bring out new trends that could be an inspiration to the Organization in the coming decade" (UNESCO, 1993, p. 1). As one can read in the referenced document, emphasis in the recommendations of the Forum was on devising:

> means enabling all people throughout the world to obtain access to all forms and levels of education within the context of lifelong education which would establish a continuum between universal primary education and forms of higher education. It should be an education that is both

> 'global' and 'comprehensive', preparing individuals for work in a world increasingly shaped by science and technology and fostering a spirit of tolerance and social harmony with a respect for nature and cultural pluralism. (pp. 1, 2)

In short, a *world system of open education* was expected to ensue.

It is not uncommon for entities, such as the Forum, working within the purview of the United Nations and its various agencies, to boldly propose and set objectives that go way beyond what has ever been attempted. I believe this to be a good thing. It keeps one alert and it is good to challenge the complacency of those who find such goals to be unrealistic. Having the opportunity, during the time I served UNESCO, of working—together with my, more than just merely enthusiastic, team—on LWF related issues was a distinct privilege. Such issues were multiple and widely varied. Meeting the demands associated with them often felt like something unattainable.[3] But nothing is unattainable for those who are determined to personally engage in bringing about a desired change. It is thus gratifying to know that others, both in and outside UNESCO, continue the work once started under the LWF banner. The Learning Development Institute is among them.

2.2 *In Search of the Meaning of Learning*

The 1999 seminar on 'Overcoming the underdevelopment of learning' marked the beginning of a lengthy process that eventually led to the book you now hold in your hands. It wasn't the seminar itself, but rather the discussions among the invited scientists, preceding the programmed conference event and following its conclusion, that set the tone. The Learning Development Institute was, at that time, still in the earliest days of its infancy, starting to explore the world of learning as it looked then, trying to identify effective inroads into making it better. How could we best overcome the 'underdevelopment of learning?' As we set out to respond to that question, we first needed to figure out what learning actually is.

Luckily, we didn't have to start from scratch. Work undertaken in the context of UNESCO's 'Learning Without Frontiers' program during the 1990s had already revealed that the frontiers of learning are multiple and much more complex than initially assumed. They have to do with more than access alone. Providing access to deliberate learning opportunities, such as those provided in the schooling context, by whatever means possible, was and remains an important goal. Advances to the attainment of that goal continue to be made, in part thanks to progress with the development of ICT[4] infrastructure around the world; increased availability of digitized learning resources, particularly OER[5]; and improved distance education practice. But more is at stake and more must be done. Finding out what we actually mean when we say that people are learning, or that they have learned something, had our particular attention in LDI's 'Meaning of Learning' (MOL) program.[6] Why? Don't we know what learning is?

The answer by most people to the question 'What is learning?' is likely to be something like 'That's what you do in school.' To find out if that is a sufficient answer or if more is at stake we carried out some research. Actually, it was quite a bit more than just 'some research' as the inquiry stretched out over a period of several years and involved organizing workshops, meetings and conference sessions with colleagues to explore the issue. In addition, we ran an extensive Learning Stories Research program, asking people of different ages with diverse experiential backgrounds, around the world, what the most meaningful learning experience was they had ever undergone in their life; why they considered it meaningful; and what conditions were present that allowed the experience to occur. It gave us the opportunity to collect and analyze hundreds of learning stories, expressed in the form of brief essays, drawings, and poetry, or delivered orally and recorded by the researcher. The stories came from a diverse spectrum of people, ranging from well-established academics to illiterate farmers living in the Andean mountains of South America. I leave it to the imagination of the reader to consider which of these groups, if any, within that spectrum might have had the more profound meaningful learning experience.

Analysis of the collected learning stories showed that for an overwhelming proportion of the respondents their most meaningful learning experiences, those that were most relevant to them in the real-life context, were situated outside the school context. One shouldn't be surprised. According to research by the *Learning in Informal and Formal Environments* (LIFE) Center at the University of Washington, Seattle, WA, we spend most of our learning life in informal settings (Bransford, Slowinski, Vye, & Mosborg, 2008). The question 'What is learning?' must therefore be addressed in a much broader—lifelong, life-wide and life-deep—perspective.

Changing the perceptions about learning is difficult in a world in which the dominant, often exclusive, emphasis remains on schooling as the preferred environment for facilitating learning (Visser & Visser, 2000). The different contexts in which we learn—of which the school must be seen as an organic part—don't exist in isolation of each other. They should be seen as mutually interacting spaces within a comprehensive learning landscape. It's the entire landscape that deserves our attention. To make things worse, despite the emphasis on schooling as the default option for societal concern with human learning, there is little creative thought about structuring the schooling context to better suit the interests, motivations and desires of the attending students. Attempts to change this require, in addition to vision, courage and persistent determination. Even so, they may still fail to flourish and endure. However successful their students and happy their teachers, they may still meet with strong resistance to change from the educational establishment and, eventually, fail (see Heckman, Chapter 5, this volume).

2.3 *An Enhanced View of Human Learning*

It should by now be clear that I value learning highly but that I am, at the same time, extremely unhappy with how it is commonly viewed. The question regarding what learning actually is has therefore long been on my mind. The results of the research we, and others, carried out, in combination with what I had learned from my experience as an educator and developer of learning systems, has helped me to rethink the meaning of learning, redefine it, and figure out why it is so difficult to change people's perceptions about it. This led to a definition of learning[7] which highlights the following four things:

- Getting better at constructive interaction with change is the essence of what it means to learn.
- Learning is a lifelong disposition rather than something one engages in only from time to time.
- The disposition referred to in the previous item does not only apply to individual people. Social entities (such as families, organizations, corporations, etc.) also learn and thus get better at interacting with change.
- Learning results from dialogue. It should be noted that we develop such dialogue along the lifespan with the human, social, biological and physical environment that surrounds us.

The formal definition itself, presented here only in a footnote so as to not interrupt the flow of reading, does not pretend to invalidate other definitions in frequent use in the context of formal education. Such definitions are geared towards measuring particular learning outcomes that occur as the result of a deliberately designed process. My point is that such formal learning is but a small part of all the learning we do along the lifespan. Most of our learning is not the result of a deliberate decision and an intentionally designed process. Rather, it is the natural consequence of being alive in specific circumstances while being disposed to have an open mind and willing to enter, metaphorically speaking, into dialogue with one's environment.

3 'Bildung' and the Scientific Mind

'Bildung' is a German term, the full meaning of which can only partially be found in single words available in the English language. It is therefore usually left untranslated and adopted as part of the vocabulary, in a way similar to what happened to the German term 'Gestalt.' Deimann and Farrow (2013) review the concept extensively in an article about open educational resources (OER) in the context of open education. They highlight the importance of Bildung as a "powerful reflective tool for approaching open education" (p. 344), using the combination of the terms self-cultivation and self-reflection as a shorthand reference to the more complex meaning that the concept Bildung has acquired in the German tradition, where it is particularly associated with the name of the philosopher Wilhelm von Humboldt, the originator of the Humboldtian model of higher education. I will be referring here specifically to Bildung as a process of becoming; a way of getting to know oneself ever better while at the same time shaping oneself, discovering and constantly clarifying one's humanity and improving one's mental and emotional faculties.

I was reminded of the rich meaning of Bildung in 2005, at the opening of our first colloquium on 'Building the Scientific Mind' (BtSM), a colloquium that later became part of the biennial series of six BtSM colloquia that ended in 2015. It was on that occasion, in 2005, that the first person who raised his hand to signal that he wanted to speak, my esteemed colleague and friend Mohsen Tawfik, remarked vehemently that "You cannot build the scientific mind."

Indeed, notwithstanding the phrase we coined in Montreal in 1999 to describe the proposed focus area of activity for our then new institute, 'Building the Scientific Mind' should not be thought of as something one does to the mind of another person. It is not like building a house. Rather, it is like a tree growing. A tree, in the setting of a natural forest, grows of its own accord. It may take centuries to shape and reshape itself. It is best left to itself and its fellow trees with which it collaboratively constitutes the forest, an organic whole of coexisting organisms that interact with each other in subtle ways, which humans have long been unaware of.

Likewise, we humans engage in a lifelong process of becoming through the 'dialogue' I referred to earlier with our human, social, biological and physical environment. This specific feature of BtSM became increasingly more pronounced and distinct as the dialogue on this matter evolved over time and around the world. The six BtSM colloquia that constituted the biennial series were deliberately organized, with most valuable local support in culturally dissimilar regions of the world straddling the different continents. It spawned local activity that followed its own path. An overview of the colloquia is available on the BtSM pages of the website of the Learning Development Institute.[8]

Over the years, since that seemingly uneventful AERA seminar in Montreal in 1999 took place, my views of what building the scientific mind entails have evolved and so have, undoubtedly, the views of those who participated in the colloquium series or who otherwise interacted with it. On the occasion of the final BtSM colloquium, held against the powerful backdrop of the rock paintings at the archeological site of Serra da Capivara in Brazil, I wrote (Visser, 2015):

> I am not sure if I have reached the end of that evolutionary process. I'm still learning, particularly thanks to discussions I have with other people. Nonetheless, I find the description I currently use helpful in guiding me now that we start working on turning theory into practice. According to that description, building the scientific mind is
>
> *a lifelong process of human development to acquire a way of being in and of the world, inspired by the heritage of the millennia-long history of the human pursuit of knowledge (scientia) for the advancement of understanding and wisdom, comprising*
> - *habits of thinking and dispositions in approaching the world, as well as*
> - *values, ethical concerns, aesthetic considerations, and attitudes, alongside*
> - *mastery of a complex array of skills and mental capabilities in select domains, with such 'select domains' depending on an individual's interests and prospective needs and desires, which are different for a carpenter, a theoretical physicist, or a musician.*

And I furthermore asked my fellow participants to note that in that compact description above the following points particularly mattered:
- I avoid using the word 'mind' as it is too easily interpreted in the Cartesian sense of the term. Instead, I have replaced it with the phrase 'way of being in and of the world.' This formulation emphasizes the intimate relationship between us as human beings and the world of which we are an inseparable part.
- I avoid the word 'science' as it is too easily interpreted in terms of its too narrow reference to the body of knowledge pertaining to disciplines such as physics, biology and chemistry. Such narrow interpretation of the word 'science' should probably be attributed to the coining by Whewell in 1833 of the term 'scientist' to designate professionals working in these areas. Consequently, building the scientific mind is then seen as a concern limited to reform in science education and relevant to those who eventually become scientists.

- I refer to mastery of skills not as an autonomous goal of the building of the scientific mind but as an important means in the process of shaping the envisioned 'way of being in and of the world.'
- I also refer to the building of the scientific mind as a continual process, which, just the same as in the case of learning, evolves along the lifespan.

Writing today I should have added a fifth item, expressing the concern that 'building' should be interpreted as 'Bildung,' in the full sense in which that concept evolved in the German tradition. The B in BtSM definitely refers to Bildung.

I conclude this segment with a graphic representation of the complex multidimensional nature of the scientific mind as it evolved thanks to the decade-long BtSM dialogue, which involved several hundred participants around the world (Figure 1.5).

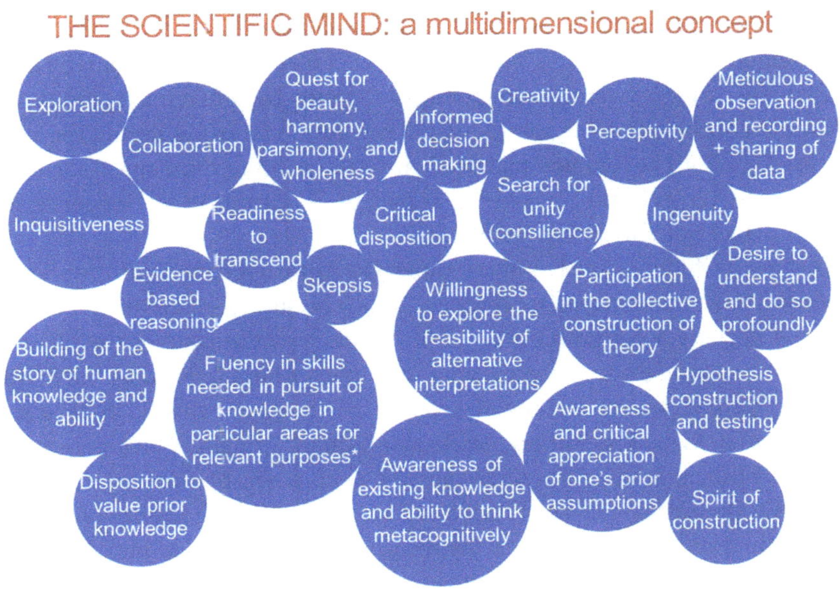

FIGURE 1.5 Graphic representation of the multidimensionality of the scientific mind

4 The Enduring Challenge: Safeguarding Sustained Intelligent Life on Earth

We live in interesting times, an era in which our survival as a species is at stake. The challenge it poses is of a kind and extent we have never met before, asking the utmost of our determination to meet the challenge, both in the short term of the coming decades and in the much longer term if we manage to take appropriate measures in the short term. It requires wisdom; profound understanding of our reality and recognition of our role in shaping it; awareness of the values we share; creativity to invent; skill and prowess in conceiving and engineering solutions that work and interact harmoniously in addressing a complex array of problems; and the courage to look ourselves in the eyes, determined to no longer fool ourselves or look the other way, but instead to dramatically change course and invent new futures (see also Mayor, Chapter 14; this volume). Here is a picture of what I want you to have a look at.

We, members of the species *Homo sapiens*, are relatively new on planet Earth, where life forms first appeared 4.28 billion years ago, not long after the formation of our planet itself, 4.54 billion years ago. Our species emerged, according to recent research by Hublin et al. (2017) and Schlebusch et al. (2017), more than 300,000 years ago. However long ago that may seem when measured against the standards of human timeframes, it is an ephemeral event against the backdrop of the 13.82 billion years it took the universe to evolve since the Big Bang. Yet, our impact has been enormous, particularly as a consequence of the invention of agriculture (dating back to circa 10,000 BCE) and more specifically since the industrial revolution (starting during the 18th century CE). Both developments have benefited us in the short run—short in evolutionary terms—but consequences may be disastrous in the longer-term perspective of human existence on Earth.

The era in which we live has been coined the Anthropocene by atmospheric chemist Crutzen and biologist Stoermer (2000) because human impact alone has now become a determining factor in influencing the state of the planet. In fact, it has brought us close to what multiple authors refer to as the Sixth Mass Extinction (see for instance Leakey & Lewin, 1996; Barnoski et al., 2011, Kolbert, 2014, Ceballos, Ehrlich, Barnosky, García, Pringle, & Palmer, 2015; Ceballos, Ehrlich, & Dirzo, 2017; Van der Valk et al., 2018). Mass extinctions—a total of five known so far—have occurred naturally over the history of life on Earth, but none of them has been the result of the actions of a single species. This is different now. The responsibility rests with us. We have, unwillingly and unknowingly, been responsible for creating the problems for which we now have the responsibility to solve them.

Huge transformations in human behavior are required, first of all to address the current serious threats to our survival, and then to ensure sustained intelligent life on planet Earth. The transformations in question depend highly on the human ability to understand deeply, to understand in the flesh, what it means to be human on an ordinary planetary satellite, of an ordinary star, located in some insignificant corner of one particular galaxy among the more than 100 billion (10^{11}) that are known to exist in the universe. The scientific mind, as defined in this chapter and elucidated in the writings of my colleagues, the contributing authors to this book, must become an essential and defining feature of human existence in the Anthropocene if sustained intelligent life on planet Earth is to endure.

Notes

1 Details about the seminar, including names of the participants and their written contributions, are available at the webpages of the Learning Development Institute (1999) as well as those of UNESCO (1999).
2 As of the time of writing, an archived copy of the LWF website (the first ever website created within the United Nations system) can still be accessed at http://www.unesco.org/education/lwf/
3 To get a glimpse of the nature and variety of LWF related activities see the 'Learning Without Frontiers' Documents and Papers' page at http://www.unesco.org/education/educprog/lwf/lwf_docs.html. A special issue of the *UNESCO Sources* magazine on Learning Without Frontiers is available at http://unesdoc.unesco.org/images/0011/001112/111215E.pdf
4 Information and Communication Technology.
5 Open Educational Resources (see for instance http://unesdoc.unesco.org/images/0021/002158/215804e.pdf).
6 See http://learndev.org/MoL.html
7 Published in 2001, see Visser (2001). The definition reads as follows:
Human learning is the disposition of human beings, and of the social entities to which they pertain, to engage in continuous dialogue with the human, social, biological and physical environment, so as to generate intelligent behavior to interact constructively with change.
8 See http://learndev.org/BtSM%202005-2015.html

References

Barnosky, A. D., Matzke, N., Tomiya, S., Wogan, G. O. U., Swartz, B., Quental, T. B., … Ferrer, E. A. (2011). Has the Earth's sixth mass extinction already arrived? *Nature, 471*, 51–57.

Bransford, J., Slowinski, M., Vye, N., & Mosborg, S. (2008). The learning sciences, technology and designs for educational systems: Some thoughts about change. In J. Visser & M. Visser-Valfrey (Eds.), *Learners in a changing learning landscape: Reflections from a dialogue on new roles and expectations.* Dordrecht: Springer.

Ceballos, G., Ehrlich, P. R., Barnosky, A. D., García, A., Pringle, R. M., & Palmer, T. M. (2015). Accelerated modern human-induced species losses: Entering the sixth mass extinction. *Science Advances, 1*(5), e1400253.

Ceballos, G., Ehrlich, P. R., & Dirzo, R. (2017). Biological annihilation via the ongoing sixth mass extinction signaled by vertebrate population losses and declines. *Proceedings of the National Academy of Sciences* (Advance online publication). doi:10.1073/pnas.1704949114

Chandrasekhar, S. (1987). *Truth and beauty: Aesthetics and motivations in science.* Chicago, IL: The University of Chicago Press.

Crutzen, P. J., & Stoermer, E. F. (2000). The 'Anthropocene.' *Global Change Newsletter, 41*, 17–18.

Daston, L., & Galison, P. (2007). *Objectivity.* New York, NY: Zone Books.

Deimann, M., & Farrow, R. (2013). Rethinking OER and their use: Open education as building. *The International Review of Research in Open and Distributed Learning, 14*(3), 344–360. Retrieved from http://www.irrodl.org/index.php/irrodl/article/view/1370/2542

Farmelo, G. (Ed.). (2002). *It must be beautiful: Great equations of modern science.* London: Granta Books.

Harari, Y. N. (2015). *Sapiens: A brief history of humankind.* New York, NY: HarperCollins Publishers.

Harman, O. (2018). *Evolutions: Fifteen myths that explain our world.* New York, NY: Farrar, Straus and Giroux.

Hublin, J.-J., Ben-Ncer, A., Bailey, S. E., Freidline, S. E., Neubauer, S., Skinner, M. M., … Gunz, P. (2017). New fossils from Jebel Irhoud, Morocco and the pan-African origin of Homo sapiens. *Nature, 546*, 289–292.

Kolbert, E. (2014). *The sixth extinction: An unnatural history.* New York, NY: Henry Holt and Co.

Leakey, R., & Lewin, R. (1996). *The sixth extinction: Patterns of life and the future of humankind.* New York, NY: Anchor Books.

Learning Development Institute. (1999). *AERA Symposium.* Retrieved from http://learndev.org/aera.html

Maeda, J. (2013). Artists and scientists: More alike than different. *Scientific American.* Retrieved from http://blogs.scientificamerican.com/guest-blog/2013/07/11/artists-and-scientists-more-alike-than-different/

Merleau-Ponty, M. (1964). *L'oeil et l'esprit* [Eye and mind]. Paris: Gallimard.

Pais, A. (1998). Paul Dirac: Aspects of his life and work. In P. Goddard (Ed.), *Paul Dirac: The man and his work* (pp. 1–45). Cambridge: Cambridge University Press.

Proctor, M. (2014, May). *An interview with Martin Rees, astronomer royal. King's Review.* Cambridge: King's College. Retrieved from http://kingsreview.co.uk/articles/an-interview-with-martin-rees-astronomer-royal-4/

Robinson, D. (2014). Hume on causation. In *Reid's critique of Hume* (Podcast series of eight lectures, Lecture 3). Oxford: University of Oxford.

Schlebusch, C. M., Malmström, H., Günther, T., Sjödin, P., Coutinho, A., Edlund, H., ... Jakobsson, M. (2017). Southern African ancient genomes estimate modern human divergence to 350,000 to 260,000 years ago. *Science, 358*(6363), 652–655.

Shotter, J. (2005). Goethe and the refiguring of intellectual inquiry: From 'aboutness'-thinking to 'withness'-thinking in everyday life. *Janus Head: Journal of Interdisciplinary Studies in Literature, Continental Philosophy, Phenomenological Psychology, and the Arts, 8*(1), 132–158.

Snow, C. P. (1961). *The two cultures and the scientific revolution* (The Rede Lecture, 1959). Cambridge: Cambridge University Press.

Turok, N. (2012). *The universe within: From quantum to cosmos.* Toronto: House of Anansi Press Inc.

UNESCO. (1993). *Consideration of the results of the deliberations of the ad hoc forum of reflection.* Paris: UNESCO Executive Board (142nd session). Retrieved from http://unesdoc.unesco.org/images/0009/000958/095846Eo.pdf

UNESCO. (1999). *AERA contributions 1999: A work in progress.* Retrieved from http://www.unesco.org/education/lwf/doc/aera1999.html

Van der Valk, T., Sandoval-Castellanos, E., Caillaud, D., Ngobobo, U., Binyinyi, E., Nishuli, R., ... Guschanski, K. (2018). Significant loss of mitochondrial diversity within the last century due to extinction of peripheral populations in eastern gorillas. *Scientific Reports, 8*(1), 6551.

Visser, J. (2001). Integrity, completeness and comprehensiveness of the learning environment: Meeting the basic learning needs of all throughout life. In D. N. Aspin, J. D. Chapman, M. J. Hatton, & Y. Sawano (Eds.), *International handbook of lifelong learning.* Dordrecht: Kluwer Academic Publishers.

Visser, J. (2015, August 17–21). *Two things commanding attention.* Brief paper contributed to the Sixth Advanced International Colloquium on Building the Scientific Mind, São Raimundo Nonato & Serra da Capivara, Piauí, Brazil. Retrieved from https://www.researchgate.net/publication/283328115

Visser, J., & Visser, Y. L. (2000, October 25–28). *On the difficulty of changing our perceptions about such things as learning.* Paper presented at Presidential Session on 'In Search of the Meaning of Learning' at the International Conference of the Association for Educational Communications and Technology, Denver, CO. Retrieved from https://www.researchgate.net/publication/259811620

INTERMEZZO 1

The Missing Piece

Jan Visser

Reflections inspired by Robert Greenberg: 'Context is everything and everything has meaning'

It was a mammoth task to develop a vision of what it means, involves and implies for human beings to engage in a lifelong process of shaping their mental and emotional faculties in ways that allow them to interact constructively with a world in constant change. It has taken no less than ten years to create such a vision, requiring the involvement of more than 300 people, scientists (including several Nobel Laureates) and non-scientists alike, pertaining to dissimilar cultures from around the world. The result: A compact, though complex, definition and a graphic overview of major characteristics and features that, when present, testify to evidence of the scientific mind in action (see Visser, Chapter 1, this volume).

The decision to produce a co-authored volume on the issue of Building the Scientific Mind (BtSM) as a lifelong human undertaking was one of the outcomes of the fifth BtSM colloquium held in Lembang, Indonesia, in May 2013.[1] Robert Greenberg—music historian, composer, prolific online teacher, and much more—was among the first authors we approached in November 2013 when we were in California. The invitation to meet him at home was warm and encouraging, even though BtSM was new to Dr. Greenberg's universe. "Your project sounds fascinating and I would be pleased to discuss it with you," Bob wrote. And he continued: "You might consider ... coming to my house in the Oakland Hills: I can offer a fine dinner, the best of martinis (if you are so inclined), and we could have the afternoon and/or evening for conversation." There was not the slightest hesitation on our part to accept this kindest of invitations. It clicked on first sight between us as soon as we arrived. The conversation was inspiring, the martinis uplifting and the dinner enriching.

Half a year later we had the chapter that follows. 'Context is everything and everything has meaning,' the title read. I had a mild shock. What has that to do with shaping the scientific mind? The question remained on my mind during the years that followed. It sensitized me to looking for context and clarifying how it mattered whenever I saw the scientific mind in action, in me, and in others. Looking back at the original diagram about the multidimensional

 | DOI: 10.1163/9789004416802_002

nature of the scientific mind, presented in Chapter 1, 'awareness of context and recognition of its meaning' stands out as the most important missing piece of the ever incomplete puzzle of an evolving concept of the scientific mind. That's why Greenberg's chapter appears up front in this book.

Note

1 See http://learndev.org/BtSM2013.html

CHAPTER 2

Context Is Everything and Everything Has Meaning

Robert Greenberg

1 Let Me Tell You a Story …

We are made to tell stories. Stories assert order and find meaning in the otherwise chaotic events that shape our lives and world. Stories create myths, paradigms and metaphors that allow different people living at different times and places to tap into the collective and, having done so, find greater meaning in their own lives and personal experience. Indeed, storytelling—the progressive organization of information over time—lies at the heart of almost every human activity. Teaching is story telling. The manner in which we relate the events of the day to our family at dinner time is storytelling. Performing and composing a piece of music is storytelling.

Teaching music demands the telling of a number of distinct yet intersecting stories: the story behind a piece of music (the life and times of its composer); the actual bricks-and-mortar content of the piece of music (music analysis); and the reason-to-be, the meaning of the piece (the story it tells, be it abstract, metaphoric, or programmatic).

This chapter intends to discuss the nature of stories and storytelling and how that applies to the creation, content, and teaching of music. But first, an exercise in intellectual and artistic futility: what I was *taught*.

2 What I Was Taught

When I was an undergraduate and graduate student in music composition in the 1970s and early '80s, I studied in departments still dominated by post-War musical modernist composers.

Such composers created music based on pre-determined formulas of various sorts. They believed that such processes rendered their music intellectually pure, autonomous, and incorruptible; true only to itself, defined only by the formulas used to create it, and thus free from external influences, expressive meaning, nationalistic content, and any attempt to tell a metaphorical story: in brief, those elements that were perceived as having created the preconditions for Nazism and Fascism.

 | DOI: 10.1163/9789004416802_003

High-minded though their intentions, the music composed by such modernists was (and remains) nearly impossible to listen to. It is music about its compositional process, music in which pitch, rhythm, and articulation are entirely objectified. Like most conceptual art, the substance of such a musical composition lies in its idea—its *formula*—rather than in its actual execution, meaning its actual sound and the story it would tell. The primary goal of such music is to obliterate everything except the present moment. And while such musical nihilism might reflect an expressive agenda, it does not reflect any ongoing expressive meaning.

Some modernists actually tried to explain away the importance of sound and meaning and storytelling in their music. For example, in 1959, the well-known composer Ernst Krenek revealed as much about his own post-War alienation as he did about his music when he wrote:

"Actually, the composer has come to distrust his inspiration because it is not really as innocent as it was supposed to be, but rather conditioned by a tremendous body of recollection, tradition, training, and experience. In order to avoid the dictations of such ghosts, he prefers to set up an impersonal mechanism which will furnish, according to premeditated patterns, unpredictable situations. The audible results of these procedures are incidental" (cited in Taruskin, 2005, p. 41).

It is a shameless confession: Krenek admits that what his music sounds like—and thus presumably means—*is incidental to the point of the piece.*

Such composers approached music history and analysis from precisely the same point of view. In what amounted to a musical version of the 'New Criticism,' musical compositions—be they by Bach, Beethoven, Brahms, Babbitt or the Beatles—were discussed as discreet objects related only to themselves. Analytical language had to grow from the piece under examination; terminology that suggested cultural norms or period styles could not be employed. For example, when I was an undergraduate student of music analysis, it was forbidden for us to use such ubiquitous structural terms as 'sonata form,' 'minuet and trio form,' and 'rondo form;' which was the equivalent of forbidding a student of anatomy the use of the words 'heart,' 'lung,' and 'kidney.'

However, most problematic for the modernists was the issue of expressive content: of meaning. For these pseudo-scientists employing a self-created, pseudo-scientific analytical language in their attempt to render music an informational science, expressive content had no place. Expressive content was irrelevant because—subjective as it is—it could not be quantified. Since it could not be quantified it could not be discussed objectively and was therefore unworthy of scientific inquiry.

Equally irrelevant to the understanding of music were the lives and personalities of the composers as well as the cultural, religious, social, and economic environment in which composers lived and worked. For the modernists, there was no admissible back story that might shed light on the style, substance, and decisions made by the composer. There was only the music.

The modernists reserved their greatest contempt for program music, that is, instrumental music that seeks to tell explicit stories and/or paint pictures in strictly musical terms. Such music, with its recourse to descriptive effect, was seen as the refuge of the compositional scoundrel. Uttering the names of composers of such music—Franz Liszt and Richard Strauss, for example—was an open invitation to private castigation and public humiliation.

This, then, is what I was taught.

3 What I Have Learned

In the thirty years since I left school, I have learned that virtually every modernist view just discussed is qualitatively, quantitatively, and categorically, absolutely wrong.

In fact, *context is everything* and *everything has meaning*.

Context. Composers write music. Composers are people, people who live, work, mate and pay taxes in the same world as we do. Like people living at any place and at any time, the lives of composers are indelibly shaped by the physical, economic, spiritual, and cultural environment in which they live. As individuals with varying life experiences and genetic predispositions, they react to their environment differently. They thus create music that can be differentiated from others.

The post-World War II, atomic-age modernists wanted to render music a science. But music is not a science; it is a language art. Science, according to my Oxford English Dictionary, encompasses "the systematic study of the structure and behavior of the physical and natural world through observation and experiment.[1] Science examines and quantifies; art conveys impressions. Scientific observation has determined that Earth's moon has an equatorial radius of 1738.14 km. That number is *singularly correct* and *absolute*; no other number assigned to the equatorial radius of the moon is valid. However, there are hundreds, even thousands of pieces of music that were inspired by or somehow evoke the moon, each of them equally valid. Art is as varied and as inexact as the people who create it.

Musical style—meaning the generalized 'sound' of the music of a particular time and place—is a function of the cultural and economic environment (the 'context') of that time and place. The manner in which individual composers go about working within (or against) the prevalent stylistic norms of their time is a function of their own artistic personality and life. But it is the environment—the context, the time and place—that shapes a composer's compositional language and expressive content and not the other way around. That this is true is demonstrated by a simple 'what if': what if Ludwig van Beethoven had been born in 1685 (the same year as Johann Sebastian Bach) and not, as he was, in 1770? Would Beethoven's music still have sounded the same? The answer is: of course not. The 'sound world' of Beethoven's music was shaped by his predecessors, his contemporaries, and his personal environment (an environment that permitted radical experiment in a manner quite new in Western art). Had he been living and working 100 years earlier, Beethoven's music would have sounded completely different. Thus: the demands of the environment create prevalent style while the individual composer manipulates that style according to his or her particular artistic impulses.

Context is everything and *everything has meaning*.

Meaning. I would suggest that almost everything we do has personal meaning. We can't help it; the need to create order and structure (and therefore meaning) in the face of chaos is not just an artistic impulse but a basic survival instinct, one shared by us all. That this is true can be demonstrated by the following activity. Give each of 10 people a small bulletin/cork board and 100 thumbtacks. Ask those good people to stick the tacks in the boards. Wait 10 minutes and then observe the results. Certain of the resulting designs will feature straight lines; others curved lines; some will contain distinct shapes and even pictorial imagery while one or two might very well seem abstract or even random in their design. Each design (and non-design) is the result of personal choice and reflects something of the life and spirit of its creator. Each design (and non-design) has *meaning* to its creator, who has invested something of herself in creating it. Each design *tells a story*.

A composer puts notes on paper. Those notes are then taken up by performers who bring the music to life (or maim it in a bad performance; another discussion for another time). The considerations that take place during the compositional process are seemingly endless, as a composer must not only constantly monitor the musical intent and content of a piece but consider as well the needs of the performers and the music's impact on an audience in performance.[2] Like those thumbtacks in the cork board, musical notes do not go on paper without being considered and reconsidered two, three, four, five times! Composing is as purposeful an act as sticking tacks in a board; as creating a great sauce; as writing a novel; as designing a building, with the result being that a degree of meaning informs almost every note.

Context is everything and everything has meaning.

4 What I Teach

A musical composition is the sum product of three distinct and explicable elements: one, the cultural, economic and spiritual milieu of a given time and place; two, the life, personality, and musical priorities of a particular composer; three, the actual musical content of a particular piece of music.

It is the symbiosis of these three elements that determines the meaning of a piece of music. And subjective though meaning might be, I will establish, later in this chapter, that illuminating meaning—be it absolute, metaphoric, autobiographic, programmatic, etc.—is the most important task of a teacher of music. In order to describe the meaning of a piece of music and the symbiosis that determines that meaning, a successful teacher of music must be able to tell a good story.

5 Storytelling

In his book *The Gates of the Forest*, Elie Wiesel (1966, n.p.) wrote:

> God made Man because He [God] loves stories.

Presumably created in God's image, we Homo sapiens love stories as well. For tens of thousands of years before the invention of writing (which happened fairly recently; about 5000 years ago), our ancestors sat around their fires telling stories about hunts, heroes, and lovers; about gods and goddesses; about the creation of the sun, moon, and stars.

Such shared stories came to anchor entire cultures in the guise of religions and national myth. But even more, stories contain and illuminate the very fabric of our existence: through analog and metaphor we discover ourselves in stories and, as storytellers, our individual experience is rendered universal.

Our species has been blessed/cursed with the understanding that we live, constantly, at the edge of chaos and that in the end each of us will die. In the same way our brains organize tiny bits of light into coherent images, so our brains superimpose narrative structures on the events, the chaos, and the terminal mystery of our lives.

Perceived this way, storytelling—like language and its sibling, music—is not something we're merely taught to do but something we are wired to do. Evolution is, after all, ruthlessly utilitarian, and as such I would suggest that like language, storytelling (and for that matter, music and making and engaging with art in general) is a survival skill: a powerful mechanism that allows us to attribute pattern and meaning to what might otherwise be considered random information. Stories allow us to communicate more effectively; they intensify and render meaningful what might on the surface seem to be unrelated experiences; they personalize information and allow us to vicariously achieve wish fulfillment, and even find escape from the regularity of our own lives in a manner beneficial to our survival.

At the heart of most any story lies conflict: in a word, a *problem*. In a literary story, we encounter characters (with whom we do—or don't—identify); in a piece of music we encounter principal and secondary *themes*. In a literary story the characters encounter problems (which in a piece of music can be projected in many different ways, from thematic fragmentation to harmonic manipulation, etc.). The characters in a literary story (as well as the themes in a musical composition) ultimately attempt to extricate themselves and resolve their problems, with varying degrees of success. This tri-partite design lies at the heart of almost all stories: character introduction, conflict, resolution. In other words, beginning-middle-end; exposition-development-recapitulation.

(I am aware that creators and consumers of modernist literature will scoff at this hopelessly conventional view of storytelling. But I would point out that conventions do not spring fully formed from the ether. Rather, they evolve and only become conventional once they have proved their utility over a period of time.)

Regarding conventions and the convention, permit me this excursion. Back in the 1970s and 1980s, it seemed apparent to a certain segment of the compositional community that the future of music lay entirely in electronics: in synthesized and digital sounds. I was asked by any number of fellow young composers how I could possibly continue to compose for antique instruments like the violin and piano, which were, according to my interlocutors, ancient technologies developed in patronymic, imperialistic, phallocentric communities with no claim to the modern world except our conventional dependence upon them. Putting aside the superb repertoire of music already created for such instrumental dinosaurs, I would first point out that they evolved not just over decades but centuries. For example, the first piano was built in 1700, in Florence. However, the fully modern grand piano did not come into being until the 1860s; it was built by Steinway & Sons in New York. And while improvements have been made since then, they are incremental and do not reflect a major design change. So: does this make the modern concert grand piano a dinosaur, a thing-of-the-past that has outlived its usefulness? No it doesn't. Like the fork, the toothbrush, and my electric Panini maker, it has achieved a near perfect balance of form and function, of beauty and utility. Like our sex organs, which also evolved long ago, the piano (and the violin, etc.) is no more antiquated than the imaginations of the people who use it. It is a techno-device, the utility of which is in no way diminished by its age. Strictly synthesized/digital electronic music, however, has today been banished to a tiny, backwater niche of the new music community. It is a sort of music that failed to evolve, that failed to create the new paradigm its adherents predicted for it 30 years ago.

So. There is nothing inherently wrong with the conventional. On the contrary, certain sorts of conventions—be they dramatic devices, character archetypes, or harmonic structures—represent a shared, coherent point of departure between an artist and her audience; the proverbial box beyond which an artist cannot possibly go if the box is not first perceived. So let us not be too quick to reject so-called conventions as dead weights anchoring us to the past. Like our opposable thumbs, some conventions have evolved into things indispensable.

For what it's worth, I find most modernist literature about as readable and meaningful as the instruction booklet that came with my first VCR.

6 How I Teach: Creating a Narrative

Narratives are good. However, if we seek to project absolute, unvarnished truths (if such things actually exist), narratives can, admittedly, sometimes be bad.

Facts—perceived truths—are sometimes victim to narratives. The tired cliché often rings true: 'Don't let the facts get in the way of a good story.' However, facts, as isolated bits of presumably accurate information, only take on cumulative meaning when they are incorporated into a narrative story. It is from so-called fact that we need to extract the substance—the story—that integrates them into a coherent narrative.

Here are some facts.

- In the fall of 1777, Mozart and his mother stayed in Mannheim.
- Mozart fell in love with Aloysia Weber.
- Mozart and his mother then went to Paris.
- Mozart's mother died in Paris.
- Mozart returned to Mannheim and proposed marriage to Aloysia.
- She said no; Mozart was upset.

Let us now extract the story from those facts.

In October of 1777, three months shy of his 22nd birthday, Mozart and his mother left Salzburg for Paris, where the Mozart clan believed their boy genius would find permanent and profitable full-time employ.

On the way, Mozart and his mother Anna Marie stopped off in Mannheim in southwestern Germany with the intention of making some money and contacts before moving on to Paris.

Mozart's mother Anna Marie was thrilled with her son's success in Mannheim. She wrote home to her husband Leopold that:

> [Wolfgang] has so much to do, that he really doesn't know whether he is standing on his head or on his heels. What with composing and giving lessons he hasn't had time to visit anybody. (Solomon, 1995, p. 141)

Well, that wasn't entirely true, because in fact Wolfgang was regularly visiting the house of a particular citizen of Mannheim. Her name was Aloysia Weber. She was a 16 year-old singer, the daughter of a local musician named Fridolin Weber, and Mozart was in love with her. In short order, he wrote his father, informing him that he would be staying in Mannheim, where Aloysia's father would help him to find work.

Mozart's father Leopold went ballistic; the twin prospects of being replaced by Herr Weber and of being deprived of Wolfgang's future income as a result of marriage were unthinkable. He commanded his son:

> Off with you to Paris! Soon! Find your place among great people! (Solomon, 1995, p. 143)

So: not for the first (or last) time, Mozart did what he was told. He and his mother left Mannheim, and arrived in Paris on March 23, 1778. However, forced to part from the girl he loved, increasingly aware that his family's hopes and future were riding on his shoulders, Mozart did what most post-adolescents would do under the circumstances: rebel. And he rebelled by hating Paris.

Then catastrophe struck. On July 3, 1778 Mozart's mother died of typhus in their Paris apartment.

Wolfgang was traumatized and did the best he could. For a 22 year-old who had rarely had to manage his own affairs, to oversee, in a foreign city, the illness, death and burial of his mother was terribly difficult.

Sadly (but predictably), Mozart's father blamed him for his mother's death; it was yet another tawdry episode in their tawdry, co-dependent relationship. Leopold commanded that his son return home to Salzburg, and with no other prospects in the offing, Mozart had no choice but to comply.

Disheartened, angry, and still grieving for his mother, Mozart set out for Salzburg. Alas, he had one last kick-to-the-groin to sustain before arriving home. You see, Mozart stopped in Mannheim where he proposed marriage to his beloved Aloysia Weber. Unfortunately for Mozart, his absence had not made Aloysia's heart grow any fonder for him. She was, accordingly, as cold as spumoni and blew off Mozart and his proposal. Mozart, to his infinite credit, sat down at a piano and with Aloysia still in the room sang a bawdy old song,

'Leck mir das Mensch im Arsch, das mich nicht will' ('The one who doesn't want me can lick my ass').

Good boy.

It's a great story, one based entirely on facts. It also gives us vital insight into Mozart's personality and provides a context for Mozart's break from his father, which occurred some 16 months after his return to Salzburg.

In talking about an actual piece of music, there are two complementary stories to tell: the backstory of a piece and the actual musical content of the piece, from which a teacher will then divine the meaning of the work.

For example. On June 14, 1787 Mozart completed the single strangest work he ever wrote, the so-called *Ein musikalsicher Spaß* (*A musical joke*). The piece contains endless examples of compositional amateurism and bad taste, many of them very subtle: bizarre harmonic progressions, banal melodies, pointless virtuosic passages, incomplete phrases, and so forth. Neither does Mozart spare the players; they are shown to be bumbling fools, particularly in the disastrous conclusion of the fourth and final movement, during which all the apparently 'wrong notes' are carefully notated.

Taken by itself, in reference only to itself, *Ein musikalsicher Spaß* is an exercise in bad taste; a singularly nonsensical curiosity in Mozart's otherwise storied compositional oeuvre.

But in truth, nothing exists outside of its context and therein lies quite a story.

Mozart's father Leopold died in Salzburg on May 28, 1787, at the age of 67. In Vienna, Wolfgang was informed almost immediately of his father's death.

The next day Mozart wrote the following letter (Hildesheimer, 1983, p. 203) to his friend Gottfried von Jacquin:

> Dearest Friend,
> Please tell Herr Exner to come at nine o'clock tomorrow morning to bleed my wife.
> I send you herewith [the sonata] and the sacred song. Please … give the sonata to your sister with my compliments and tell her to tackle it at once, for it is rather difficult. Adieu.
> Your true friend, Mozart
> PS: I inform you that I have received the sad news of my most beloved father's death. You can imagine the state I am in.

Actually, given the conversational, even cheerful tone of his letter, it is difficult to imagine the state Mozart is in. And this letter, with its rather muted 'outpouring' of grief, is pretty much the only documentary evidence we have of Mozart's reaction to his father's death. His dealings with his sister over his father's estate were entirely businesslike, without a hint of grief.

Now, Mozart had a pet bird—a starling. Three days after his father's death, Mozart's bird dropped dead in its cage. Mozart's grief over the bird was such that he wrote a twenty-four-line eulogy in its memory. One does not need to be a psychiatric professional to see this for what it was: an epic example of suppression and transference.

For Wolfgang, insult was soon added to the injury of his father's death. You see, it was only after his father's death that Mozart, reading his father's will, discovered that he had been almost entirely disinherited because he had decided to marry, without his father's consent, five years before. Instead, Mozart's sister, the ever-obedient Marianne, inherited almost everything, including Wolfgang's own scores, which had been left with his father for safekeeping. The irony is that Leopold's assets—which were considerable—had been purchased with money earned by the child Mozart during his various European tours. By being excluded from Leopold's will Mozart was denied his share of money that he himself had earned. What a swindle!

Mozart's bitterness immediately played itself out in his music. The first music he composed after his father's death was not some sort of elegy or memorial to Leopold. No, what he wrote was the so-called *Ein musikalsicher Spaß*, a pre-PDQ Bach-type musical parody that purposely does wrong all those things that Mozart's father—as his composition teacher—taught him to do correctly.

'A musical joke' is Mozart's monument to his father. In this context, its musical crudity and bumptious humor take on a dark, bitter, and ironic cast. Taken in context, the meaning of the piece becomes clear, an autobiographical 'meaning' that far transcends merely the notes on the page. Mozart was a son yearning for his father's approval, but he was also a proud man and a great artist who carried scars and harbored rage borne from a lifetime of exploitation. All of that—and more—is what 'A musical joke' is all *about*. Without the context afforded by a story, that meaning cannot be understood.

7 Expression and Meaning in Music

Music, like any sensual (sense based) experience, must in the end be understood in its own terms, which means that we must *listen to it*. Thus, it is the responsibility of any teacher to guide his/her charges through the musical and expressive complexities of a score by carefully selected listening. However, before that can be done, a teacher has to decide first—for herself—what a given piece of music is, indeed, about: whether the piece is to be discussed as an abstract work of art, related only to itself; whether it functions metaphorically; or whether it depicts an explicit program. However a teacher chooses to interpret and teach a work, that teacher must first do her homework. She must know what the historical literature says about the piece; she must know how and where the piece fits into its composer's life and output; and she must understand how the piece works compositionally: thematically, structurally, and harmonically.

That last job is a big one. It means understanding a work as a musical composition and not just as a historical document. This does not mean analyzing the piece using one (or more) of the methodologies so favored by twentieth century music theorists. No; it means breaking a piece down into its component parts and asking—always—what was the composer's intent: why these notes; why these harmonies; why this structure. It means thinking like a composer, something many teachers of music are simply unprepared to do.

I must open a Pandora's box by broaching the issue of practitioner versus observer: who is better prepared to divine compositional intent in a given piece of music: a trained composer or a trained music historian/musicologist?

In my opinion, the greatest musicologist of the last 60 years—Joseph Kerman (1924–2014)—was great for two reasons. One, his prose had the same elegance, depth, drama, humor and beauty as the music he was writing about and, two, *he thought like a composer*. My parochialism clear, I would suggest that that is the highest praise that can be given a practitioner of musicology, a field dominated by historiography (the study of the methodology of historians and the development of 'history' as a discipline) and what Joseph Kerman (1994) himself calls "quasi-scientific scholarly research in music" (p. 19).

My assertion that Kerman's greatness as a musicologist stemmed from not having actually thought like one might strike some as being unfair, like criticizing cats for not being more like dogs. But since musicologists are almost always those faculty assigned to teach music history and literature courses, from Music 101 (toons for goons) through graduate pro-seminars in music literature, it is germane and fair.

It is, for me, a truism that only a practitioner can understand, at both a gut and intellectual level, the craft, thought, process, and spirit that goes into a particular activity. Show me a good batting coach and I'll show you someone who was a good hitter himself. Dance teachers and choreographers are, themselves, dancers. A violinist cannot diagnose subtle flaws in an oboist's technique; those flaws must be diagnosed by another *oboist*.

Only someone willing, able, and empathic enough to engage with a piece of music in a manner approaching that of its own composer can sense that composer's compositional intentions and thus divine, at both a gut and intellectual level, what that piece means.

The art and craft of musical composition takes years—decades—to acquire. One must learn how to 'push notes around' by learning the crafts of melody writing, counterpoint, harmony, form and orchestration. A young composer must learn to imitate historical models as she searches for her own voice. A professional composer brings to bear all she knows and feels while she writes. A professional composer plans and executes dramatic scenarios; pushes and eases up on the expressive pedal as necessary; constantly balances and develops her original intentions based on discoveries made in the moment. In the end, she must trust instincts that have been honed over decades of writing.

Now, of course, this is not meant to say that composers are intrinsically better teachers of music than non-composers. No, I can assure the reader that pedagogic ineptitude is even more rife among my fellow composers than among our musicological brethren. What it does say, however, is that along with knowing something of a composer's life and time, having a degree of insight into what a composer actually thinks and does while composing is a necessity if someone wants to *understand* what is going on in a particular piece of music.

8 Say It with Music

That music is a language is indisputable. Like spoken language, it consists of sound in time or, if you prefer, time defined by sound. Like spoken language, music employs varying levels of pitch and speed, rhythmic declamation, and relative loudness and softness to provide expressive cues as to what it might be saying. However, unlike spoken/written languages, music lacks semantic content: musical syntax does not feature a vocabulary in which stable, specific meaning is provided by *words*.

The seemingly self-evident idea that spoken language is a relatively specific entity is not accepted by everyone. For example, in an attempt to assert that spoken languages are as non-specific as music, the eminent American musicologist Leo Treitler (2011) writes: "the tools of language have been so blunted by a combination of rhetorical and doctrinal motives that language has been drained [of its] potency and versatility" (p. 7). This is nonsense; a perfect example of an academic's need to dismiss the self-evident in order to build a 'theory' that is only tangentially related to perceivable reality. However blunted language has (or has not) become, it is still—*relative to music*—more literal: more information specific than music.

A game proves the point. Assume a composer has set to music these words for voice and piano:

> Don't the moon look lonesome, shining through the trees?
> Don't the moon look lonesome, shining through the trees?
> Don't your heart feel lonesome when your baby packs up to leave.[3]

Imagine a performance, though instead of the singer, a flute plays the vocal part. Following the performance, ask the members of the audience what the piece meant. They will almost certainly perceive the generalized mood, because in all probability the composer will have employed certain musical conventions that represent sadness and loneliness. But will the audience be able to identify such literary specifics as a 'lonesome moon' and someone's 'baby packing up to leave' without actually having heard the words? Of course not; that degree of informational specificity can only be transmitted by words, not music alone.

Musical syntax is not absolute and thus the meaning of a particular passage of instrumental music will vary from listener to listener. However, as we just observed, this 'variability of meaning' will stay within a fairly restricted range. Another example: the jagged, turgid, C Minor opening of Beethoven's Fifth Symphony of 1808 might evoke a variety of meanings, but all within a fairly restricted expressive range: anger, rage, violence, ferocity. I would suggest that anyone who hears the opening of Beethoven's Fifth Symphony as describing 'a delightful frolic in the park' be medicated immediately.

While the presence of words in vocal and choral music does indeed provide absolute literary meaning, it is still the music that imbues vocal music with its deeper meaning, meaning that often transcends the words. That this is true can be easily demonstrated. Think of a favorite song, one that evokes a special time and place in life. Recite the words as poetry, without the music. Then sing the music, without the words. Where lies the expressive depth, the evocative expressive essence of the song? In its music. Every time.

The power of music to transmit expressive information beyond the power of words alone has been described since antiquity.

The Roman educator Quintilius (Marcus Fabius Quintilianus) advocated the inclusion of music in the teaching of oratory, claiming that an orator could co-opt the power of music by learning to use its inflections:

"Give me the knowledge of the principles of music, which have the power to excite or assuage the emotions of mankind. We are told that Pythagoras, on one occasion, when some young men were led astray by their passions to commit an outrage, calmed them by ordering the piper to change her strain [in a certain way], while [the Stoic philosopher] Chrysippus selects a special tune to be used by nurses to entice their little charges to sleep" (Wiess & Taruskin, 1984, p. 14).

The revolutionary theologian and lapsed priest Martin Luther was an accomplished musician who firmly believed that music was capable of changing the hearts of people and the face of nature. In a letter dated October 4, 1530, he made his feelings plain when he wrote:

> Next to theology there is no art that is the equal of music, for she alone, after theology, can do what otherwise only theology can accomplish, namely, quiet and cheer up the soul of man. (Atlas, 1998, pp. 511–512)

By the nineteenth century, instrumental music—ephemeral, non-tactile, and seemingly boundless—had come to be considered the ultimate art form. According to E. T. A. Hoffmann (1776-1822), the most influential music critic of his time, instrumental music:

> discloses to man an unknown realm, a world that has nothing in common with the external sensual world in which he leaves behind him all definite feelings to surrender himself to an inexpressible longing. (Strunk & Treitler, 1998, p. 1193)

It was this perception of overwhelming expressive power of music that led the English essayist and critic Walter Pater (1839-1894) to assert "All art constantly aspires to the condition of music" (Grout & Palisca, 1988, p. 659).

In his essay *Poetry and Drama* (1950), the writer, poet, essayist (and Nobel Laureate) T. S. Eliot spoke directly to the power of music to articulate expressive meaning beyond the power of words.

> It is the function of all art to give us some perception of an order in life, by imposing order upon it. The painter works by selection, combination, and emphasis among the elements of the visual world; the musician, in the world of sound. It seems to me that beyond the nameable, classifiable emotions and motives of our conscious life ... there is a fringe of indefinite extent, of feeling which we can only detect, so to speak, out of the corner of the eye and on which we cannot completely focus ... This peculiar range of sensibility can be expressed by dramatic poetry, at its moments of greatest intensity. At such moments, we touch the border of those feelings which only music can express. (cited in Kerman, 1988, p. 5)

The power of music to project meaning is indisputable (or nearly so; more in a moment).

How music manages to project extra-musical expressive information is another matter entirely. The most easily perceived (and least disputable) such projections are pictorially literal: that is, when musical sounds are used to imitate nature. The twittering of birds (as in the first movement of 'Spring' from Vivaldi's *The Four Seasons*); the squeal-song of whales (as in George Crumb's *Vox Balaenae*); the pitter-patter of raindrops (as in the third movement of Beethoven's Symphony No. 6) all indisputably evoke natural sounds and thus project literal programmatic meaning.

Music can also evoke extra-musical meaning through metaphor: musical gestures that evoke meaning through association. For example, for hundreds of years, it has been conventional to evoke death by using moderate-to-slow descending minor or descending chromatic bass lines. Such gestures became conventional not on a whim or merely through habit; the step-wise descent marked by its contour is a physical analog for the descent into the grave.

But by what affective reasoning can we say that a passage of music is, for example, literally mournful in mood? It has been argued (endlessly) that music—in itself, by itself—is incapable of evoking any such mood, and that only by becoming habituated to certain musical conventions over the years have we come to identify certain expressive moods with certain musical gestures. The brain-game used to support this assertion describes a New Guinea aboriginal hearing the second movement funeral march of Beethoven's Symphony No. 3 for the first time. Lacking any cultural commonality, would the aboriginal be able to identify the movement as being a funeral march? Even more, would she be able to identify the music as being mournful in mood?

The answer is, I think: unlikely. But in fact, the brain-game asks the wrong question. No one ever claimed that the second movement of Beethoven's Symphony No. 3 was 'absolutely mournful.' Rather, the movement contains musical elements that render it identifiably 'mournful' relative to the other music of its time.

For example. Its slow tempo (relative to our resting heartbeats) creates a comparatively more contemplative, less physically active state. The thrumming low strings clearly mimic drum roles, which—along with the slow, steady beat—literally depict a march. The music's setting in minor (C Minor) creates a quantifiably more complex harmonic environment than a like passage in major, a complexity that we have come to identify as being darker in tone than one in major. The relatively low registral compass in which the instruments play generates a heavy, lugubrious sensibility which, when coupled with the slow tempo creates a powerful musical analog for weariness.

The point: the mournful, funeral march-like mood of the second movement of Beethoven's Third Symphony is the cumulative result of many different factors, the artistic whole greater than the sum of its conventional parts. (As observed earlier in this chapter, conventions only become conventions because they play a useful role over a long enough period of time *to become conventional.*)

9 The Inherent Challenge of Describing Music with Words

For the teacher and/or writer of music, the greatest single challenge in talking about musical meaning is finding the right words with which *to talk about music*: how to describe something in words that is, by its nature, beyond the power of words to describe. For the modernists, this was not a problem at all; the fact that musical expression/meaning could not be quantified rendered it unworthy of discussion. But for the rest of us this remains an issue.

"Verbal dialectic is powerless to define musical dialectic in its totality" (Stravinsky, 1970, p. 123). Thus spoke the great twentieth century composer Igor Stravinsky when he delivered the Charles Eliot Norton Lectures at Harvard in 1939–1940. No doubt inspired by the august academic environment around him, Stravinsky said rather pompously something both simple and profound—namely that words cannot fully describe the experience of music.

In this Stravinsky is absolutely correct.

However, we should note that Stravinsky was not always so correct. In his autobiography, written rather prematurely in 1936 (he lived another 35 years, until 1971) Stravinsky wrote this about musical expression and meaning:

> *I consider that music is, by its very nature, essentially powerless to express anything at all, whether a feeling, an attitude of mind, a psychological mood, a phenomenon of nature, etc. Expression has never been an inherent property of music. That is by no means the purpose of its existence. If, as is nearly always the case, music appears to express something, this is only an illusion and not a reality. It is simply an additional attribute which, by tacit and inveterate agreement, we have lent it, thrust upon it, as a label, a convention—in short, an aspect which, unconsciously or by force of habit, we have come to confuse with its essential being.* (1962, pp. 53, 54)

In 1961, having been criticized for that statement for 25 years, Stravinsky offered this update:

> The over-publicized bit about expression (or non-expression) was simply a way of saying that music is supra-personal and super-real and as such beyond verbal meanings and verbal descriptions. It was offhand and annoyingly incomplete, but even the stupider critics could have seen that it did not deny musical expressivity, but only the validity of a type of verbal statement about musical expressivity. I stand by the remark, incidentally, though today I would put it the other way around: music expresses itself. (Stravinsky & Craft, 1962, p. 101)

Whatever. As one of those admittedly 'stupider' critics I will assert that Stravinsky is still wrong, because much—if not most—music expresses something much greater than just the sum of itself.

10 Teaching Meaning in Music

In my humble opinion, the key to teaching meaning in a given piece of music is to find and identify the story line in that piece of music.

Clearly, different sorts of music, created at different times by different people will tell different sorts of stories. Opera, song, choral music and most religious music are by their nature vocal. The stories they tell are contained in their words. For me, this is the easiest music to teach because the essential meaning of such a work is contained in its words. My job is to demonstrate how music distills, deepens and intensifies the meaning of those words and the feelings behind the words.

When it comes to divining meaning and finding a story line, instrumental music poses an entirely different set of challenges.

Instrumental music—as a tradition unto itself—came into existence during the seventeenth century. Yes, instrumental music was composed before the seventeenth century. But almost all such works were arrangements of vocal music or dance music. It wasn't until the seventeenth century, with the standardization of string instruments, tuning practices, and the emergence of the major/minor system of tonality, that composers had the syntactical tools necessary to create music that could mean something *absent of words*.

It is beyond the scope of this chapter to describe the evolution of instrumental music, so the following statements will have to suffice:

'Narrative instrumental music' is instrumental music that embodies the *operatic impulse*. The operatic impulse is the inclination to compose music that tells some sort of a story; music that employs, either explicitly or implicitly, the structural devices and affective states that were originally the province of opera.

Opera (which was invented in Florence around 1600) is the ultimate narrative music; its reason-to-be is to tell a literary story intensified by music. This narrative, *operatic impulse* was adapted to the world of instrumental music during the Classical and Romantic eras, periods spanning roughly the years from 1740 to 1900, during which opera and instrumental music evolved side-by-side in a constantly reinforcing, synergistic loop.

As in any narrative art, in narrative instrumental music different sorts of passages play different expressive roles. Themes (usually, but not always memorable melodies called 'tunes') are the 'characters,' the 'principal players' in a musical narrative. Multiple themes within a single movement create contrast and the potential for conflict. Thematic variations deepen and extend the character of a theme; development (that is, the fragmentation and/or metamorphosis and/or harmonic manipulation of a theme (or themes) creates action sequences and dramatic confrontation; bridges transit between themes; introductions create a musical environment appropriate to the thematic material that follows.

Some musicologists claim that instrumental music cannot be considered narrative. For example, according to Carolyn Abbate (1991), music cannot express past tense and thus cannot be narrative, since narrative—by Abbate's definition—is "a tale told later, by one who escaped to the outside to tell the tale" (p. 53).

Alas, this is an example of a musicologist thinking like a musicologist and not like a composer, with the consequence being that Ms. Abbate is incorrect. Instrumental music reflects on the past all the time. When a composer repeats or varies a previously heard theme, it is quite often in the spirit of backwards reflection. As for her definition of 'narrative,' twentieth and twenty-first century literature clearly demonstrates that tense is not a gauge for determining whether something is or is not narrative.

I am not claiming that musical narrative is a precise analog to literary narrative. But I will claim that the vast majority of composers plan and execute scenarios in their works, and that having an inkling of what that often *narrative* scenario represents is the very first job of the performer, analyst, and teacher.

11 Putting It All Together

A teacher of music must first create a historical case—a context—for the cultural role of music and the sound (the style) of music in a given community. That teacher must then place an individual composer within his historical milieu, seeking, always, to understand his external life, personality, and something of his interior emotional landscape. Next, in teaching a particular piece of music by a particular composer created at a particular time and place, a teacher must illustrate, through selected listening, the content and structure of the music under study: its theme(s) and the compositional processes employed in the piece (repetition, variation, development, etc.). Finally, having accomplished all of the above, a teacher must divine and communicate what in her opinion the piece is intended to *mean*.

The process just described constitutes precisely what a conductor does when preparing a piece of music for performance. A conductor will immerse himself in the life and times of a composer in order to understand, as best as he can, that life and those times. A conductor will then study and analyze the work to be performed, with the final intention of divining what the work *means*. Only then is a conductor prepared to stand before an orchestra, there to rehearse and finally perform the work.

Just as different listeners will have different ideas regarding the meaning of a piece of music, so the members of an orchestra will have a variety of opinions about a work under performance. However, like a conductor, it is my job to decide what a piece of music means and then, like a conductor, project that meaning to my listeners.

A professional orchestra consists of roughly 85 of the most talented, opinionated, and—for better or for worse—potentially hostile people you will find anywhere. The members of such an orchestra will have a wide variety of opinions about any work under performance. For those players to be willing to put aside their own impulses and instead, follow those of a conductor, the players have to *trust* that conductor. While trust can be generated in many ways—through charisma, empathy, and energy—no single element is more important than competence. A conductor must be able to convince the women and men of an orchestra that she knows exactly what she's doing: that she has done her homework, that she has a cogent vision of the meaning of the music under performance and the ability to clearly communicate that vision to the orchestra in rehearsal and performance.

Trust born from this sort of conductorial competence allows the players to buy into the conductor's vision/interpretation of a work whether or not they are in complete agreement with that vision.

So it is with teaching as well. No matter what his discipline, a teacher must confer form upon information by creating a story: a cogent vision of the material under study that resonates with and builds upon his students' own life experiences. Students do not follow their teachers blindly, and the days of rapping their knuckles with a ruler in order to brutalize them into doing so are long passed. Students of every age—and as the father of four children I know this to be a fact—hate having their time wasted by ill-considered teaching and will react poorly when faced with such.

There is a contract, then, between a conductor and an orchestra and a teacher and his students. It is the conductor's/teacher's responsibility to exhibit competence by mastering his material, comprehending its larger meaning and then presenting it in a coherent narrative. Provided this is done, it is the musicians'/students' responsibility to trust that narrative, not blindly, but to the degree that they can subsume it into their particular experience and, in this way, make it their own.

12 What I (Now) Know

We are hardwired to make and listen to music. Far from being merely a form of entertainment, music takes over where words leave off, deepening experience, intensifying meaning, and coordinating group activity. We have evolved to make and listen to music. Generally, but accurately speaking, evolution does not waste its time and energy on idle pursuits.

With the advent of music notation some 1200 years ago, Western music diverged from most other world music cultures and became, over time, a self-expressive art form practiced—increasingly—by literate musicians who notated their music.

This self-expressive impulse has caused the style—the sound—of Western music to change constantly over the last 1000 years, as composers have sought ever-new musical means to reflect the changing world around them. Because the great bulk of this music was notated it has survived to become a repertoire, one that can be studied, taught, re-performed and re-interpreted.

If we are to understand the stylistic content and expressive meaning of a piece of music, we must know something of the time and place that saw its creation and something of the composer—the person—who made it. If we are to understand the story the music tells, we must also know the story behind its creation.

For those proponents of the 'New Criticism' and modernism, who would contend that a piece of art is relevant only to itself—a singular object bereft of metaphoric meaning or cultural context—we must feel no small bit of pity. By denying the interconnectedness of art with life, they are subscribing to a worldview of utmost loneliness. We, however, will not be drawn into such a nihilistic worldview. Instead, we will celebrate the interconnectedness of human activity, and remember always that context is everything and everything has meaning.

Notes

1 See https://en.oxforddictionaries.com/definition/science
2 The reader may be interested in this regard in pianist Imogen Cooper's May 2013 Oxford lecture on *The hidden power of the re-creative process in music*, available at https://www.youtube.com/watch?v=UQoDm_LPqeA
3 Johnny Ray's complete lyrics can be found at http://www.metrolyrics.com/sent-for-you-yesterday-lyrics-johnnie-ray.html. A captivating YouTube rendition is available at https://www.youtube.com/watch?v=MKIVmWIKHOI

References

Abbate, C. (1991). *Unsung voices: Opera and musical narrative in the nineteenth century*. Princeton, NJ: Princeton University Press.

Atlas, A. W. (1998). *Renaissance music*. New York, NY: W. W. Norton.

Grout, D. J., & Palisca, C. V. (1988). *A history of western music* (4th ed.). New York, NY: W. W. Norton.

Hildesheimer, W. (1983): *Mozart*. New York, NY: Vintage Books.

Kerman, J. (1988). *Opera as drama* (New and rev. ed.). Berkeley, CA & Los Angeles, CA: University of California Press. (First published in 1956, New York: Alfred A. Knopf)

Kerman, J. (1994). *Write all these down*. Berkeley, CA & London: University of California Press.

Solomon, M. (1995). *Mozart: A life*. New York, NY: Harper Collins.

Stravinsky, I. (1962). *An autobiography*. New York, NY: W. W. Norton.

Stravinsky, I. (1970). *Poetics of music in the form of six lessons*. Cambridge, MA: Harvard University Press.

Stravinsky, I., & Craft, R. (1962). *Expositions and developments*. Berkeley, CA & Los Angeles, CA: University of California Press.

Strunk, W. O., & Treitler, L. (1998). *Source readings in music history*. New York, NY: W. W. Norton.

Taruskin, R. (2005). *The Oxford history of Western music* (Vol. 5). Oxford & New York, NY: Oxford University Press.

Treitler, L. (2011). *Reflections on musical meaning and its representations*. Bloomington, IN & Indianapolis, IN: Indiana University Press.

Weiss, P., & Taruskin, R. (1984). *Music in the Western worlds: A history in documents*. New York, NY: Schirmer Books.

Wiesel, E. (1966). *The gates of the forest*. New York, NY: Schocken Books.

INTERMEZZO 2

Life Starts Long before Its Beginning

Jan Visser

Reflections inspired by Emily Vargas-Barón: 'From preconception to preschool: The foundation of the scientific mind'

Life is short. Really? Our biological make-up is such that we cannot sustain ourselves for, roughly, more than 120 years. For many of us life is much shorter. We often think of it in terms of multiples of ten. I have 80 years to look back on. Long ago I passed the traditional milestones of 30 and 50 without giving them any particular thought. Those anniversaries were 'just another day in my life.' However, I traversed the threshold into the ninth decade of my life hesitatingly, respectful of its deeper meaning. Should I? Could it be avoided? How do I live with the prospect of an uncertain number of decades, if not just years, left?

I have long been aware how much of my life has taken shape against the backdrop of not only the events I experienced along its course,[1] but also of the life that preceded my biological existence; that of my parents and grandparents, and those that went before them. Only more recently it occurred to me that life doesn't end, either, with its biological demise. There is life after life, and it's not the mythical 'afterlife'—with which we try to console ourselves—in a world that isn't ours. It's here and it's then, after we are gone.

We are part of a continuum that starts, as I see it, at the time, long ago, when our universe came into being. The material world became more and more articulated, and gradually more conscious and reflective of itself, after rudimentary biological lifeforms emerged on the surface of our planet and in the depth of its seas. It may, in a not yet foreseeable future, touch upon intelligent forms of life elsewhere in the universe; it does already interplay with intelligent lifeforms—which we refer to as 'artificial'—that are of our own making.

I met Emily Vargas-Barón for the first time face-to-face some 25 years ago in the course of my work for UNESCO's *Learning Without Frontiers* initiative. We had known each other well before that time because of the interests we share in contributing to a better world. For Emily that meant, in the first place, ensuring that the context in which conscious life comes into being and undergoes its early development are optimally present everywhere in places around the world. For many, such care for Early Child Development (ECD) starts at birth of the new human being. Not in Emily's thinking. It doesn't start at conception

 | DOI: 10.1163/9789004416802_004

either. Her chapter thus focuses not only on infancy and early childhood, it is equally concerned with the stages of life that are commonly referred to as gestation and lactation as well as with the conditions that prevail long before conception. They all matter greatly for the awakening and early growth of the scientific mind in the newborn.

Note

1 I highlight the importance of 'situational awareness' in this regard in a recent contribution to a Festschrift for an 80-year old colleague and friend of mine, available at https://www.researchgate.net/publication/325312929

CHAPTER 3

From Preconception to Preschool: The Foundation of the Scientific Mind

Emily Vargas-Barón

1 Introduction

To create this unique and profound book, Jan Visser and his colleagues spent 17 years exploring ways to build the scientific mind. Synthesizing their results, Jan prepared the following multidimensional definition of the scientific mind:

> The development of the scientific mind is:
> A lifelong process of human development to acquire a way of being in and of the world, inspired by the heritage of the millennia-long history of the human pursuit of knowledge (*scientia*) for the advancement of understanding and wisdom, comprising:
> - habits of thinking and dispositions in approaching the world, as well as
> - values, ethical concerns, aesthetic considerations, and attitudes, alongside
> - mastery of a complex array of skills and mental capabilities in select domains, with such 'select domains' depending on an individual's interests and prospective needs and desires, which are different for a carpenter, a theoretical physicist, or a musician.

In essence, scientific research has demonstrated that each of us develops our unique 'scientific mind' in very different ways based on the health, nutritional and psychosocial status of our parents; our development in the womb; outcomes of our births; the parenting skills of our mother, father, relatives and caregivers; plus the genetic and cultural heritages we receive from our parents and families.

- From birth onward, how do we become who we are?
- What impels us to learn, to explore and to imagine?
- How do we form our values and moral principles, attitudes and behaviors? And might they be—to some extent—innately within us?

© KONINKLIJKE BRILL NV, LEIDEN, 2020 | DOI: 10.1163/9789004416802_005

- How much of 'who we are' is the product of our ancestors, and how much do we develop as a result of our explorations, our home and community environments and life experiences including conflicts and challenges as well as love and support?
- To what extent are our parents and extended family members important in our formation?
- Or are the influences of our schools, external communities, organizations, and cultures ultimately more important than our parents and immediate caregivers?

These questions have perplexed ancient and modern minds alike, from Socrates to Piaget, and we continue to refine our responses as we explore the new and fascinating world of neuroscience. Indeed, some questions may never be answered …

This chapter addresses some of my own early experiences, followed by discussions on preconception and prenatal education and care. If there is one thing we have learned about child development during the past 30 years, it is that we must give far greater priority in all world regions to improving preparation for pregnancy, promoting good fetal development and early parenting skills to achieve better birth outcomes. I then discuss infancy, including early moral decision making, as a prelude for reviewing major barriers to good child development, including those that are still prevalent in many low- and middle-income countries and in some populations of upper-income countries.

After recognizing the dire challenges we still face, I identify and discuss five major priorities for conducting national policy planning to overcome these barriers in order to build an enquiring, scientific mind in children who often live in challenging situations. Finally, I attempt to weave the strands of the discussion into a series of recommendations for parents and caregivers regarding ways to promote each child's scientific mind while also ensuring their full responsibility in society.

∵

From childhood, I was steeped in philosophical writings on science and in many dimensions of the scientific method. This was due in large part to having one grandfather who was a medical researcher, and the other, a philosopher, physicist and electrical engineer—one of the first international development leaders. My family abounds with scientists and educators, from my great-grandfather, who was a minister of education and renowned educational leader, to one grandmother trained in medicine and the other in teacher education and the development of school systems. When approached to write this chapter,

I viewed it as a major challenge but also a way to bring together some of the disparate strands of my ancestors and my life to help, perhaps, some parents to give their children a good start in life and a disposition to scientific thinking.

From my earliest moments my father, Dr. Aníbal Vargas-Barón, a professor of Romance Languages and Literature, encouraged me to discover my world, learn children's songs and poetry, and develop my imagination—my fantasy life. At a very early age, my mother, Ingeborg von Raven, an inveterate reader, taught me to love to read—primarily in English. My scholarly Colombian father taught me to delight in reading children's literature in Spanish and French. As a young child I voraciously read a wide array of books in a headlong rush to discover knowledge by myself. My explorations varied from the fascinating notebooks of Leonardo da Vinci to my grandmother's amazingly advanced primary school textbooks on botany and zoology. I was encouraged to create new worlds in my imagination and to talk about them with my parents. I delighted in discovering nature, from tempting fate by observing and playing around mounds covered with slithering rattlesnakes warming themselves in the sun, to finding and observing fascinating salamanders in mountain streams, and digging clams in Puget Sound.

Perhaps the most delightful activities of my earliest years were the building of 'huge' cities created out of shaped hardwood blocks of all sizes, Lincoln Logs, toy people and animals, cars, and masses of sod, grasses and colorful weeds, which I hauled into our parlor from outside ... until each 'city' would fill the entire room. My parents thoughtfully tiptoed around and over my creations as though they were of singular importance. And I do believe they thought my cities were important, for they were wise and loving parents. Once each elaborate and magnificent city had been fully created, my three to four-year old self would stand back in joy and wonderment at my achievement. And then, and only then, once the city had been fully created, would I regretfully take it apart. The parlor was transformed back into its former self and became an empty pallet for the building of a newer, grander and even more marvelous city ... I must have driven my parents wild! But I do remember that my mother had the slightest of smiles as she wound her way around my creations.

And so this primitive activity of exploration and creativity continued through my early years ... and well beyond. Only now, instead of the cities of my dreams, I conduct global research projects and help countries in all world regions create policies and plans for improving services to help ALL of their young children achieve their full potential.

I believe there is a direct thread from the parenting styles of my mother and father to the life I chose to lead. And for their forbearance and wisdom, and for their nurturing love, they have my eternal gratitude.

I profoundly believe—and now the marvelous field of neuroscience has confirmed—that from the vantage point of the cycle of human life, brain development is so rapid and so all-absorbing during the first three years of life that it truly sets the basic parameters, indeed the neural pathways, for the 'scientific mind' during the foundational period that extends from preconception and gestation to birth, infancy and early childhood.

We now know that from a glossy, apparently undifferentiated and multipurpose mass at birth, the infant's still fetal brain begins rapidly to develop its basic sections and neural pathways, including most surprisingly one for her or his theory of mind and social cognition, a section of the brain called the right temporal-parietal junction (Saxe & Kanwisher, 2003, pp. 1835–1836). Indeed, infants' brains expand by 88% during the first year of life, and each major section is developed to conduct essential functions. At the same time, neural pathways proliferate and connect the sections of the brain to ensure adequate functional development (Deen et al., 2017).

We are only at the beginning of the exploration of the development and capacity of infants' brains. My daughter-in-law, Rebecca Saxe (2016), was the first to use fMRI machines to study infant brains. For a window into her views regarding the scientific method, studies of infant brains, and motherhood, I suggest you watch her TEDx presentation on 'Baby brains: Unlocking our humanity.'

We do not know how to maximize human brain development, although from child development, anthropological and neuroscience research, we have a beginning understanding what may constitute 'good practice' in parenting in various world cultures.

We do know that we must begin with preconception education and care for the future mother and father, and continue with high-quality prenatal education and care. After birth, continuous early childhood development and early childhood intervention services are essential to support parents in their critically important roles of guiding the development of their infants. This work includes ensuring good transitions from nurturing homes to inclusive centers for early childhood development and care as well as inclusive preschools.

We have learned that early childhood is not only 'prologue;' it also establishes the parameters of our abilities for lifelong learning in formal, non-formal and informal educational settings. We know that the parents of children who are very low in birth weight, have fewer than 37 weeks of gestation or have developmental delays, disabilities or another type of fragile status at birth, will immediately require special early support and early childhood intervention services. Indeed, "… the effects of poor neonatal health on adult outcomes are … set very early" (Figlio, Guryan, Karbownik, & Roth, 2013, p. 1). Specifically, low birth weight is found to have long-term negative effects on cognitive development that are very difficult to reverse (ibid.).

During the subsequent three years, infants and toddlers who are identified later to have developmental delays, disabilities or behavioral conditions, such as autism spectrum or attention deficit and hyperactivity disorders, will require early childhood intervention services that are more intensive, individualized and somewhat more costly than general early childhood development services and support.

Since 2000, rapid and very exciting research in the fields of neuroscience, child development, early childhood intervention, learning psychology, education, health, nutrition, hygiene, child rights and protection have been helping us discern effective ways to ensure all children will attain their full developmental potential. Alison Gopnik has written extensively about the infant scientist, focusing especially on their openness to learning and exploring (Gopnik, Meltzoff, & Kuhl, 1999; Gopnik, 2009). Indeed, enquiry is an essential part of developing the scientific mind.

We are now able to use this growing evidence base for improved early childhood policy planning (Vargas-Barón, 2005, 2016). In addition, evaluation research is helping us refine program initiatives that are based on evidence.

> *Considering the available evidence, one critically important dimension of each child's development includes the presence and active engagement of supportive parents and caregivers to unlock and build on the inborn potential of the infant brain that in turn enables the flourishing of the 'scientific mind.' I am convinced that all children are born with a development potential and an interest in enquiry and learning that is far beyond our current understanding.*

2 The Role of the Preconception Period

In the 1980s and 90's, we believed that to improve birth outcomes, such as reducing rates of low birth weight, pre-term birth and birth anomalies, it was important to ensure all future mothers received good prenatal education and care. While this is always important, we now know that essential elements of the human development process begin long before conception. Preconception education and care of prospective parents should begin in adolescence and be reinforced once couples begin to think about conceiving a child (Vargas-Barón, 2009b).

We must focus on the mother and father many years or at least several months before conception. The genetic conditions, environmental situations, daily behaviors, and the physical, emotional and economic status of the mother and father all establish the parameters for gestation, infant status at birth and her or his developmental potential.

Realizing that birth status and the future development of each child are highly dependent on this early phase, nations are investing increasingly in services for preconception education, health care and nutrition.

Specifically, I find that nations are expanding investments to:

- Assess genetic history and risk screening of the mother and father before pregnancy;
- Review the nutritional status and especially micronutrient levels of the mother and father, including levels of folic acid, iron, calcium, vitamins A, B12, D, iodine and other micronutrients that can affect brain development;
- Reduce the weight of obese women because maternal obesity often affects pregnancy and later child growth, development and health;
- Educate parents about the avoidance of certain prescription drugs and substances that are injurious to the fetus;
- Improve the dental status of the mother, including caries that can cause systemic infections affecting the fetus;
- Provide information about risks related to maternal and paternal age; for example, older fathers are correlated with the occurrence of Down syndrome, and pregnant young adolescents have higher levels of pre-term and low birth weight infants than older mothers;
- Reduce levels of stress, anxiety and/or depression in both mothers and fathers because both are highly correlated with the occurrence of depressed infants and young children;
- Ensure the timely provision of essential immunizations for future mothers before pregnancy;
- Avoid environmental toxins that appear to be associated with some cases of autism, developmental delays and allergies;
- Prevent chronic illnesses, such as diabetes, HIV/AIDS, and mosquito-borne diseases, including malaria, Zika and chicungunya, that can cause developmental delays and high levels of child morbidity and mortality;
- Identify high-risk parents with low levels of formal education, income, wealth, social status, and access to essential services that are frequently correlated with higher levels of developmental delays and health, nutritional and educational challenges;
- Expand maternal and paternal access to services for health, nutrition, education and social protection essential to preventing developmental delays, child morbidity and mortality; and

- Provide family and parenting support to help ensure good child development (ACOG, 1994; Atrash, Johnson, Adams, Cordero, & Howse, 2006; Boulet, Parker, & Atrash, 2006; Comerford, Moos, & Curtis, 2006; Lee, 2009; Offenbacher, 1996; PHAC, 2008; Vargas-Barón, 2009b).

No consensus yet exists regarding the extent to which genetics influences child development (Borowski, Ramey, & Power, 2002). However, in some cases genetics is of key importance, such as childhood disability due to a hereditary disease. In other cases, research reveals parent-child relationships are more important than genetics in maximizing infant and child development. Essentially, genetics, epigenetics, the nurturing roles of parents, and early stimulation are exceedingly important in child development, as are family and community environments (Black et al., 2016). At this point, I believe that for the foreseeable future, the relative impacts of genetics, epigenetics, parenting and environmental influences will continue to vary from child to child and from culture to culture.

Good practice includes combining preconception services seamlessly with early and continuous prenatal education and health care and subsequently with neonatal parent education and support. Research has shown that education, nutrition and health care during the preconception period can help future mothers and fathers give birth to healthy and full-term infants (ACOG, 1994; Denny, Floyd, Green, & Hayes, 2012). High-risk and adolescent mothers can especially be helped to improve their birth outcomes. For example, preconception care has been shown to be very effective in reducing the incidence of low birth weight (less than 2,800 grams) and pre-term infants (less than 37 weeks) as well as lowering the rate of the birth of infants with low Apgar scores, developmental delays and disabilities (Atrash et al., 2006). Research has found that, "Above all, health care providers and health agencies must understand that early prenatal care is too late for many women and babies. In particular, some evidence-based interventions recommended for implementation during pregnancy could be more effective and beneficial if implemented before conception" (Atrash et al., 2006, p. 10:S9).

3 Prenatal Education and Care

In health planning, it is largely accepted that early and continuous preconception and prenatal education and care services improve birth outcomes and also greatly reduce costs for health and nutrition care as well as early childhood intervention services for infants with developmental delays, malnutrition and disabilities. However, in countries throughout the world, I find inadequate services for preconception and prenatal education and care. This situation seriously limits fetal development, thereby greatly reducing children's opportunities to develop well.

> *Infants and toddlers with developmental delays and inadequate early stimulation engage less in exploratory activities that are essential for developing the scientific mind.*

I have found that developmental delays due to inadequate stimulation and nurturing care are exceedingly challenging to overcome after the age of thirty-six months. Some but not all low- and middle-income nations still lack the will to invest in young children when their resources for them, including services, professionals, training programs, are few and/or of poor quality. In many countries, some preconception and prenatal education and care services exist but their provision is inequitable in terms of access, usage and quality. Well-to-do families secure essential services through paying for them; however, the poor usually cannot afford them. In addition, subordinate ethnic and linguistic groups are often excluded from essential preconception, prenatal and early parenting support services, thereby widening even more the gaps between groups.

The IT revolution promises to help equalize this situation—to some degree. Cell phone applications for prospective parents, pregnant women, parents, caregivers and professionals are appearing rapidly around the world in many languages. Important learning modules and research results are now reaching isolated and underserved groups as never before. Global, regional and national information networks are using social media including webinars, articles and URL references to inform and educate families and agencies in far-flung corners of the world in a matter of hours or days. Quality control is a concern; however, most of the early childhood information systems I have reviewed are of surprisingly high quality.

Much research has demonstrated the importance of providing monthly prenatal education and health care visits. Many nations follow the minimum guidelines of the World Health Organization for the provision of only four prenatal clinic visits. Although those health care systems usually allow more visits for pregnant women with high-risk conditions, nonetheless high rates of infant mortality, miscarriages, infant mortality, low-birth weight and preterm births in many low- and middle-income countries reveal the general inadequacy of limited prenatal health and nutrition care. In many nations I have found a virtual absence of prenatal education in spite of efforts to provide it.

As noted above, in many countries services for comprehensive preconception and prenatal health care and education have not been fully designed, implemented, monitored and evaluated. However, these services are increasingly being included in the national policies and strategic plans of low- and middle-income countries for early childhood development, health, and nutrition. As a result, improved systems for preconception and prenatal education and care are increasingly being designed and implemented in more of these countries.

There is reason to hope that the major movement for early childhood policy planning and implementation will lead to rapid and measurable improvements in birth outcomes and early childhood development by 2030, if not by 2025.

4 Child Development in the Womb

In addition to the impacts of biological and environmental insults and inadequate services on growing fetuses and fragile newborns, it is important to consider recent neuroscience research regarding *in utero* child development for improving early perceptual, language and cognitive development.

Children are born learning because learning begins in the womb.

Brain development and learning continues at a frenetic pace immediately after birth, and especially during the period from birth to 12 months, and then continues at a slightly slower pace from 12 to 36 months. After the third year, brain growth slows further, and neural pathways are "pruned" and strengthened to fit the child's needs. The early years constitute the phase of most rapid human learning and exploration as synapses proliferate (Deen et al., 2017).

Considerable neuroscience research, clearly illustrates that fetuses begin to learn *well before birth* (Shonkoff & Phillips, 2000). Neuroscientists have discerned that cognitive functions begin about three months before birth, and perceptual and language development begin at or about minus two to three months. Undoubtedly, future research will reveal even earlier learning activities that occur *in utero*.

The health, nutrition and psychological statuses of the mother directly impact the development of the fetus. In addition, each infant is born already knowing his or her mother's voice and the voices of close family members. They hear music, and they already know the cadences and range of sounds of their 'mother tongue'—their own mother's speech.

Infants are born ready to continue learning. Therefore, it is up to parents and caregivers to build upon the rich foundation of learning already established during the gestational period and the early moments of each infant's life. Increasingly, well designed early childhood policies and strategic plans include provisions for continuous services to promote positive parenting and early learning.

5 Infancy and Intergenerational Commitment

Maureen McClure, an eminent professor of international education, asserts that an essential part of early learning is related to parents' innate intergenerational commitment (Verbal communication, 25 July 2017). Indeed, extensive literature has been devoted to early socialization (sociology), enculturation (anthropology), early childhood development (developmental psychology), and brain development through parent-child interaction (neuroscience).

Anthropologists have found that much of early learning is based on an implicit, and often overlooked, intergenerational commitment that is intrinsic to each culture (Herskovits, 1948). Unfortunately, modern urban societies do not always recognize or reinforce the importance of intergenerational commitment to engendering good parenting and child development.

To develop the scientific mind well, purposeful attention must be given to early learning processes that enhance intergenerational commitment.

It is generally accepted that to learn well, children must be healthy, well nourished, stimulated, nurtured and loved by their parents and consistent caregivers. They must be free of trauma as well as a host of negative environmental impacts. Their days should be well structured, and at all times, they should feel happy, secure and appreciated.

When awake, alert and well fed, infants and toddlers are usually ready to engage in rewarding relationships and positive interactions with those who love and care for them. As virtually any observant parent can attest, well cared for infants are not passive learners. They actively seek to interact with their parents and caregivers, and they enjoy exploring their environment as long as they feel secure. They reach out to stimulate others in order to obtain what they want.

Indeed, I find that infants and toddlers develop best when their parents and caregivers also seek to stimulate them in gentle but purposeful ways, such as lightly touching the tips of their fingers, introducing a new activity or engaging in a peek-a-boo game. Parents of virtually all cultures observe their infant's reactions, and then, following their baby's lead, they continue to interact with her or him. Infants and toddlers constantly learn from such activities. To them, all play and exploration is learning and all learning is play. As 'little scientists,' infants discover 'what works,' to achieve expected reactions. They also learn from observing totally unexpected results. From their earliest weeks and months, infants actively seek to understand both 'cause' and 'effect'—many times each day.

Before infants and toddlers develop expressive language, they learn many basic concepts through their experiences and especially through interactions with their parents and others in their midst. Perceptive parents realize that although their infant is capable of understanding a great deal of what is said to him or her, their child develops speech much more slowly. Virtually all of infants' first sounds are their attempts to speak with us; sometimes if we listen carefully we find that they appear to be babbling in complete sentences. Indeed from a syntactic point of view, that is exactly what they are doing, only as yet, we are unable to understand them.

I often hear people say, "Children need to learn how to learn." In actual fact, this is not the case. Rather, infants are born learning; therefore, it is critically important to stimulate and enhance their learning and to avoid impeding it.

6 Moral Decision-Making of Infants and the Scientific Mind

Piaget posited that children's moral decision-making did not begin until around the age of eleven (Piaget, 1932). However, the research of Paul Bloom and Karen Wynn, principle researchers of the Infant Cognition Center at Yale University, has demonstrated that infants begin to make moral decisions from four months of age onward—and perhaps even before, although as yet we are unable to measure it (Wynn & Bloom, 2013; Bloom & Wynn, 2016). They discovered that infants are able to distinguish between good and bad behavior most of the time, and that they vastly prefer good behavior. For their studies, they used moving shapes of different types, sizes and colors as well as several puppets with various behaviors. They varied their tests in multiple ways with many cohorts of infants, and they routinely found similar results.

Putting aside the great epistemological question regarding *how* this could be the case, the implications of their discovery for the development of the scientific mind are profoundly important.

> *Knowing that infants and toddlers are observing and judging our moral decisions and developing their neural pathways regarding moral decision making during infancy, the initial behaviors of parents and caregivers become even more important than had been thought.*

Infants and toddlers are learning from our conduct, and it is well known that as they mature, they display behaviors that mimic what they have observed. If children are reared in loving, virtuous and ethical ways, it may be posited that most of them will build on their initial base of 'good' morality and will become loving, virtuous and ethical. Sadly, all of us have observed how children reared in homes, communities and even cultures filled with intolerance, hatred, abuse and conflict often become so very much like their parents and others around them.

The research of Bloom and Wynn is also very important with respect to developing the scientific mind in all children. Infants' brains are "wired" for exploration, trying out activities, stimulating others, observing others, learning from responses, and building rapidly on what is learned. The challenge is to ensure this open-ended search for knowledge is sustained throughout early childhood and the full lifespan. In our interactions with our human, social, biological and physical environments children are continually making judgments and they should do so based on trustworthy evidence. In making decisions based on such judgments, young children begin to consider the ethical implications of the things they decide to do or not to do.

Of course, the building of the scientific mind is also germane to those youngsters who become professional scientists. They should develop a high level of research ethics and ethical conduct with colleagues. This could begin with each 'young scientist' becoming aware of and sensitive to the moral and ethical dimensions of his or her exploratory activities. Will little Mary blow up the tool shed to try out her homemade gunpowder? Will little Allan warn his friend that a particular experiment might be dangerous? Will Aisha ensure her research will cause no harm to the family dog, the innocent subject of her research? And later, will Mary, Allan and Aisha become fully ethical in their research work?

In addition, ethical behavior is ever more important in citizenship, community service, the workplace, and national leadership. I like to think that parents have the power to create a better society most especially by focusing on *maintaining moral decision-making capacities* that virtually all infants exhibit by four months of age. Our challenge is to reinforce and ensure the sustainability of positive ethical behaviors, attitudes and values in the next generation of children.

7 Negative Impacts on Child Development

My work has been deeply impacted by striking new research conducted during recent years on many barriers to good infant, toddler and young child development.

Challenges listed below have been found to seriously undermine the capacity of millions of young children to explore their environments, learn from others, and develop scientific thinking skills:
- Institutionalization of children from birth to 60 months of age;
- Child neglect, abuse and domestic violence;
- Traumas that cause negative stress including natural disasters, conflicts, and internal and cross-border displacement often caused by factors related to war and climate change;
- Marginalized ethnic, social class or linguistic groups;
- Malnutrition and chronic illnesses; and
- Poverty.

7.1 *Institutionalization and Deinstitutionalization*

Seminal research on the devastating impacts of institutionalization of children from birth to 36 months of age in orphanages and other separate housing arrangements has revealed that those who entered institutions with typical levels of development rapidly became developmentally delayed. Furthermore, children with disabilities who were institutionalized became even more disabled (Gamer, 2011; Nelson, Furtado, Fox, & Zeanah, 2009).

Due to poverty, substance abuse, social stigma of various types, the cost of caring for children with disabilities, violence and wars, religious traditions, social pressures, and other reasons, some parents in all world regions decide to place their infants and young children in institutions. These parents usually believe their children will receive better care, food, clothing and education in or through these institutions. Sometimes doctors, social workers, and managers of orphanages encourage parents and single mothers to give their children away. Around the world, these 'social orphans' number in the millions. In addition to social orphans, infants and children orphaned by HIV/AIDS, parental death, war and domestic violence are often placed in institutions.

Indeed, most countries have traditions promoting the creation of orphanages, and charitable giving is often focused on supporting these institutions, many of which are sponsored by faith-based organizations. Only in recent years have we come to understand the highly detrimental impacts orphanages and similar institutions usually have on early child development.

UNICEF has established a worldwide policy of promoting family placement for these children because all of them require the loving care of nurturing families. Increasingly, systems are being developed in most countries to:
- offer family therapy and family preservation services, plus cash support;
- arrange for substitute families or adoption, plus cash support, whenever needed; and

- provide early childhood intervention services to children's original, substitute or adoptive families.

These prevention services are essential to ensuring at-risk children will not be placed in institutions and will not become developmentally delayed, disabled or deeply scarred in their social and emotional development.

In addition, as young children are deinstitutionalized, social workers and child development specialists have found that virtually all of them have become developmentally delayed, disabled or atypical in their behaviors.

To enable successful family placements, nations are establishing national systems for early childhood intervention (ECI). These services include comprehensive child and family assessments and the development and implementation of individualized family service plans with new parents, legal guardians or caregivers. These plans guide the provision of structured and frequent visits in the natural environments of the child, such as the home or sometimes in day care centers, parks or community centers. Support is also given for children's transitions from ECI services to inclusive preschool and primary school education. If deinstitutionalized children receive ECI services before 36 months of age, many of them will overcome their delays. With continuous support, many formerly institutionalized children will be able to lead happy and productive lives.

7.2 *Child Neglect, Abuse and Domestic Violence*

With respect to child development, research has revealed that child maltreatment, including both abuse and neglect, alters brain development and is associated with changes in neuroendocrine and immune functions (Jaffee & Christian, 2014). Both abuse and neglect cause a child to withdraw and not cope well with his or her environment. Child abuse (physical, sexual, mental and emotional) and the observation of domestic violence are very damaging for a child. He or she ceases to explore and slowly but surely gives up on parents and caregivers as sources of love, support, stimulation and security.

> *Child abuse and neglect defeat the scientific mind whereas positive stimulation, encouragement, exploratory activities, and praise promote scientific thinking.*

New generations of parents—and especially those living in poverty in urban communities who are often also dealing with the stresses of acculturation from one culture or way of life to another—often need to learn how to positively discipline a child, structure a child's day, and ensure the child receives adequate attention, stimulation and nurturing care.

Virtually all positive parenting behaviors and skills can be easily learned. However, in spite of our many efforts, parent education and support is still inadequate in most nations. This is especially the case for parents of children from 0 to 36 months of age, just when children are most susceptible to child neglect, abuse and domestic violence.

7.3 *Traumas Causing Negative Stress*

In our increasingly complex and conflict-filled world that is already heavily impacted by the deleterious effects of climate change, many young children experience negative stress; often referred to as 'toxic stress.' Such challenges are highly detrimental to achieving balanced child development. Families with young children affected by war, community violence, natural disasters, climate change, refugee status, internal displacement, extreme poverty and domestic violence are virtually always impacted by highly negative stresses. In infants and young children, negative stressors have exceedingly deleterious impacts on brain development, releasing toxins that inhibit brain growth and development in addition to creating epigenetic changes that can be passed on to succeeding generations (American Academy of Pediatrics, Policy Statement, 2012; Lupien, 2009; Wagner et al., 2016).

Although psychosocial interventions are usually helpful at the time of maximum stress, young children often require continuous psychosocial support for many years because they tend to cycle back frequently through traumas they have experienced. Without continuous services, some of these children become increasingly withdrawn, fearful and even aggressive. Some become mentally ill.

Traumas causing negative stress must be identified and treated quickly. Left untreated or inadequately treated, stress levels not only inhibit brain growth and development but also preclude exploratory activities that help children develop skills that are essential to the scientific mind.

7.4 *The Impact of Poverty on the Growing Brain*

I consider severe poverty to be one of the most negative conditions for the development of children. Over a 45-year period, considerable research has revealed that children living in families and communities with high levels of poverty do not attain the levels of child development and educational achievement that other children, born into more favorable circumstances, attain. Children living in poverty tend to have more developmental delays and chronic malnutrition and illnesses (Black et al., 2016).

However, exceptions always exist, even in the most impoverished communities. For example, Marian Zeitlin conducted positive deviance studies in Senegal where she identified families living in poverty where some children were able to develop well in contrast to most others from the same socio-economic background. She then identified the activities these families conducted with their children, and considering their apparent beneficial impact, helped disseminate those activities throughout their communities (Zeitlin, 1996, 2003).

Although considerable research has been conducted on the negative impacts of poverty on parenting skills and child development, less research has been devoted to showing that even without improving the economic status of families, services for parenting skills and support can markedly improve child development. It may be that because most policy planners and national leaders want to lift families out of poverty, this approach has not been given sufficient attention.

Taking a poverty focus, the World Bank and other agencies have vigorously promoted cash transfers as a way to improve family standards of living and child development (García, 2012). However, research on cash transfers and conditional cash transfers has revealed that their provision rarely results in markedly improved levels of child development. For example, in Colombia, the cash transfer program, *Familias en Acción,* found that children's development was not improved by means of providing a cash transfer alone (Vargas-Barón, 2009a). Subsequently, program leaders added parenting education and support along with early child development services as conditions for cash transfers. According to verbal reports from national leaders in early childhood development (ECD), children's development did improve. However, considerably more research is needed on this topic in Colombia and elsewhere.

7.5 *Children from Marginalized Groups*

In all world regions, we find that subordinate ethnic, class and linguistic groups are often excluded from basic health, nutrition and educational services of dominant social groups or they are rendered inaccessible due to notable language and cultural differences. For example, Roma, (Traveler or Gitano) groups often live among other cultural groups in the Americas, Europe and Levant; however, their children tend to be excluded from basic services. When these children are identified as 'social problems,' they are referred to expensive protective services whose personnel often lack the requisite skills to support them in culturally and linguistically appropriate ways.

In most countries, some (but not all) children of marginalized groups exhibit higher rates of developmental delays and at-risk situations. This is usually related mainly to poverty but may also occur due to a lack of governmental support for the provision of early learning and parenting services in mother tongues. Later on, once they gain their first concepts in their home languages, they can more easily add one or several languages as languages of instruction (Ball, 2010).

> *Research has demonstrated that mother tongue-based early education and development services are essential for young children to learn well.*

In addition, many of these cultures possess rich child rearing traditions that, if reinforced and not interrupted, would give children a good start in life. Interestingly, many of the informal or non-formal teaching traditions of such cultures possess a wide repertoire of effective teaching styles, including exploratory learning, skills development, methods for making mathematical estimates, creative problem solving, and experimentation. When identified and reinforced, these locally derived and effective teaching/learning techniques help enrich generic early learning curricula based on universal aspects of child development (Verbal communication of Aye Aye Yee on the adaptation, translation and use of kindergarten materials from the Myanmar language to over 65 other languages, January, 2017).

8 Overcoming Barriers to Developing the Scientific Mind in Vulnerable Children

Abundant research has shown that vulnerable young children living in impoverished communities usually benefit the most from early childhood development services (UNESCO, 2007). To serve these children and their families effectively, attention must be given to overcoming many barriers to good child development, promoting scientific thinking, and adopting effective teaching and learning strategies used successfully in cultures around the world.

In addition to attaining general indicators on child development that have been proposed globally and regionally for achieving Sustainable Development Goals (SDG) for early childhood development, nations should also establish specific SDG indicators that will help monitor advances in overcoming barriers to good child development.

I strongly recommend national policy planners identify and give high priority to their nation's most vulnerable young children who need comprehensive, intensive and individualized services. To overcome a myriad of barriers to good child development, I propose the following five priorities for national early childhood planning to promote the development of the 'scientific mind' in young children, and most especially, in vulnerable infants and toddlers who live in poverty and other challenging situations.

Five planning priorities for services targeting young children living in vulnerable situations

I. Develop a multisectoral 'continuum of care' for vulnerable children.
II. Provide purposeful parent education and support.
III. Offer both home-based and centre-based programs of good quality for infants and toddlers.
IV. Provide universal developmental screening and early childhood intervention services.
V. Ensure access for all children to high-quality inclusive preschool education.

I. Develop a multisectoral 'continuum of care' for vulnerable children

Every nation should offer a comprehensive 'continuum of care' for all children, with priority given to vulnerable children and their families.

Every nation requires an array of continuous services for child health, nutrition, hygiene, education, protection and rights especially aimed at children living in vulnerable situations. Actual services will vary from country to country depending upon the needs and status of their children and families as well as their policy contexts and resources in terms of services, workforce, training and financial capacity.

Services will range on a *lifecycle continuum* from preconception and prenatal education and care to transition to inclusive primary school. They will include services appropriate for parents whose children are developing in a typical manner albeit in challenging situations to targeted interventions for parents and children who require additional individualized and intensive support in order to make progress and achieve their full potential.

All early childhood development services in the continuum of care should be well coordinated and linked, both horizontally and vertically, in order to provide 'seamless support' for all children and their families.

II. Provide purposeful parent education and support

Parents who learn how to parent effectively during their child's infancy not only maximize the child's development but also usually continue to parent well as their child matures.

Parenting education should begin during preconception and prenatal education services. It should continue after delivery during the brief neonatal phase when parents are very open to learning. It should be emphasized during the period from birth to 36 months and continue during and after transitions to inclusive preschool and primary school. From the first days onward, parenting and family support services should be adapted to meet family needs and the developmental status of each child.

A wide array of parent education and support services should exist in all communities, from general parenting groups, classes, and peer groups to more focused early childhood intervention and child protection services for families and children with greater needs.

Some early childhood specialists used to be fearful of advocating 'parenting education,' saying that it might imply that parents were inadequate, invade their privacy and demean their all important role in child development.

We are now aware that parent education is essential for all parents.

I am convinced that all parents need to be supported in our challenging new world. Even the most advanced child development specialists constantly seek to improve their understanding of child growth and development, informed by new neuroscience and other research. We all need to read avidly in this growing literature simply to keep abreast of new and promising discoveries.

At the same time, in our age, thanks especially to the Convention on the Rights of the Child and the Convention on the Rights of Persons with Disabilities as well as to national Children's Laws or Codes, parents' rights have come to the fore. Today's parents in all cultures cannot depend solely on the use of traditional parenting methods and practices. They must learn more about scientific findings regarding infant and child development and also be empowered to guide their children's development in creative and positive ways.

At some point, virtually all parents 'throw away the books' and advice of others. They observe their children, and do their very best to respond and stimulate them in ways they believe are appropriate to the individual characteristics, personalities and temperaments of their children. Indeed, parents of two or more children always vary their parenting styles for each child because they become acutely aware of individual differences.

Although child development sequences are generally universal, cultures differ greatly in the way they promote or reject certain parenting skills, leading to different timings in the appearance of specific skills or competencies.

In addition, some traditional skills become dysfunctional in modern society while others stand the test of time.

At this time, within virtually every culture, each inter-generational family exhibits a small cycle of changing enculturation processes. Parents and grandparents may disagree with one another, and slowly some new consensuses are reached, culture-by-culture, regarding parenting approaches, within evolving frameworks of evidence from the fields of neuroscience, child development, anthropology, and child health, nutrition and protection. These changes are affecting how parents introduce many topics, including languages, mathematics and scientific concepts to their children. The complexity of the many challenges facing parents of all cultures during this historical period cannot be overstated.

It is important to stress that parents have the ultimate right and responsibility for child rearing. They have the right to make all-important decisions for their children regarding health, nutrition and development *unless* they commit acts that are harmful for their children. Under specific national laws for child protection, when warranted, the state will intervene to protect and improve children's lives.

Also, every family unit, be it formed by a single parent, a couple or an extended family household, requires strong family and community support. In highly mobile societies, family support groups, peer groups, community agencies, playgrounds and other community/neighborhood activities are becoming increasingly essential to promoting family wellbeing and good child development.

III. Offer both home-based and centre-based programs of good quality for infants and toddlers

> Both home-based programs for infants and toddlers as well as centre-based programs should be prioritized in all nations to improve child development and reduce developmental delays.

Abundant research has shown that home visits that provide not only parent education but also training in conducting stimulating child development activities help to improve child development and overcome incipient developmental delays. By uniting community health, nutrition, child protection, and infant care home visits, parents learn how to appropriately nurture and stimulate their infants and toddlers during the period of the brain's most rapid growth (Olds et al., 2010, 2013; and other related publications).

Increasingly, ministries of education, health and protection in countries such as Bosnia and Herzegovina, Colombia, Georgia, Lesotho, Myanmar,

Rwanda, and Tunisia, are recognizing the importance of developing, coordinating and even integrating their multisectoral home-based services to achieve improved child development and prevent developmental delays.

In addition, evolving needs for child care are creating a major demand for improving child care services both in homes and in centers. Mainly because of dual parental employment from approximately six weeks after birth onward, many infants and toddlers are now being placed in the care of relatives, in centre-based child care and development programs, or in paid home-based child care. These services are variously referred to as nurseries, day care centers, child care centers, crèches or family child care homes. Such services are often most needed by the children of marginalized, impoverished and otherwise at-risk families where mothers must work to feed their children. However, the quality of such care is often abysmal.

All paid home- and center-based early child care and development services should be well regulated to ensure quality and accountability. In addition, they should be complemented by comprehensive parent education services. However, if parent education is offered at all in some low- and middle-income countries, it is often provided as a stand-alone service.

Although infant and child care centers have been established by non-governmental, faith-based and community-based organizations in virtually all countries, most of them are private and costly. Consequently, they often mainly serve social and economic elites. Very impoverished mothers manage other private child care centers for the poor. They tend to be exceedingly low in quality and serve mothers who work outside of the home.

Around the world, ministries of education and ministries of social welfare or social protection have learned about the importance of ensuring good child development from birth to 36 months of age. As a result, they are beginning to develop publicly-financed infant and toddler care and development programs, especially for low-income families. Some of these programs are of good quality; however, I have observed that few of them are well coordinated with parent education and support services.

With the rising employment of young mothers, the early development of infants and toddlers is being placed in the hands of inadequate child care centers and family day care homes. Most child caregivers are not professionally trained, and consequently, their quality varies and is usually very poor. In addition, these caregivers are almost always underpaid and unsupervised, and they are rarely monitored or evaluated. This situation is impeding the improvement of child development in countries throughout the world, and most especially in low- and middle-income countries.

Essentially, the future of most of the world's infants and toddlers who are not cared for by their parents during major parts of the day or night is in the hands of untrained persons who often prop infants up in front of a television set or—worse—ignore them altogether. Many, who were born with all expected abilities, rapidly become malnourished and developmentally delayed. This situation is very injurious to the development of children to say nothing of the development of their mental, physical, social and emotional abilities, such as those identified in this book as essential dimensions of the scientific mind. It goes almost without saying that this situation must be reversed as quickly as possible.

IV. Provide universal developmental screening and early childhood intervention services

Ensure that all infants and young children receive universal and regular developmental screening and that national systems of early childhood intervention are established in all nations.

The 2016 Lancet ECD series stated: "New estimates, based on proxy measures of stunting and poverty, indicate that 250 million children (43%) younger than 5 years in low-income and middle-income countries are at risk of not reaching their developmental potential" (Black et al., 2016, p. 1). This exceedingly high rate constitutes a global crisis and reveals a major gap in inclusive services for parents whose children live in at-risk situations, have developmental delays, disabilities or behavioral conditions.

The incidence of developmental delays and disabilities varies greatly from country to country. Overall, their rates are far higher than generally believed. For example, the American Academy of Pediatrics reports that by 24 months of age, at least 13% of U.S. children have developmental delays (Rosenberg, Zhang, & Robinson, 2008). This high level of early developmental delay has been shown to result in later developmental disabilities in over 15% of U.S. children from 3 to 17 years of age. In addition, by age of 9 to 17 years, 21% of all U.S. children are diagnosed with a mental health condition. Overall, at least 25% of U.S. children develop life-long conditions. These children require intensive and individualized services for early childhood intervention (ECI)—services that can prevent and reverse at least 70% of these delays, disabilities and conditions. Yet only 2.8% of children from birth to 36 months currently receive ECI services in the U.S. (Rosenberg, Robinson, Shaw, & Ellison, 2013).

Other nations, and especially low- and middle-income countries, have even higher rates of developmental delays and disabilities than the U.S. due principally to poor birth outcomes, inadequate health care and nutrition services, poverty, war, malnutrition, chronic illnesses, domestic violence, and other situations. However, few nations have conducted rigorous studies on the major needs and developmental levels of their children, mainly because, up until recently, they did not realize how important it is to identify vulnerable young children and reverse their delays very early in life—from birth to 36 months of age.

Well-trained and experienced ECI specialists know how to develop highly effective services. However, in each country, early childhood policy leaders must put in place accessible and effective systems of developmental screening to identify their fragile infants and young children. They must also develop nationwide ECI services that have a legal basis, program guidelines and procedures, service and personnel standards, and other essential elements.

From birth to three years of age, developmental delays should be quickly identified, prevented and reversed. If at-risk children do not receive timely ECI services, their delays usually worsen, and for most children they will become permanent. These children later perform poorly in school, repeat grades, drop out of school, and lead unproductive lives, causing high costs to society.

Every country should develop universal and regular developmental screening services—using a validated instrument—for identifying developmental delays, disabilities, behavioral problems and mental health conditions in infants and young children from birth to three, and up to five years. Children identified to have special needs should be rapidly referred to ECI services. Effective service coordination is essential to ensure all children and families referred to ECI services actually apply to enter those services where they will receive comprehensive assessments to address all areas of child development and behavior as well as family risk factors, such as parental depression or anxiety, domestic violence, parental substance abuse, poverty, etc.

Creating National ECI Systems

A nationwide and equitable policy or strategic plan plus ECI program guidelines and procedures are necessary to ensure all eligible children and families will be identified and invited to enroll in ECI services. Once enrolled, parents become members of a transdisciplinary team to prepare an Individualized Family Service Plan wherein parents establish developmental goals for their children and list the services they want to receive. Subsequently, a regular schedule of visits is provided in the natural environment of the child, usually in the home and sometimes an inclusive child care centre or preschool. ECI programs also assist parents and children to transition to inclusive preschools or primary schools. Many children simply exit ECI services once they have attained typical levels of development.

To ensure national ECI services are effective, they should be fully accountable. A national system for child tracking, monitoring and evaluation is needed to assess rates of early identification, referrals and enrolment in ECI and other community services, developmental-behavioral assessments, and child outcomes, including kindergarten readiness. Effective child tracking systems help nations adjust their services and ensure their policies are effective.

National ECI systems help ensure vulnerable children improve in their development and help to ensure they will be able to develop their scientific thinking abilities and skills, including those relevant to the building of the scientific mind, to the greatest extent of their inborn potential.

Put simply: Without national developmental screening and ECI services, from 15% to 45% of children, especially in low- and middle-income countries, will be unable to effectively build their 'scientific mind' and other faculties essential to leading a fulfilled life and contributing to the well being and productivity of their country.

V. Ensure access for all children to high-quality inclusive preschool education

Inclusive and high-quality preschool education is beneficial for all children—not just children with delays, disabilities or behavioral or mental health conditions.

Because in many industrialized and agricultural economies, often both parents must work outside of the home, preschool education services are increasingly beginning by 24 to 36 months of age, and larger proportions of younger children are now attending preschools.

As children become preschoolers, they continue to learn through constant exploration, interaction with each other, and engaging in stimulating planned and spontaneously developed activities. They may also become obstreperous, try to control others, and not share their toys as they become increasingly single-minded and self-centered. This is the period when they need gentle and loving guidance to play well with others, learn to share, and demonstrate empathy for and help others.

Children's innate capacity to explore and learn through interactions and formal and informal play is preserved if parents and preschools cultivate those activities and explorations. However, at very early stages, certain situations can inhibit children's development, including harsh parenting practices, restrictions on exploratory learning, and forcing young children to learn through rote drill and practice. Preschool education is a time for developing positive interpersonal relationships and promoting education for peace, conflict resolution skills, and above all, empathy.

The most important capacity for ensuring learning during the school years is the development of good self-regulation skills in preschool (McClelland et al., 2007). This is a critically important time for parents, caregivers and preschool teachers to build on each child's capacity for observation and moral decision-making.

Preschool education must be fully inclusive through serving children at all levels of abilities and disabilities. By learning to communicate well with children of all abilities and by forming strong friendships, all children make major progress in social and emotional development. They become more empathetic and more self-assured. Children who used to be excluded from preschools due to having different abilities, belonging to marginalized ethnic groups, speaking different languages, or having another status (such as refugees, internally displaced persons or those of separate castes) benefit greatly from inclusive education.

Similarly, according to the research of the eminent Swedish professor of child development and inclusive education, Ulf Janson, children with 'typical abilities' benefit just as much and sometimes even more than children with disabilities from inclusive education (Janson, 2001, 2007). He observed that with the support of their parents, caregivers and preschool teachers, children with typical levels of development become more empathetic with respect to other children possessing varying levels of abilities. They learn to identify and appreciate the abilities of children with manifest disabilities. They make new friends and support their friends when they meet new people. They gain new and improved communication skills. These benefits to children with typical abilities should be better known, especially in traditional cultures, preschools and schools because this should help to promote a more inclusive approach and overcome the exclusion of children with delays and disabilities from preschools, schools and society in general.

It is very important to remember that each child should have good transition experiences from home to inclusive preschool, and then to inclusive kindergarten and/or inclusive primary school. Parents and children should get to know school directors and teachers, and they in turn should visit the child's home and family. If the child has been enrolled in ECI services, then the strengths, capacities and emerging abilities of the child should be explained to his or her new teacher and resource persons with the goal of ensuring a seamless transition to happy learning and play for all children in inclusive primary school.

9 Promoting the Child's Scientific Mind While Ensuring Full Responsibility in Society

Through all stages for the development of exploratory learning, it is essential—especially in the world of information technology—that parents, caregivers and teachers focus on the primary goal of developing the scientific mind while also ensuring balanced child development through promoting good moral decision-making, language skills, physical growth, fine motor abilities in art and music, and good self-regulation skills combined with sound social and emotional development.

To the extent possible, parent education services, pre- and in-service training programs, service and personnel standards, and supervisory systems should be designed with the above core goal of balanced child development in mind.

Above all, future parents must prepare well before conceiving a child and secure the best prenatal education and care services available in their communities. They should remember that all dimensions for developing the scientific mind are already fully present in infancy and early childhood and their role is to give their child loving and nurturing care combined with appropriate stimulation and learning activities through play.

As noted, infants and young children possess early learning skills that enable them to develop much more rapidly than had previously been thought. It is not that we want them to excel and achieve more in terms of cognitive development. Rather, we should want them to fulfill their true potential in all developmental domains: physical, perceptual, language, social and emotional as well as cognitive.

As parents, grandparents, extended family members and close family friends, it is up to each of us to promote well-balanced child development as we help to lay the child's foundation for lifelong enquiry and learning.

And you might ask, *"What can I do with my baby or young child to help her or him develop a capable and well-rounded 'scientific mind'?"*

Frankly, this is where research meets art, love enables nurturing interventions for good child development, and the miracle of life reveals new possibilities. I am the fortunate grandmother of two wonderful young adults and two little ones, who are three and four years of age and love to learn. I have observed my children and their spouses as they nurtured and stimulated their young children.

The following suggestions are a composite of research findings, personal experiences, and observations of evolving parenting skills in my family and in many cultures around the world.

Give your children your time and love

– You do NOT need a computer, television or even fancy and expensive toys for your child to build her or his scientific mind. Every day, you DO need to spend as much special time as possible interacting with your baby or young child.

Stimulate your baby before birth

– Because learning begins before birth, you can help develop the language, cognitive and perceptual skills of your baby before birth by talking to him or her and playing beautiful music. Of course, being happy, eating a good balanced diet, and going to regular prenatal education and care services will also help greatly!

Build on your baby's inborn capacities

– After birth, your principle role is to release, stimulate, build on and above all not impede her or his inborn capacities.

Talk and read often to your baby and young child

– Promote language learning through talking often during each day with your baby, reading books to baby from the first days onward, encouraging babbling and early speech, and explaining what you are doing and why you are doing it—because baby understands much, much more than you realize.

Have fun with numbers

– Develop your child's numeracy skills through counting items around the house, identifying and sorting shapes, and playing with homemade jig saw puzzles, shapes and colors.

Promote enquiry-based learning

– Ask your child daily questions beginning with: Why? How? What? Which? Who? Where? When? For example: Which pants do you want to wear? The blue one or the white one?

Practice non-closure

– When you notice your baby or young child begins to lose interest in an activity—that is to say when her or his eyes begin to wander—gently end the activity, and then engage in your child's next interest. If you do this then he or she will usually enjoy returning to the original activity at a later time, perhaps on another day.

Use your listening skills

– Always try to listen attentively to your child, looking her or him in the eye and responding to comments or questions.

Avoid criticism and praise exemplary creativity or efforts

– Support your child. Never criticize attempts to do something. All attempts are good. However, avoid praising all the time. Rather, praise really good efforts as well as major successes.

Answer your child's questions

– Always answer your child's questions, and try to do so in a way that is at their level of understanding. Don't be afraid to tell the child when you do not know the answer, making it known that you too have an exploratory mind and that the two of you together may be able to find the answer in a book. (Later on, you can search together for answers online.)

Ask leading questions

– Ask your child questions about—and express your interest in—what she or he thinks, how things work, why he or she is doing certain things, and what she or he is drawing. Ask, What are the parts of a flower he or she likes? And then help him or her to find out.

Model exploratory behaviors

– Model exploratory behaviors with your child by saying, "Let's look this up in a book." "Let's go to the library to find a book about dinosaurs." "Let's look that word up in the dictionary." "Let's go outside, find leaves to figure out what types of trees we have in our neighborhood." Let's go talk with our friend the scientist (the pharmacist, the farmer) and ask her or him for an answer to your question." Later on, you will want to begin online searches with your child.

Make good use of playgrounds

– Should you lack a nearby playground for toddlers and preschoolers, be sure to advocate for and help to build a safe and stimulating play structure because all children benefit greatly from playgrounds. In the playground, accompany your child and talk with him or her at all times to encourage the exploration of new spaces, the development of physical strength, good balance and new skills. Help your child to learn how to take calculated risks and safely climb and descend different types of structures.

Follow your child's interests and share appropriate resource books

– Share and read together some adult-level books about interesting topics, such as animals, fish, airplanes and cars. Ensure your child has at least five books, perhaps homemade, and take him or her frequently to a library (or to someone who has books to share) to select and read books. If children select their own books, they will want you to read to them, and soon they will want to read themselves. Give children opportunities to read about the lives of scientists and others who interest them.

Do learning adventures together

– Help your child create learning adventures by asking what he or she wants to do, encouraging exploration, and observing the results of his or her initiatives. From this beginning, soon your child will begin to plan learning adventures by themselves. Parents are often invited to join in!

Do beginning-level scientific experiments

– Encourage your child to guess what will work and what will not work, and then support explorations to discover the best way to do things or to find out what does not work. Except for when the child's safety or health is at risk, try not to impede the child's natural exploration to find things out through experimentation or careful observation.

Create order and identify patterns

– Ask your child to make order out of apparent chaos and identify patterns, forms, sequences, colors, textures, and sizes of objects.

Build with blocks and shapes

– Play with blocks and shapes to create buildings, cities and other places that interest your child and sit back to enjoy and admire her or his creativity!

Begin to help your child state objectives for what she or he wants to find out

– Encourage your child to build habits of enquiry by stating his or her hypotheses or objectives, and then—when and if you are asked—help your child achieve her or his stated objectives.

Stimulate your child's fantasy world—and creativity!

– Above all promote your child's fantasy world by playing with puppets, creating theatres, playing dress up, and talking with dolls, animals or childhood figurines. This will help your child think about alternatives and will promote that essential quality of the scientific mind: Creativity.

Encourage other parents to build their children's scientific mind

- Consider doing some of these activities in front of other parents with their children. Without telling them what to do, they will begin doing the same types of activities. By modeling good parenting skills, you will encourage them to become more involved with their children's good development. Ultimately, you will contribute to helping all children in your community to achieve their full potential.

You can do all of these activities and many more during your usual daily care giving routines, with your infant or young child, such as dressing, changing, feeding, playing, bathing and preparing for bedtime. If you do them conscientiously, you will truly help your child develop her or his scientific mind.

I conducted these activities with my own children, and I am so proud of their creativity, scientific achievements, and wonderful contributions to our world.

And quietly, I think my parents would be pleased to know that I remember their devotion to my brother Robert and me, and that the example of their wise teachings, guidance and support continue on through their grandchildren and great grandchildren …

References

American Academy of Pediatrics. (2012). Policy statement: Early childhood adversity, toxic stress, and the role of the pediatrician: Translating developmental science into lifelong health. *Pediatrics, 129*, e224–e231.

American College of Obstetricians and Gynecologists. (1994). *Exercise during pregnancy and the postpartum period* (ACOG Technical Bulletin, No. 189). Washington, DC: ACOG.

Atrash, H. K., Johnson, K., Adams, M., Cordero, J. F., & Howse, J. (2006). Preconception care for improving perinatal outcomes: The time to act. *Maternal Child Health Journal, 10*, S3–S11.

Ball, J. (2010). *Enhancing learning of children from diverse language backgrounds: Mother tongue-based bilingual or multilingual education in the early years: Literature review*. Paris: UNESCO.

Black, M. M., Walker, S. P., Fernald, L. C. H., Andersen, C. T., DiGirolamo, A. M., Lu, C., … Grantham-McGregor, S. (2016). Advancing early childhood development: From science to scale: Early childhood development coming of age: Science through the life course. *Lancet.* http://dx.doi.org/10.1016/S0140-6736(16)31389-7

Bloom, P., & Wynn, K. (2016). What develops in moral development. In D. Barner & A. S. Baron (Eds.), *Cognitive development and conceptual change* (pp. 347–364). Oxford: Oxford University Press.

Borowski, J. G., Ramey, S. L., & Power, M. B. (Ed.). (2002, 2012). *Parenting and the child's world: Influences on academic, intellectual, and social-emotional development* (Monographs in Parenting Series). Mahwah, NJ: Lawrence Erlbaum Associates, Inc.

Boulet, S., Parker, C., & Atrash, H. K. (2006). Preconception care in international settings. *Maternal Child Health Journal, 10*, S29–S35.

Boyle, C. A., Boulet, S., Schieve, L. A., Cohen, R. A., Blumberg, S. J., Yeargin-Allsopp, ... Kogan, M. D. (2011). Trends in the prevalence of developmental disabilities in US children, 1997–2008. *Pediatrics, 127*(6). Retrieved from http://pediatrics.aappublications.org/content/127/6/1034

Comerford, F. M., Moos, M. K., & Curtis, M. (2006). The history of preconception care: Evolving guidelines and standards. *Maternal Child Health Journal, 10*, S43–S52.

Deen, B., Richardson, H., Dilks, D. D., Takahashi, A., Kiel, B., Wald, L. L., ... Saxe, R. (2007). Organization of high-level visual cortex in human infants. *Nature Communications, 8*, Article No. 13995. doi:10.1038/ncomms13995

Denny, C. H., Floyd, R. L., Green, P. P., & Hayes, D. K. (2012). Racial and ethnic disparities in preconception risk factors and preconception care. *Journal of Women's Health, 21*(7), 720–729. doi:10.1089/jwh.2011.3259

Figlio, D. N., Guryan, J., Karbownik, K., & Roth, J. (2013). *The effects of poor neonatal health on children's cognitive development* (NBER Working Paper No. 18846, JEL No. I14, I20, I24). National Bureau of Economic Research. Retrieved from http://www.nber.org/papers/w18846

Gamer, G. N. (2011). Earlier is better: Children thrive best in families. In *Early childhood development: What parliamentarians need to know and do*. Geneva: UNICEF.

García, M., & Moore, C. M. T. (2012). *The cash dividend: The rise of cash transfer programs in Sub-Saharan Africa*. Washington, DC: The World Bank.

Gopnik, A. (2009). *The philosophical baby: What children's minds tell us about truth, love, and the meaning of life.* New York, NY: Farrar, Straus and Giroux.

Gopnik, A., Meltzoff, A. N., & Kuhl, P. (1999). *The scientist in the crib: What early learning tells us about the mind.* New York, NY: William Morrow & Company.

Hair, N. L., Hanson, J. L., Wolfe, B. L., & Pollak, S. D. (2015). Association of child poverty, brain development, and academic achievement. *JAMA Pediatrics*. doi:10.1001/jamapediatrics.2015.1475

Herskovits, M. J. (1948). *Man and his works: the science of cultural anthropology.* New York, NY: Knopf.

Hueston, W. J., Quattlebaum, R. J., & Bentch, J. L. (2008). How much money can early prenatal care for teen pregnancies save: A cost-benefit analysis. *Journal of the American Board of Family Medicine, 21*(3), 184–190.

Jaffee, S. R., & Christian, C. W. (2014). The biological embedding of child abuse and neglect, implications for policy and practice. *Social Policy Report, 28*(1), 36.

Janson, U. (2001). Togetherness and diversity in preschool play. *International Journal for Early Years Education, 9*, 135–143.

Janson, U. (2007). Preschool cultures and the inclusion of children with disabilities. *Japanese Journal of Special Education, 44*, 405–422.

Johnson, K. et al. (2006). *Recommendations to improve preconception health and health care – United States. A report of the centers for disease control/ATSDR preconception care work group and the select panel on preconception care.* Atlanta: Centers for Disease Control.

Lee, E., Mitchell-Herzfield, S. D., Lowenfels, A. A., Greene, R., Dorabawila, V., & DuMont, K. A. (2009). Reducing low birth weight through home visitation. *American Journal of Preventive Medicine, 36*(2), 154–160.

Lupien, S. J., McEwen, B. S., Gunnar, M. R., & Heim, C. (2009). Effects of stress throughout the lifespan on the brain, behaviour and cognition. *Nature Reviews Neuroscience, 10*, 434–445.

March of Dimes. (2007, 2009). Retrieved from http://www.marchofdimes.com

McClelland, M. M., Cameron, C. E., Connor, C. M., Farris, C. L., Jewkes, A. M., & Morrison, F. J. (2007). Links between behavioral regulation and preschoolers literacy, vocabulary, and math skills. *Developmental Psychology, 43*, 947–959.

Nelson, C. A., Furtado, E. A., Fox, N. A., & Zeanah, C. A. (2009). Developmental deficits among institutionalized Romanian children—and later improvements—strengthen the case for individualized care. *American Scientist, 97*, 222–229.

Offenbacher, S., Katz, V., Fertik, G., Collins, J., Boyd, D., Maynor, G., ... Beck, J. (1996). Periodontal infection as a possible risk factor for preterm low birth weight. *Journal of Periodontology, 67*(10 Suppl.), 1103–1113.

Olds, D. L., Holmberg, J. R., Donelan-McCall, N., Luckey, D. W., Knudtson, M. D., & Robinson, J. (2013). Effects of home visits by paraprofessionals and by nurses on children: Follow-up of a randomized trial at ages 6 and 9 years. *JAMA Pediatrics.*, E1–E8.

Olds, D. L., Kitzman, H. J., Cole, R. E., Hanks, C. A., Arcoleo, K. J., Anson, E. A., ... Stevenson, A. J. (2010). Enduring effects of prenatal and infancy home visiting by nurses on maternal life course and government spending in American Medical Association. *Archives of Pediatrics and Adolescent Medicine Journal, 164*(5).

Piaget, J. (1932). *The moral judgement of the child.* London: Kegan Paul, Trench, Trubner and Co.

Public Health Agency of Canada. (2008). *Preconception care, family-centred maternity and newborn care: National guidelines*. Toronto: Public Health Agency of Canada.

Rosenberg, S. A., Robinson, C. C., Shaw, E. F., & Ellison, M. C. (2013, January). Part C Early intervention for infants and toddlers: Percentage eligible versus served. *Pediatrics, 131*(1). Retrieved from http://pediatrics.aappublications.org/content/131/1/38

Rosenberg, S. A., Zhang, D., & Robinson, C. C. (2008). Prevalence of developmental delays and participation in early intervention services for young children. *Pediatrics, 121*(6), e1503–e1509. Retrieved from http://pediatrics.aappublications.org/content/121/6/e1503

Saxe, R. (2016). *Baby brains: Unlocking our humanity*. TEDxCambridge talk. Retrieved from https://www.youtube.com/watch?v=PDefVd8DRxM

Saxe, R., & Kanwisher, N. (2003). People thinking about thinking people. The role of the temporo-parietal junction in "theory of mind." *Neuroimage, 19*, 1835–1842. http://doi.org/10.1016/S1053-8119(03)00230-1

Shiono, P., & Behrman, R. E. (1995). Low birth weight: Analysis and recommendations. *The Future of Children, 5*(1), 4.

Shonkoff, J. P., & Phillips, D. A. (Ed.). (2000). *From neurons to neighborhoods: The science of early childhood development*. Washington, DC: National Academies Press.

UNESCO. (2007). *Education for All global monitoring report: Strong beginnings, early care and education*. Paris: UNESCO.

U.S. Department of Health and Human Services. (1999). *Mental health: A report of the surgeon general*. Rockville, MD: U.S. Department of Health and Human Services, Substance Abuse and Mental Health Services Administration, Center for Mental Health Services, National Institutes of Health, National Institute of Mental Health. Retrieved from https://profiles.nlm.nih.gov/ps/access/NNBBHS.pdf

Vargas-Barón, E. (2005). *Planning policies for early childhood development: Guidelines for action*. Paris: UNICEF, UNESCO, ADEA, CINDE and Red Primera Infancia.

Vargas-Barón, E. (2009a). *Going to scale: Early childhood development in Latin America, designing successful and sustainable ECD programs with national-level coverage*. Washington, DC: The World Bank.

Vargas-Barón, E. (2009b). *Guide for presenting "Prepare for a healthy pregnancy: A preconception booklet."* Washington, DC: The RISE Institute.

Vargas-Barón, E. (2016). Policy planning for early childhood care and education: 2000–2014. *Prospects: Quarterly Review of Comparative Education, 176*, 4–26. doi:10.1007/s11125-016-9377-2

Wagner, S. L., Cepeda, I., Krieger, D., Maggi, S., D'Angiulli, A., Weinberg, J., & Grunau, R. (2016). Higher cortisol is associated with poorer executive functioning in preschool children: The role of parenting stress, parent coping and quality of daycare. *Child Neuropsychology, 22*(7), 853–869. http://dx.doi.org/10.1080/09297049.2015.1080232

World Health Organization. (2005). *World health report, 2005: Make every mother and child count*. Geneva: World Health Organization.

Wynn, K., & Bloom, P. (2014). The moral baby. In M. Killen & J. G. Smetana (Eds.), *Handbook of moral development* (2nd ed., pp. 435–453). New York, NY: Psychology Press.

Zeitlin, M. F. (1996). *Child care and nutrition: the findings from positive deviance research* (Nutrition Monograph Series No. 27). Ithaca, NY: Cornell International Nutrition Program.

Zeitlin, M. F. (2003). *Causeries développées au cours d'un programme en Déviance Positive en nutrition et éveil en forme de dialogue maïeutique entre l'animateur et les mères, Nder Positive Deviance Program.*

INTERMEZZO 3

Touching the Encountered World

Jan Visser

Reflections inspired by Arvind Gupta: 'Saga of a small science center'

Touch is a newborn's primordial mode of inquiry. When a newly born baby awakens to the world it immediately starts exploring it. Reaching out to touch and feel things, or to elicit a response from the environment, is among its first acts. Gopnik, Meltzoff, and Kuhl (1999) describe this primeval encounter with an unknown, yet to explore, world in the moving first paragraph of their book *The Scientist in the Crib* as follows:

> Walk upstairs, open the door gently, and look in the crib. What do you see? Most of us see a picture of innocence and helplessness, a clean slate. But, in fact, what we see in the crib is the greatest mind that has ever existed, the most powerful learning machine in the universe. The tiny fingers and mouth are exploration devices that probe the alien world around them with more precision than any Mars rover. (p. 1)

This innate ability to explore the encountered world never leaves us of its own accord. It is basic to our instincts, as the sculptor knows. It is essential to our survival in the wilderness, our natural home, of which we live in exile. Most importantly, it's the bridge to sustaining our childhood along the entire lifespan, as long as we keep touching the world. Parents usually recognize this behavior as a positive sign of the well-being of their child (my parents certainly did), and in kindergarten it is usually at least allowed, if not encouraged. However, it is largely absent in the formal school setting, from the elementary level upward. In fact, the structure of formal education is counter to further nurturing the probing instincts of its attending students. Real learning in school is supposed to be academic, not tactile. And that's how one of our most essential bridges to the world gets destroyed, for many permanently.

It need not be that way. Starting from scratch and following one's instincts as a teacher while allowing students to do the same, entirely different models emerge. I had this experience while working as a science educator in newly independent African countries where one wasn't yet under the pressure of existing regularities. They had not yet had time to develop.

 | DOI: 10.1163/9789004416802_006

In such circumstances one naturally turns to using 'trash' as raw material for constructing 'toys' to play with and learn the principles by which nature works. The pleasure to do so is contagious. It should pervade school systems around the world as it is such a forceful way for the scientific mind to take shape. Arvind Gupta's lifetime work may make it happen—if we want.

Reference

Gopnik, A., Meltzoff, A. N., & Kuhl, P. K. (1999). *The scientist in the crib: Minds, brains, and how children learn.* New York, NY: William Morrow and Company, Inc.

CHAPTER 4

Saga of a Small Science Center

Arvind Gupta

1 The Setting

Pune is the cultural capital of Maharashtra, India. In the past 150 years, this city has produced a string of social reformers, whose ideas and actions have had a deep impact on India's development. These luminaries include, among others, Mahatma Phule (1827–1890)—first to raise the voice of the subjugated, poorest castes; Gopal Krishna Gokhale (1866–1915)—Gandhi's political mentor; the firebrand nationalist Bal Gangadhar Tilak (1856–1920); Maharshi Karve (1858–1962)—pioneer of women's education; and Hamid Dalwai (1932–1977)—founder of Muslim Satyashodhak to stimulate reform in his community. It is fitting for such a historical context that further initiatives to reform and enhance social and individual life continue to emerge. In this chapter I describe one such development, which culminated in the establishment of a small science center in the development of which it has been my privilege and pleasure to play an active role. Following is a brief overview of how and why the initiative evolved.

2 The Beginning

Crucial to the development of the Science Center was the prior initiative by renowned astrophysicist Jayant V. Narlikar[1] to create, in 1988, the Inter-University Center for Astronomy & Astrophysics (IUCAA) within the Savitribai Phule Pune University. Professor Narlikar is India's most celebrated scientist, having won the second highest civilian award—Padma Vibhushan. Two decades earlier he had won the Kalinga Award—UNESCO's highest recognition for science popularization. As part of his research institute, his dream was to set up a Children's Science Center. Why? To offer children at an early age the opportunity to experience the joy, and develop the love, of science. A large donation from Smt. Sunita Deshpande—the wife of his late friend P. L. Deshpande, a cultural icon in India—turned the dream into reality. The children's Exploratory Science Center or *Muktangan Vidnyan Shodhika* (MVS) was inaugurated in 2004 by Professor Yashpal, a fellow Kalinga Awardee who

 | DOI: 10.1163/9789004416802_007

FIGURE 4.1 Building toys from trash to build the scientific mind

shared Jayant Narlikar's vision of science. Right from its inception, IUCAA had an active public outreach program. Every second Saturday there was a popular science lecture/demonstration for some 1000 students and teachers from over a 100 schools from Pune.

My own academic and engineering background is from the Indian Institute of Technology Kanpur in Uttar Pradesh with many years of grassroots experience in popular science. Like many idealists in the 1970s, I had left a well-paying job to plunge into social change working for two decades with various People's Science Movements. In 2003, I was invited by Professor Narlikar to become the first coordinator of the Science Center, initially for a period of six months, but ultimately I spent 11 years at the Children's Science Center. I was soon joined by Vidula Mhaiskar, a microbiologist with four years PostDoc experience at Stanford, and by Ashok Rupner, who brought in a decade of experience of working in a Mobile Science Lab. This was all made possible through funding from the philanthropic Tata Trust. This core group continued for almost a decade, with support from occasional part-time volunteers who pitched in for short periods of time.

India is a large country with a population of more than 1.3 billion people. The education system is highly segregated with badly run government schools for the poor and private well run schools for the rich. Irrespective of the school, all children learn science by rote—mugging up definitions and formulae. The rich schools have well-equipped science labs with burettes, pipettes, and test tubes. But all this fancy equipment often lies unused, locked in cupboards and laden with a layer of dust. Most teachers come from a mutilated system where they have never 'dirtied their hands with experiments.' So, they prefer the age-old chalk-and-talk method.

3 The Hoshangabad Science Teaching Programme (HSTP)

The 1970s were full of positive political turmoil—student uprisings, civil rights and anti-war movements. In this decade the environment and women's movements made rapid strides. Sensitive scientists asked fundamental questions: "Should we work for human welfare or design bombs and missiles?" The defining political slogan was "Go to the people, love them, live with them, start from what they know, and build on what they have." It was against this backdrop that the 'Hoshangabad Science Teaching Programme' (HSTP) emerged, initiated by Anil Sadgopal, to revitalize science learning in village schools where there were no labs.

This pioneering program started with 16 schools but soon spread to over a thousand village schools in Central India. The HSTP drew its inspiration from the Nuffield Science Program in England. It emphasized the 'discovery method' and 'learning by doing.' Many progressive and creative scientists from colleges and universities across India lent support to this program. The HSTP was a wonderful partnership between a local group (a Non-Governmental Organization) and the state government. Teachers and children actively participated in designing low-cost experiments, based on local, easily accessible materials. It unleashed mass creativity and experimentation. The credo was, "think differently; look at your milieu critically." So, children substituted 'dissecting needles' with long Acacia thorns plucked from local bushes. When the indicator 'phenolphthalein' could not be bought, a teacher dissolved a tablet

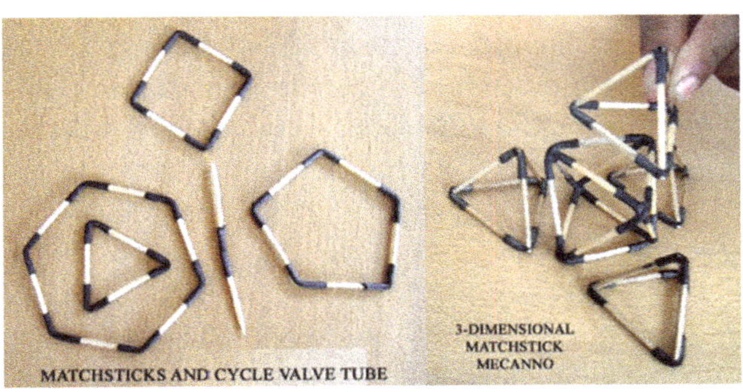

FIGURE 4.2 Two-dimensional and three-dimensional matchstick assemblages

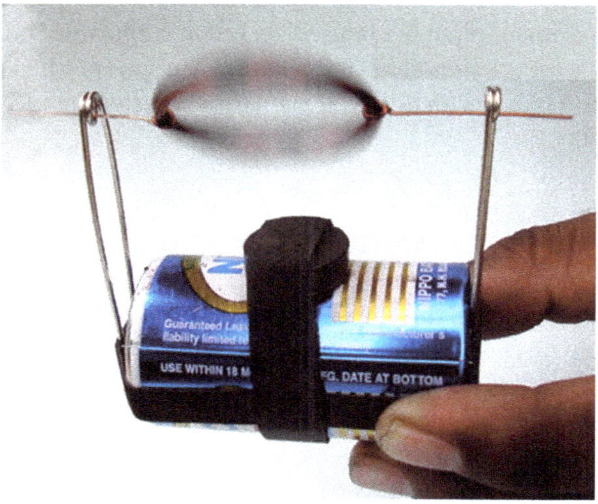

FIGURE 4.3
Simple DC motor

of the commonly available laxative "Vaculax" (which contained 0.2 mg of phenolphthalein) in water to conduct a successful experiment. Bicycles are ubiquitous on village roads. The pieces of bicycle valve tube (a thin black flexible tube) were used as 'joints' and matchsticks as members to make an array of two- and three-dimensional structures (Figure 4.2). A simple, working model of the DC motor (Figure 4.3) was devised using a torch battery, magnet and insulated copper wire! These experiments captured the imagination of children. With everyday, local materials it was easy for the children to relate science to their daily, real-life affairs. Science could be done without lab coats and outside the standard labs.

The 1990s saw the World Bank (WB) sponsored District Primary Education Program (DPEP) pushed across India. The DPEP was touted as a panacea for India's educational ills. The entire political class smelt quick dollars in this program. The home grown, low-cost Hoshangabad Science Teaching Program—with its vast experience, was the only real threat to the WB sponsored DPEP. So, in 2002 the state government shut down the HSTP! This conclusively demonstrated that any creative science program can only survive under the political support of the state government. Once the political umbrella is withdrawn the program will collapse and die.

4 A Typical Day in the Centre

The HSTP taught us many 'scientific' and 'political' lessons which were incorporated in the Science Activity Center at IUCAA, Pune. Two days a week we conducted free workshops for school children and teachers. Any school could

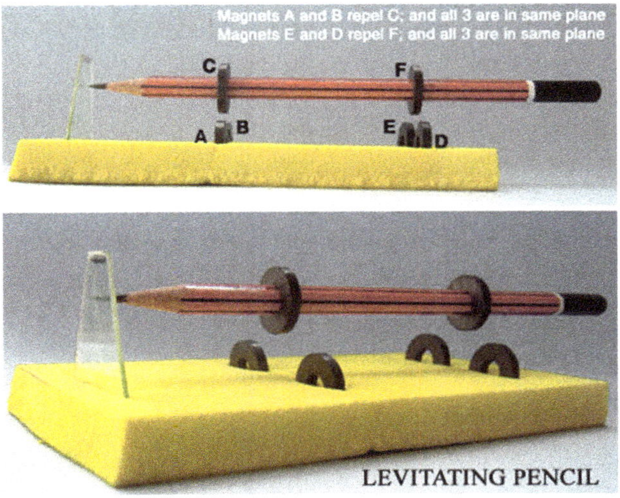

FIGURE 4.4
Levitating pencil

register and bring 50 children to the Science Center for an amazing 3-hour experience. During these sessions children could see many improvised, working science models, made from plastic bottles, Tetra Pak cartons, matchboxes, broomsticks, plastic straws, etc. (see e.g. Figure 4.4.). The children would also make many things with their own hands. For example they would make the Straw Flute, Sprinkler, and a simple Electric Motor; fold a Cricket Cap from a newspaper and make a dynamic Spinner from a broomstick. The children took home their creations. For once they got a taste and feel for 'good' science. These workshops became vastly popular. Schools had to be booked two months in advance.

The Electric Motor and Generator are part of the class ninth science curriculum. Children mechanically mug-up Faraday's laws of Induction to pass exams. This way their knowledge is very superficial. There is no deep insight or understanding. So to assemble an electric motor from scratch, and make it work, was for children an unforgettable experience. You could see the joy of discovery 'sparkling' in their eyes. Over the years thousands of children also made the 'Syringe Generator.' For this they wrap 1000-turns of insulated copper wire on the plastic barrel of a disused syringe to make the coil and then attach the coil ends to a LED. Finally they place a strong Neodymium Magnet inside the syringe barrel. On shaking the magnet the LED glows! The moving magnetic field produces a small current in the coil and lights up the LED. For the first time they produced current! It helps them relate theory to practice. This empowers children. A girl from a poor unelectrified village is likely to exclaim: "Today I light one LED; someday I will light up my village!"

We conduct workshops on only two days a week. What do we do on other days? We read, listen to inspiring TED talks, great music and meet many inspired people. Every day we design new experiments from trash. Modern society produces mounts of junk. We use Tetra Pak cartons, plastic bottles, newspapers, cardboard cartons, caps, old pens, and other throwaway stuff. Recycling and environmental sensitivities are inbuilt in our designs. We always ask, "Can we replace this expensive material with something low-cost or throwaway stuff?" Today we have 50 models made from waste Tetra Pak material, and over a 100 for which we use old plastic bottles! These models can be made by the poorest children who can never afford expensive science kits! Our message is: pick up rubbish—and we will help you make beautiful fun toys. Most school science has outdated curricula that are totally out of sync with societal needs. In India we are never far away from abject poverty. We want our science toys to help children in Asia, Africa and Latin America. We wish to reach out to the poorest children on earth!

5 Main Activities

The ethos of the Science Center is people centric, to help children who need it most—especially children from the oppressed section who study in Government and Municipal schools. For seven years we ran a twice weekly Science Club in the only school located in 400-acre Pune University Campus—the Vidyapeeth Shaala.

What else did we do? We document science models and toys in simple but profusely illustrated books. Today we have 20 activity books with beautiful line drawing. Most books have been translated and published in five or six Indian languages and have gone into multiple editions. The Fondazione Montessori—the Montessori Foundation in Italy—is publishing an anthology from our various books in Italian for the children there. The city of Pune has an enormous appetite for science. Its people have abundant energy to do meaningful things. The Center gets many visitors and some of them contribute significantly. We seize this wonderful opportunity and build upon the goodness of people. They have translated many books on education, science, math and the environment in the local language Marathi. In ten years over a 100 world class books were translated in Marathi. Educational classics like *Summerhill* by A. S. Neill; *Free at Last* by Daniel Greenberg (Sudbury School); *Totto-chan* by Tetsuko Kuroyangi; *Teacher* by Sylvia Ashton Warner; and *Danger: School* by Paulo Freire's group. In science we have the *VSO Science Teachers and Math's Teachers Handbooks*, the *UNESCO Sourcebook for Science Teaching*, along with a series of 36 books by the great Isaac Asimov: *How did we find out about … Black Holes, Germs, Vitamins, Antarctica, and Solar Energy* etc. These lucid, wonderful books on the history of science were written by the world's greatest science writer. Marathi was the first language in which this entire series was translated, followed by Telugu and Hindi.

In 2003 the website www.arvindguptatoys.com was set up. Initially, all our own science activity books were uploaded. Gradually other books were scanned and added. Slowly the corpus of books grew. Today the website hosts a large collection of 5,500 books in Marathi, Hindi and English. Many of these books have been out of print for years and cannot be bought. Every day 20,000 books get downloaded for free! It just shows the hunger for good books among our people.

From 2005, a young and dynamic volunteer Monil Dalal from India's premier National Institute of Design interned with us during summers. He documented all our teaching aids in sequential photographs. In keeping with the dictum that "a picture speaks more than a thousand words," we labeled, captioned and uploaded these photographs. The photos help children make science models. The photos are easier and clearer to follow than just written instructions.

In India the National Council for Education Research and Training (NCERT) is the apex body which designs curriculum. We collaborated and made over 125 videos on simple science models with the TV wing of the NCERT. These programs were dubbed in many Indian languages and beamed on the National TV channel *Doordarshan*. India has hundreds of TV channels. But *Doordarshan* has the largest reach and is viewed by millions in far flung villages.

There is a dearth of good educational material in the regional languages—especially in Hindi—the language spoken in 5 states. Close to 500-million people speak Hindi but there is acute scarcity of good books on science, math, the environment, and children's books. We have helped translate over 175 books in Hindi. This includes classic books like *Chemical History of a Candle* by Michael Faraday; *Soap Bubbles* by C. V. Boys; *The Autobiography of Charles Darwin*; and *The UNESCO Source Book of Science in the Primary School* by Harlen and Elstgeest, *Romping in Numberland* by P. K. Srinivasan and many other world class books. Most children in India do not have access to public libraries or supplementary books. Apart from school text books they cannot afford to buy other books. So, the digitized books made available online for free have helped many.

A major breakthrough took place in May 2009. A young social activist, Sourabh Phadke, taught us how to make short videos using a digital camera. He enthusiastically told us, "It's quite easy. Shoot short clips on the digital camera, weave them in Movie Maker—which came with Windows X software, and then overlay your voice using a small mike. Then upload the videos on YouTube." The training took less than half an hour. Soon Ashok internalized the process. In seven years he has single-handedly shot 970 videos. In June 2009, we produced and uploaded our first video—*Balancing Nails*. This 1-minute video showed how to balance a dozen nails on the head of one vertical nail. Since then we have never looked back. In April 2016—six and a half years later—we have 7000 videos in 20 different languages with a viewership of more than 47-million! Google (owners of YouTube) said that our channel was the second largest channel of educational videos from India! This was indeed a compliment considering that all this was done with minimal resources. Our room in the Science Center measures just 400 sqft (37 m^2)!

These videos have been dubbed in many languages by volunteers. Prabhakar K. Nanavati—a people's science activist—has dubbed 900 science videos in Kannada. Today children in far flung villages of Karnataka can access these videos in their native language Kannada through mobile phones. Another wonderful volunteer was Breezy Ocana Flaquer. Breezy came as a Post Doc student and visited our center during her first week in Pune. She originally hails from the Dominican Republic. As a sensitive person—coming from a relatively poor country—she saw immense possibilities in these Toys from Trash. She dubbed over 300 videos in her native language Spanish. Spanish is spoken in over 26 countries throughout the world. Today millions of children in Latin America have seen the Toys from Trash videos in Spanish. We have over 200 videos in Chinese, another 150 in Russian.

However, we have not been lucky with several Indian languages. We have just a few videos dubbed in Bengali, Oriya and Punjabi. We wish our videos were dubbed in Nepalese. Nepal is a poor country, struck by a massive earthquake in 2015. The videos in Nepalese would help thousands of children and teachers in Nepal. Similarly, the videos in Bengali would not only help learners in West Bengal but also in Bangladesh.

As of January 2016 we have 930 videos in English. These videos have been divided into 22 broad categories like Air, Water, Math, Electricity, Magnetism, Paper Toys etc. There are some novel categories like Toys from Trash and Pumps from the Dump. We have designed and documented over 30 different kinds of Pumps—which can inflate a balloon, or throw water 15-feet away, sprinkle water, etc. All these pumps use simple film can bottles, bicycle tubes, straws, bicycle spokes, etc. Children start with simple pumps and graduate on to difficult ones. In the process of assembling and making them work children get a very good 'feel' about valves and all else which goes into making a good pump.

Do children really learn science with these toys? Teachers/parents often lament that children get so engrossed in these projects that they do not pay much attention to their textbooks. Where is the science in all these toys? Will they help in children scoring high marks, or in clearing stiff entrance tests? The kids have loads of fun, but are they really learning science? Libertarian pedagogues have asked these questions for centuries. Play is primordial and children are born to play. Good learning takes place in an atmosphere of fun and joy. Great pedagogues like Erik Erikson, Lev Vygotsky and Jean Piaget sincerely believed that children's play is very serious business. Summerhill's founder A. S. Neill said, "I would rather see a school produce a happy street cleaner than a neurotic scholar."[2] Children who play remain happy. They do not invent missiles and guns when they grow up. Happy people do not hurt or kill others. We sincerely believe in making children (especially the poor) happy by helping them make their own toys. Because there will be no peace on earth unless there are happy children!

5.1 *Epilogue*

Arvind Gupta and Vidula Mhaiskar quit the Science Center in December 2014. Arvind Gupta quit because of his medical condition. Vidula became the Director of the Garware Children's Center in Pune. Ashok Rupner is the new coordinator of the Science Center. Work continues, as there is much left that remains to be done.

Notes

1 A 22-minute Walk the Talk interview with Professor Narlikar is available at https://www.ndtv.com/video/shows/walk-the-talk/walk-the-talk-with-renowned-astrophysicist-professor-jayant-narlikar-372830

2 https://archive.org/stream/Summerhill-English-A.S.Neill/summerhill_djvu.txt

INTERMEZZO 4

The Worthwhile Struggle to Overcome Inertia

Jan Visser

Reflections inspired by Paul Heckman: 'Expanding the dialogue: Challenging the mental models of schooling through indigenous invention'

Reform in education at a level beyond tinkering with the existing systems is hard to accomplish. Yet, there are courageous attempts worth knowing. They can serve to inspire. They are therefore worth to be studied and learned from. Some such attempts may be considered successful and seen to endure amid an otherwise inert and sometimes hostile environment. Others may live only a short life and go under. Yet others survive longer thanks to the fact that they develop and remain 'in hiding.' From an evolutionary perspective these incidental deviations from the norm may be considered mutants. Without mutants there would be no evolution. We must be happy and thankful that they occur and value their fragile, often ephemeral, existence.

Through a, so far unfunded, research effort we try to collect their narratives. It's a slow process because of the lack of funding, but we take advantage of other activities with which our Counter Narratives Research (CNR) can be combined. It started in the Fall of 2015 when we visited, by invitation, a school in West Sacramento, California. The school was populated by underprivileged youths, eager to learn by following their own interests, and motivated teachers, willing to change learned habits in the process of indigenous (re)invention. In addition, they had an inspiring and dynamic school principal as well as moral and intellectual support from progressive elements in the School of Education at the University of California at Davis. At WestSac, as the school was commonly referred to, there was no school bell, nor were there other mechanisms that could mark timeframes within which particular learning objectives were to be achieved and tested. There was no set curriculum either. Students felt clearly at home. They felt no urge to rebel against the system, which was of their own making. They learned, driven by their own interest, touching the world they encountered, and did so to satisfy their curiosity and for the pleasure to find things out, rather than to distinguish themselves in competition with their peers. They were forthcoming in voluntarily talking about their experiences and had no problem to reveal to us, or their fellow students, aspects of their own being, such as having dyslexia or ADHD, which they had to cope with.

 | DOI: 10.1163/9789004416802_008

Dialogue among students, teachers, principal and the key person involved on the part of UC Davis, Paul Heckman, was an essential part of what made the school tick.

Sadly enough, the success of the school was not to endure for reasons explained in Heckman's chapter, which is aimed at expanding the dialogue around what is implied by challenging existing mental models and focusing on indigenous invention as a way forward.

There is much to be learned from WestSac's brief life as it was, as well as from its eventual demise and Heckman's post-mortem analysis. We lost Paul Heckman to cancer while this book was in press.

CHAPTER 5

Expanding the Dialogue: Challenging the Mental Models of Schooling through Indigenous Invention

Paul E. Heckman

1 Introduction

In his landmark book, *The Culture of the School and Problem of Change*, Seymour Sarason (1982, 1996) describes the staying power of the culture of the school in maintaining the status quo despite untold attempts to alter that culture. Tyack and Cuban (1995) in their book on the history of reform identify this phenomenon as the 'grammar of schooling'—emphasizing how the ideas and practices in schools remain the same despite a century of work to fundamentally alter them. This intractable nature of schooling continues to thwart educational change (Payne, 2008). Nonetheless, and, fortunately, as responsible educators, we persist.

This chapter presents a school reform effort that builds upon the lessons of the past in its approach to the reform process and focuses on addressing the persistent lack of engagement and achievement among students from economically poor and disenfranchised neighborhoods and families. Turning this situation around was at the heart of the establishment of the West Sacramento Early College Prep Charter School (WSECP). Following is the story of what happened. It identifies successes and gains for students and teachers, and illuminates lessons for reformers.

The city of West Sacramento is adjacent to the city of Sacramento, California. This city, located between the Sacramento River and the deep water Port of Sacramento, is home to a highly diverse population including recent immigrants from Mexico and Ukraine. This diversity has been a challenge for the schools. With a major university nearby and resources from a neighboring community college, a partnership was built. The University of California, Davis's School of Education; Sacramento City College; and the West Sacramento Unified School District worked together to establish a charter school for middle and senior high school students serving students from these diverse and economically challenged neighborhoods.

 | DOI: 10.1163/9789004416802_009

On August 22, 2007, the West Sacramento Early College Prep Charter Middle and Senior High School opened its doors. A news release by UC Davis, highlighted the following hopes with respect to the anticipated results:

- UC Davis has joined with Sacramento City College and the Washington Unified School District in West Sacramento to launch an innovative charter school.
- The school will provide an exceptional opportunity for historically underserved students to achieve the promise of a postsecondary degree …
- The school will offer small classes, tutoring, community service, caring relationships and engaging school work and investigations. All students will have the opportunity to build strong relationships with their teachers, college student mentors and college professors.

Bringing together these resources held incredible potential, as well as some age-old challenges. As the Associate Dean of the School of Education and someone who has worked in educational reform, I knew these issues well and was committed to utilizing what had been learned in previous work. From this perspective then West Sacramento Early College Prep (WSECP) provided an opportunity to address the disparities in opportunities and benefits faced by poor students and students of color in California. One of the primary goals for this project was to collaboratively create and research solutions to the troubling high school drop-out rate, which hovered at over 50 percent or higher among the most underserved populations in the state.

A larger foundation of ideas also lay beneath these hopes. These ideas would guide colleagues who were involved in the school's development in a process called 'Indigenous Invention' (Heckman & Montera, 2009). Through weekly, sometimes daily work in groups and individually with the teachers, students, and the principal of the school, alternative learning environment and experiences were created. The process was far from simple and far from linear.

This chapter discusses the ideas underlying Indigenous Invention. It will analyze what happened and what was accomplished as the school engaged in this process. I will provide examples of how Indigenous Invention was implemented in practice, such as through 'dialogues' between teachers and the principal. The examples will highlight the ideas that the educators examined in these 'dialogues' at the school. Of fundamental importance to the way of working was that these ideas challenged the established regularities and mental models of schooling (see Tyack & Tobin, 1994, p. 454). The various examples illustrate the newly created dimensions of the school (Sarason, 1972) that arose in this Indigenous Invention process. As change processes are often contested, the examples will also discuss the tensions that arose as a result of the new practices and structures that were introduced—tensions which ultimately affected the continuity of the innovation.

2 Indigenous Invention: Challenging Regularities

When the school opened in September of 2007, 'mental models of schooling' guided educators' and students' thoughts and actions in classrooms of the school. The educators at the school, like most educators, had guides in their minds for the arrangement and practice of school. So did the students. Everyone knew how to do school. These memories, like mental maps, came from prior experiences in schools, as students, and then later through the role of educators (Lortie, 1975).

A question and two sub-questions guided the Indigenous Invention process. The main question asked was: Can the process of Indigenous Invention provide an opportunity for local educators to challenge the regularities and create practices consistent with current knowledge on learning? This question led to two sub-questions.

The first sub-question sought to establish whether a school—especially one serving the economically poorest students and families—can successfully promote pupil success (in spite of the odds).

Since the publication of the Coleman (1966) report half a century ago, the research literature has documented the links among economic poverty, color, and student success (which is usually measured through academic achievement). Unfortunately, the subsequent fifty plus years have seen little change in the stubborn relationship between poverty and educational performance in the United States. Despite numerous federal, state, and local educational initiatives, which have sought to disrupt this predictable relationship, children and young people from economically and socially disenfranchised backgrounds continue to generally perform less well in school.

Various explanations have been offered in the literature about the lack of school success and educational attainment of children and youth who live in economic poverty. For example, economically poor families do not 'invest' in their children by providing supplemental educational programs, activities, and resources to the same high degree that wealthy and upper middle class families do. Similarly, the educational accomplishments of parents affect the resources that children bring with them when they come to school. Health and food disparities also influence children's ability to learn in school and the extent to which they will acquire the 'readiness' skills for succeeding in school.

The students at the West Sacramento Early College Prep Charter School (WSECP) in many ways were no different from those that the literature suggested were doomed in terms of educational success. Due to their families' socio-economic status, 75% of WSECP's learners qualified for free or reduced meal prices; 13% were English Learners (ELs)[1]; 20% were classified fluent-English-proficient; and 52% of students came from families with little formal education and no experience in post-secondary institutions.

Our premise was not to deny the influence of these factors on the well-being of children and youth, but to question whether other factors less recognized could be more influential in their school success. In particular, we felt strongly that the research evidence that we held about children's knowledge and family assets challenged these explanations. One research understanding had to do with children and youth knowing a lot. And so we wanted to make sure that children's own knowledge would bring value to the learning process. Another research understanding acknowledged that regardless of wealth and education, families have accumulated knowledge that they utilize. This knowledge, too, would be a valuable asset in children's learning. I return to these two concepts below in the discussion of the example about children knowing a lot and the way that insight guided the Indigenous Invention process.

We also believed that it should be possible to put the vast amount of time that children spend in school in classrooms (six to seven hours a day during a 180-day school year) to better use. We believed the same with respect to the funding. The United States of America (USA) spends an average of $11,009 per year for each student, and while these resources may not be sufficient, they are important and can be dedicated to enhancing the educational experience. Similarly, we were strongly convinced that during the time that children spend in school, what happens in the classrooms—both in terms of thoughts and actions—has the potential to promote children's success.

The starting point in the Indigenous Invention process was therefore to focus on how the thoughts, actions, and learning conditions in classrooms and the schools could be used to encourage student and teacher learning. Doing this would mean bringing about change in the processes of schooling and challenging mental models underlying the regularities of schooling.

The overwhelming evidence that education reform can be very difficult (see below) led us to the second sub-question. In essence this question queried whether in this school the 'regularities' (the established ways of working) of schooling could be removed, and new structures and practices created in their place.

The challenges of replacing the regularities of school are discussed at length in a now somewhat dated, but unfortunately still very relevant, 1990 article in the *Educational Researcher* by Larry Cuban. The article entitled "Reforming, Again, Again, and Again," discusses at length the difficulty school reformers have encountered in replacing the regularities of schooling. Cuban notes that, "... another generation of reformers is fighting against a technical, subject-centered form of instruction expressed in mastery learning, measurement driven curricula, and bookkeeperlike accountability" (p. 4). This "common core of academic knowledge" forms part of the foundation for the regularities of schooling and the difficulties in reinventing them (Sarason, 1996).[2] They have persisted throughout the past century and a half in schools in the USA (Goodlad, 1984; Jackson, 1968; Sirotnik, 1983).

One example of a regularity is that teachers usually do most of the talking in classrooms. Students sit and listen. Teachers expect students to give back the words, ideas, and thoughts which reflect what was expressed by the teacher. Assessments follow and reflect a similar format—short answer questions with correct answers. The teacher and texts are viewed as the source of knowledge and information. The student is the recipient of that information. That information is in turn assumed to become knowledge in the minds of the students.

Breaking with these 'regularities' was the focus of the efforts in the Indigenous Invention process at WSECP. A key part of our efforts in the school was to try to replace these predictable structures and practices—which manifested themselves in boring didactics, seatwork, and tests—with other ways of working more in line with these premises and the research ideas described below.

We decided to do this in a way that was different from the usual efforts in school reform. We realized that reform is often driven by experts and designers of educational innovations who bring new practices and structures to the school which seek to replace the regularities of schooling. The recipients of the innovations (teachers and pupils) are encouraged (or told) to implement these new ways of working with fidelity. Thus, the assumption behind these efforts at school reform is that teachers follow the rules for these innovations and put the new practices or structures into action inside of their classrooms or throughout the school in a manner detailed by the designers (Goodlad, 1975).

Like didactics in classrooms, replacement practices and structures coming to teachers from outside often do not have that intended effect (Berman & McLaughlin, 1978). The transmission of information about new thoughts and practices to be implemented assumes that teachers or students will understand the presented information, embrace its meaning, and take new actions based on what they have been told.

David Olson and Jerome Bruner (1996, p. 16), borrowing from Gilbert Ryle, called this "knowing that." Paul Duguid (2005) summarizes the differences between Ryle's the learning of 'that' and the learning of 'how' in this way:

> ... the two aspects of knowing [that and how] are complementary, knowing how helping to make knowing that actionable. They are not, however, substitutable: Accumulation of know *that* does not lead to knowing *how*. Know *that*, we acquire in the form of explicit, codified information. By contrast, "we learn *how*," Ryle argues, "by practice." (p. 111)

The emphasis in schooling, for both students and teachers, has been on knowing *that*, which limits knowing *how* and *learning*.

Teachers' own experience with schooling (during their 12 to 16 years in school as children and students) plays a firm role in their socialization as educators. These early experiences in schools form the predictable practices and structures of schooling that teachers are likely to enact later in their own teaching profession.

As Dan Lortie (1975) and Phil Jackson (1968) noted many years ago, learning about the practice of schooling happens in a process of socialization and enculturation in schools. The culture of school (Sarason, 1982) heuristic is the foundation for Lortie's and Jackson's premise that processes of socialization and enculturation guide teachers' views about the regularities of schooling.

Educational scholars, like Sarason (1982), suggest that the culture of schools is a key heuristic for understanding educators' commitment to regularities of structures and practices across schools in the USA. Other organizational researchers also suggest that 'organizational culture' is at the center of the ongoing organizational characteristics. Gelfand, Lim, and Raver (2004) argue, "that as individuals are enculturated through socialization in a particular sociocultural context, they develop cognitive maps of how various individuals, groups, and organizations are answerable and accountable to one another" (p. 137).

Spillane, Reiser and Reimer (2002) refer to such cognitive maps as 'schemas.' Schemas arise from memories and reflect the individuals' own experience with the practices and structures of schooling. These then remain in the mind.

> Research on schemas also stressed the mental representation of dynamic processes, called mental models. An important finding in this research is that people construct intuitive models from their experiences apart from formal instruction and use those models to envision a situation,

> essentially 'running' the model to make predictions about its causes and outcomes. (p. 395)

This is especially true for educators. Educators coming into the cultural setting of schools bring with them their mental models of schooling, and find themselves immersed in the regularities which they experienced in their own schooling. Their schemata quickly make sense of these predictable classroom practices and structures of schooling. They fit their expectations to these contexts. They do not question and wonder much about them and the details of classroom or school actions in which they find themselves. They 'do' schooling in the same way as it was done to them. Looking across the educational system in the United States, it then becomes clear why the regularities have remained steadfast, despite calls to alter them.

3 Indigenous Invention: Using Dialogues to Change Schemas and to Change Learning

Indigenous Invention encourages questioning the regularities of schooling, and not accepting them as normal. This is why Dewey prioritized the scientific method and inquiry for education and classrooms and schools (Schön, 1992). It is also why the book in which this chapter appears focuses on '*understanding*' as the ultimate goal for using the Scientific Mind.

In the Indigenous Invention process that WSECP implemented, teachers learned to question their everyday actions. They, thereby, inquired into the daily life of their classrooms and school. They became researchers of the context in which they worked.

Dialogues were central to this. Such dialogues—essentially inquiry discussions between teachers and other staff in the school—served as a forum for teachers to identify, describe, and discuss the problematic aspects of the regularities of schooling that they themselves were enacting in their classrooms. The dialogues encouraged teachers to ask why classroom actions happened.

For example, one dialogue began with teachers describing the typical practice of providing letter grades for student papers or class work. The discussion brought out that there was variation in the grades that students received. This led to the question, why grades? In responding to that question, teachers offered their explanations. Essentially, they argued that grades are necessary because students only do the classroom work and complete assignments if they receive grades as a reward. The discussion also highlighted that some students received high grades, while others received low grades. Students, who lacked—in the teachers' views—the necessary motivation, received the poor grades. To discourage low or missing motivation, teachers were convinced that they had to punish the low or unmotivated students with a low grade, and reward the more motivated students with higher grades.

Subsequent dialogues followed on the same topic and took place over several weeks. These dialogues went further in examining the explanation that grading drives motivation. It led to teachers questioning to what extent the explanations offered about grades and motivation had 'warrants' (Buchmann, 1992). That is, was there really strong research evidence in support of this relationship? Did the explanation make sense? Were there alternative ways of motivating students?

The weekly dialogues provided a time and opportunity for teachers and the principal in the school to examine this kind of issue. They also provided an opportunity to actively seek out alternative explanations and bring these to the table. The dialogues were thus key in helping uncover alternative explanations for students who lacked motivation, or for those students with low or high motivation.

For example, the participants in this particular dialogue agreed to read chapters from Daniel Pink's (2009) book, *Drive*, in preparation for each of the dialogues. Pink challenges the usual reward and punishment relationship which was held by many members of the dialogue. He also shows the importance of students' interests and control. The reading brought out evidence that led to the construction of new meanings by the teachers. It also informed actions by teachers to do things differently based on their learning and explorations from the dialogues.

The next section of this chapter discusses three examples of efforts to challenge established schemata and to fundamentally change what happens in schools. In these examples, we moved beyond learning about the various instances of 'that.' Instead we moved toward practicing alternative explanations that arose from the dialogue. This meant action, experimentation and trying out new invented practices and/or structures in classrooms and the school.

4 Putting Indigenous Invention in Practice—Three Examples

In this section three examples of Indigenous Invention illustrate the powerful changes that can arise when questioning of established ways of thinking and doing becomes possible.

In the first example below, educators at WSECP questioned whether it was true that students lacked knowledge and that this was the reason for the deficits (poor performance). They decided to work from an alternative explanation, namely that students had a lot of knowledge and many interests and that focusing on these assets would encourage them to do better in school.

The second example shows what happened when teachers questioned why students were assigned to classes by age. As part of this process, teachers explored and talked about the history and shortcomings of the graded school and grade levels. This led them to try out multi-age, inter- and intra-class grouping arrangements of students.

The third example involved teachers questioning the established way of teaching which focuses on transmitting content to students—what Olson and Bruner called the 'that's,' which were discussed earlier in this chapter. By questioning and exploring this topic, teachers came to better understand the importance of students' thinking, cognition and doing. The outcome of this was a stronger focus by teachers and students on problem-based investigations. Students' own questions, interests, and prior knowledge directed these investigations.

4.1 *Example 1—Knowing a Lot by Paying Attention to Students', Teachers', and Communities' Prior Knowledge and Interests*

One development that happened in the Indigenous Invention process at WSECP had to do with teachers and students beginning to understand that each person in the school had a mind full of ideas and knowledge. This was true despite many of the students living in families and neighborhoods with a great deal of economic poverty. The idea that students know a lot was not understood when the school opened and the dialogues began. Teachers became aware of this in the course of their investigations into the reasons behind students' lack of success.

In the dialogues, teachers noted that the students' lack of success happened because the students were missing the knowledge and skills necessary to succeed with the given material. For the teachers, the problem arose because students had failed to adequately learn school material in prior schooling experiences. They were also failing, according to the teachers, because their home and neighborhood did not provide the necessary background knowledge for school success. This 'deficit' explanation held sway in classrooms where there was a predominant feeling that students did not know much, and that the lack of knowledge was the result of the students' poverty, backgrounds, community, and families (Garcia & Guerra, 2004). We had to talk about the details of these explanations, question them, and entertain alternative explanations and warrants for students' lack of success.

For example, the discussion in dialogues highlighted that educational settings model education and learning on what is expected primarily from upper-middle class and white students. As a result, all students are expected to know particular stories and details of certain phenomena in the world that are accessible to this group. This expectation is encouraged in families and communities through books and newspapers, the nature of family events, and the language used by individuals and groups in these various communities. When students do not possess these attributes, then educators and others usually conclude that these students have 'deficits.' Deschenes, Cuban, and Tyack (2001) discuss this historical process of placing 'labels' on students who lack the expected prior experiences:

> Labels also reveal just how embedded categorization and construction of difference are in the structure of schooling. Over time, as the institutional structure of schools changed and as bureaucracies evolved and expanded, so did the way educators and reformers thought of students they served. But the fact that there were students labeled as failures remained constant. (p. 529)

In the Indigenous Invention process at WSECP, those involved in the process questioned these ideas and labels. Through the discussion in the dialogues, such labels came to be seen as unhelpful. Teachers sought more useful and powerful conceptions for thinking about students who were not succeeding. For example, teachers read and discussed literature that highlighted that students bring to the classroom minds full of knowledge and that there is enormous potential in these full minds. Daniel Pink's (2006) other book, *A Whole New Mind,* aided these discussions.

The initial views of teachers were that students did not have full minds and their interests did not matter. Pink's (2006) work challenged this, focusing instead on the importance of what students have themselves learnt, and what meanings they have made of this, the critical place of joy, and creative thinking, as opposed to right answers, routines, and standardization. Pink's work also challenged the rational linear process of lesson planning and testing that was being pressed on the teachers and students by the policy context in which WSECP operated. Like James March (1972), whose work the teachers also read, Pink called for being playful and creative. For March and Pink being linear and rational had to be challenged.

The teachers also read Paul Churchland's (1996) research about the human brain. We discussed his research that all brains, except those of individuals with severe brain injury, have about 100 trillion modifiable synaptic connections. And, even with injuries, minds and brains greatly adapt. We also identified and discussed the importance of children and youth's mental plasticity

(Meltzoff, Kuhl, Movellan, & Sejnowski, 2009). These facts are so for all children and youth in various families and communities—wealthy and not so wealthy.

To test out this idea of a mind full of knowledge, teachers returned to their classrooms and actively sought to engage students in an open discussion which lasted several days. The discussion focused on students' insights of what they knew. In the process, teachers found out that the students knew a lot about many different areas of their lives. Often, these areas did not come to the attention of the teachers because the students had so few opportunities to talk, and in particular to talk about their knowledge, experiences, and backgrounds, and about their families' stories and experiences.

The other dimension that emerged from these investigations was the recognition that important knowledge exists in the families and communities of students. Work by N. Gonzalez, Moll, Tenery, Rivera, Rendon, R. Gonzales, and Amanti (1995), for example, has shown that Latino households, especially those in economic poverty, hold vast "funds of knowledge" (p. 445). Earlier work by John Kretzmann and John McKnight (1993, 1996) had found similar evidence of knowledge resources among families in Cabrini Green—a no longer existing public housing complex—in the late nineteen eighties in Chicago. McKnight's and Kretzmann's work of 'asset mapping' revealed that individuals and community-based organizations in a neighborhood shared knowledge and skills. Some community members assisted others who had no or limited mobility by providing, transportation to do grocery shopping or make physician appointments. Many others in the community helped out in similar ways to address challenges of fellow community members. We used the dialogue to discuss these examples and the ideas of assets and resources. Through these discussions teachers became more aware of these factors in the local neighborhoods surrounding WSECP. It now made sense to them that families have assets and knowledge and that students gain these assets in their family and community settings.

As the ideas about the children's knowledge and family resources started being acknowledged, there was also an emerging understanding of the importance of students' knowledge. We found that, when encouraged, students would be vocal about what they knew and about their interests. They had questions that interested them, and were puzzled by certain things. And these interests and questions varied among the students. There was not just only one set of questions or interests that would motivate and keep students' focus. Through the process of dialogues and changing classroom practices, teachers gradually embraced this variation. They started actively seeking students' knowledge and assets by asking students about what they knew, their interests, and what they had questions about, and by helping them to further develop such knowledge.

4.2 *Example 2—Breaking away from Established Ideas about What Learners Need to Learn at What Age: The Experience of Multi-Age Grouping Practices*

As teachers started working from and with students' knowledge and assets—working in ways that were fundamentally different from what they had done before—another regularity of schooling entered into the picture, and into the dialogue. This regularity concerned the established practice of working from grade level designations and expectations—something which in many ways appeared to be constraining learning.

So the dialogues then turned to grade levels, grade level material and skills, and the assignment of students to classes according to grade level. That meant challenging and questioning the graded school and graded curriculum structure.

In the first year of the school, only sixth and seventh grade students were enrolled. Each subsequent year, a new grade level was added, as the school grew into a middle and high school. Families and their children made the decision to attend this school knowing that it would eventually have students aged 11 through 18 years of age—or, as the school mental model would state, students in grades 6 through 12.[3]

The grade level dialogue initially looked at who the students were that were in the sixth and seventh grade classrooms. The students made clear to their teachers that they wanted to attend this school because they were not doing well in their previous school. But, they wanted to succeed. The teachers, students, and their parents were motivated by the hopes that had been expressed when the new school was established, namely that it would provide an opportunity for students to succeed, complete high school, and go onto post-secondary education.

The teachers realized that the students had not done well in their previous schooling and that they were not meeting the grade level expectations. For the teachers the students' lack of success reflected the fact that they did not know much at all and that they had little prior knowledge of school material or middle class experiences. Because of these shortcomings, the teachers worried that the students would not do well on the California State Tests which they had to take each spring. The success of the school would be judged by these test results. The accountability rhetoric that was occurring outside of the school community in the newspapers and legislature also raised questions about the prospects for student success.

In addition, the dialogues revealed that parents, students, and School Board members held tightly to these conceptions of grade level success. All had a mental model of schooling based on assumptions about grade level performance and content—sixth graders had to focus on particular sixth grade content and activities, and seventh graders had a respective focus on seventh grade materials and skills and knowledge. And if successful in 'learning' grade level material and skills, students would be promoted at the end of the school year to the next level. And, ideally, they would also score average or above on state assessment tests.

Questioning the idea of grade level was part of the discussions in our dialogues. My own mentor John I. Goodlad along with Robert Anderson co-wrote a book, *The Nongraded Elementary School* (1963). It grew out of Goodlad's dissertation study in 1948. In that study, like a number of subsequent studies, Goodlad explored the promotion and non-promotion issue (Raudenbush, 2005; Shepard & Smith, 1990). He found that students who were retained did no better than those that were promoted. However, those who were retained had a much lower sense of self, and their chances of dropping out increased significantly. Sadly, subsequent studies largely supported these findings, and the belief in the importance of not promoting those who do not do well continues today, with deleterious results for many retained students (Jimerson, 1999).

Another point that also helped move away from grade level content and skills and grouping practices had to do with our discussion of the history of the graded school. According to Frederick McClusky, the idea of grade level arose in 1848 in Quincy, Massachusetts and before that in Cincinnati, Ohio (McClusky, 1920). In those cities, the Industrial Revolution and its assembly line approach dominated. The production of goods and products in these settings reflected a standardized process of adding pieces of material to a product on an assembly line until a final product was produced at the end of the assembly line. In 1848, this model served as a guide and a metaphor for thinking and designing the curriculum and its graded school format. David Cohen and Jay Mehta (2017) have suggested that the graded school and graded textbooks became a way for schools to manage large numbers of students as enrollments in elementary and secondary schools increased over the decades.

The colleagues at WSECP found this history perspective particularly interesting and agreed that the strategy of holding pupils back made little sense. As a consequence of these deliberations, the teachers decided to mix students of different ages in the same group. The mixed-age grouping practice, despite some early discomfort, proved to be very successful. It showed that older students were more attentive to pro-social norms, and more self-regulated, and that they urged younger students to do the same. Teachers also saw improvements in student self-management, and therefore spent less time regulating students in these mixed age groups. As actual experiences with multi-age grouping increased, teachers encouraged larger age ranges in inter- and intra-class groupings. This resulted in multi-age groups which could go as far as covering the full range of 11 years to 18 years, although mostly, the groups would include 11 to 14 years of age, and 14 through 18 years. The issue of promotion and failure to meet grade level expectations diminished as students succeeded in these multi-age groupings.

4.3 *Example 3—Problem-based Investigations and Practicing Like the Practitioners in the Field*

This third example involves students engaging in problem-based investigations. Students' own questions, interests, and prior knowledge were central to these efforts. This approach challenged the conventional notions that it is academic and curricular standards that determine what pupils should learn, based on pre-determined and standardized lists of content.

The focus on problem-based investigations grew out of dialogues around the examinations at WSECP and how that related to the content of learning and to the curriculum. This discussion also reflected the issues examined in the second example above, about grade level and expectations.

The dialogues revealed that the lists of content to be learned mimicked similar curricular lists that had been drawn up more than a century ago. It appeared that little had changed since then. In fact, the standardization of the high school curriculum and structure dates back to the turn of the 20th century. Tyack and Tobin (1994) show, in *The "Grammar" of Schooling*, that the Carnegie Foundation, the College Board, and prestigious university presidents had joined hands at the turn of the 20th century to standardize the high school curriculum and structure:

> It was also necessary to develop a standard measurement of time and credit for each subject—the Carnegie unit—and to demand that a college require at least fourteen of these units. The Foundation did not stop there: it also went on for eight pages specifying in great detail the content of units in subjects like English, mathematics, Latin, Greek, foreign languages, history, and science. Thus, they standardized not only time and credits but gave pride of place to traditional academic subjects that were increasingly being organized into high school *departments*—another key element in the grammar of secondary schooling. (p. 461)

The resulting lists, then and now, are based on the idea that knowledge and information is acquired in a linear and slowly expanding manner. That reality probably justified the curricular decisions of the graded school curriculum in the eighteen hundreds—a list of limited knowledge and skills and content for each grade level to be covered in nine months. Students learning those facts in the mid-eighteen hundreds would then finish the required years (six or more years) of schooling with a rudimentary understanding of key areas of human knowledge and having a basis for being a citizen in a democracy. Later these lists became more extensive and the expectations of an informed and active citizenry more intense.

The dialogues led the teachers at WSECP to question the idea of these lists and of teaching students the material on these lists. Both the lists and the teaching of these lists constituted one of the regularities of schooling. Many teachers felt that presenting information (what we have called the 'that's) to students through a lesson, textbook page, or chapter and then requiring students to give that information back to the teachers in the form of a test or an assignment did not work well for the students.

One of the problematic aspects of the above regularities involved the 21st century explosion of information and knowledge, which has led to the characterization that we live in an Age of Information and Knowledge (Gleick, 2011). Much of what happened many centuries ago was never recorded or digitally archived. Yet, today, performances, or speeches, or even events that happen in an elevator or on a street corner can easily be stored on digital devices or recordings and then become information and sources for knowledge. For example, "In biology, the federally funded Human Genome project will store thousands of bytes for each of the several billion genetic bases" (Frawley, Piatetsky-Shapiro, & Matheus, 1992, p. 57). Examples like these testify to the rapidly exploding information and knowledge of this age (ibid.). This kind of discussion helped us rethink and take new actions regarding the usual lists of material of the curriculum of the school to be learned.

Another aspect in support of an alternative conception of today's world and the curriculum involved Thomas Friedman's (2005) assertion about the world in which we live being a flat world. His metaphor of flatness grew out of the realization of the world being a global world. It is interconnected among its people, knowledge bases, and shared ways of thinking. John Seely Brown and Richard Adler (2008) also provided us with the following added ideas about that world:

> Thanks to massive improvements in communications and transportation, virtually any place on earth can be connected to markets and can become globally competitive. But at the same time that world has become flatter, it has also become "spikier"; the places that are globally competitive are those that have robust local ecosystems or resources

> supporting innovation and productiveness. A key part of any such ecosystem is a well-educated workforce with the requisite competitive skills. And in a rapidly changing world, these ecosystems must not only supply this workforce but also provide support for continuous learning and the ongoing creation of new ideas and skills. (p. 17)

The regularity of a graded school curriculum and lists of content, which was founded in the mid-eighteen hundreds, differs from the one these writers accurately portrayed of today's world. The expectations for the graduates of schools in the old world had to do with filling industrial era jobs after a few years of schooling. In the new 'flat' and 'spiked' world, our expectations are that students will graduate from high school, go on to post-secondary programs, and be active, creative, and problem-solving graduates and citizens.

Through dialogue, we concluded that the curricular content was not responsive to this 'flat' and 'spiked' world. An alternative curriculum would have to embrace the explosion of knowledge and an ever-changing world. It would also have as its foundation our students' experiences, prior knowledge, and their communities' household knowledge, and it would prepare students for becoming the democratic citizens of the 21st century.

Our attention went away from lists of content to be learned to Bruner's ideas of promoting 'practicing.' In essence, practicing meant engaging with the subject matter in the manner that subject field specialists would do, so approaching history as a historian, or writing as a writer. To do so we identified criteria for practicing and thinking in each of the subjects and used them as guides for the students' investigations.

For example, in history the criteria which were identified encouraged students to "determine the relevance and credibility of sources by distinguishing the difference between fact and opinion, understanding different points of view and perspectives, identifying bias and prejudice." These were the criteria for practicing being a historian. Criteria for being a writer highlighted the importance of "generating ideas for writing from their personal reading, past experiences, observations and creative thoughts." And, for being a biologist, one of the key criteria was the importance of "formulating questions that can be answered through experimentation and observation" (WSECP, 2011).[4]

To encourage the use of these criteria and student investigations, we spent a considerable amount of time reading and distilling what we could learn from the literature, for example, in the field of cognitive research, from the research on out of school learning, and from studies on motivation (Heath, 2014; Strobel, Kirshner, Donoghue, & McLaughlin, 2008). This led us to identifying six conditions for student learning (the six circles in the graphic below,

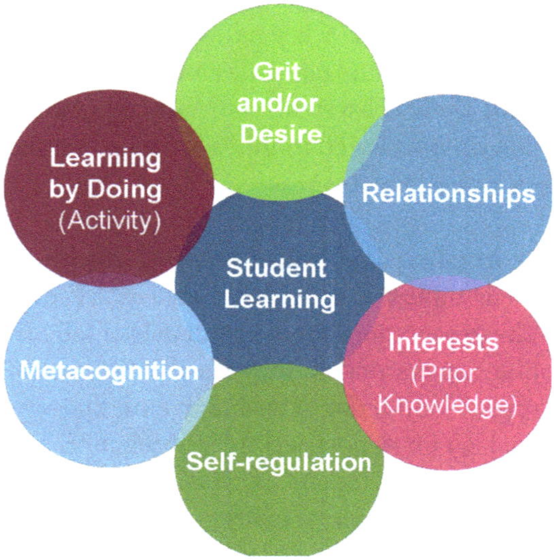

FIGURE 5.1
Six conditions for student learning

the center circle being the learner) which would guide teachers' and students' efforts to encourage investigations investigations (Figure 5.1).

Students were to make sure that the investigation focused on an idea or question they were interested in and had prior knowledge about. They were not just finding a topic like 'the stars and planets' but instead they sought and identified a compelling question—something they were wondering about—to guide their investigation. Teachers encouraged the students to think about and identify their 'wonderings.' And if students found themselves less focused on what they were investigating, the teachers would encourage students thinking (metacognition) and heighten their interests by relating what they were investigating to phenomena about which they knew a lot from their experiences.

In the resulting processes, student investigations became all about doing and experiencing many phenomena (touching moving and transforming materials), reflecting what Dewey called the sense of experience (Dewey, 2007). We also introduced public performances that constituted a form of assessment (Scardamalia, Bransford, Kozma, & Quellmalz, 2012). They occurred periodically, usually once a quarter. Students presented their investigations to each other and their teachers. Sometimes parents and community members also attended these events. At these public discussions students would describe and share their investigation. They would explain the question that guided the inquiry, the research literature that they reviewed, the evidence they found, the data collected, the analysis that led them to their findings, the conclusions drawn from their findings, and if they were to do another similar investigation the lessons learned about undertaking their investigation. Fellow students, teachers, parents and other community members actively discussed the criteria for the relevant subject/field and how the students' efforts reflected these. The discussions also highlighted the conditions for learning, and how those conditions aided the students in the investigation and in the learning process.

These performances elicited uplifting conversations. Parents, students, teachers and community members usually left those discussions with a realization that the students were learning important ideas and drawing worthwhile conclusions about the subject field from their investigations. The students clearly expressed their thoughts, showed the power of their metacognitive skills, and expressed their understandings about such things as science, writing, and history.

5 The Power of Indigenous Invention to Alter the Mental Models of Schooling

The examples above reflect the new ideas and practices that arose at WSCEP. These ideas emanated from the dialogues and were followed by actions in the classrooms. In essence, through the Indigenous Invention process, these actions challenged the usual regularities of schooling that had previously characterized schooling for students as well as teachers.

Students showed excitement about being able to express and explore their own questions. They responded positively to the learning conditions in the school. These conditions encouraged them to identify questions and investigate their own interests and queries. Many areas of investigation had personal meanings for them. For example, one student studied Parkinson's disease. She wanted to understand the nature of the disease and the various treatments that might mitigate its symptoms to help her father who had this disease. What she learned in the process allowed her to help her father sing using a karaoke machine, bringing great joy to him and the family. When her parents went to see the doctors, she went along and served as the spokesperson and advocate for her father. In this way, the expertise she gained from her investigations helped the family and her father. And, in the process, she gained insights about the practice of science.

These newly created features of schooling enabled students who had previously not succeeded in school, to succeed. The public assessment performances demonstrated that students were developing insights and actions. Most importantly, the students graduated from high school and many students went on to other post-secondary accomplishments. Graduates went on to matriculate and are now graduating from UC Davis, UCLA, and UC Santa Cruz,

TABLE 5.1 Grades obtained by a selection of WSECP pupils in courses at Sacramento City College

Grade	A	B	C	D	F
	42	28	11	4	1

for example. Others went to Sacramento State, some directly from WSECP. Yet others attended community college and some went into the military. If not all, almost all of the students who had started in the school from middle school and completed high school at WSECP have graduated and been successful in post-secondary educational settings.

Students excelled at many other areas. In 2013, for example, students at WSECP simultaneously took courses at Sacramento City College. They achieved excellent grades as shown in Table 5.1, further demonstrating that the students had gained knowledge and skills in the subject fields.

Teachers were enthusiastic about the student accomplishments. In interviews teachers reported gaining a sense of personal and professional satisfaction out of the dialogues and the Indigenous Invention process. Many teachers found the process exhilarating. Teachers referred to the intellectual and personal challenges and opportunities as moments of joy. Teachers enjoyed the freedom and opportunity to build and create the school and be the authors of what was happening in their classrooms. As might be expected, some also expressed feelings of frustration with certain parts of the process because of the ambiguity associated with the process of questioning and of exploring and inventing new concepts and ways of working. In dealing with the ambiguity and recognizing the accomplishments collegiality also grew.

The accomplishments that have been noted throughout this chapter were important to teachers, students, the principal, and me. All of us were directly a part of the school's daily development. By inquiring into the regularities of schooling, we understood them in a new way that enabled further inquiry and invention. The indigenous process had influence on those of us most directly involved in inventions—teachers, principal and students, and myself. The indigenous process is local and had influence on altering the mental models of those most directly involved in the daily invention, the inventors—students, teachers, and the principal.

Researchers have determined that the area that is most difficult to alter in schooling, and the one that is at the same time central to learning, concerns the interactions between students and teachers (Tyack & Cuban, 1995). It is clear that the Indigenous Invention process impacted the interactions between teachers and students, contributed to students' success, and altered the educational attainment trajectories for these students.

6 Further Challenges: Altering the *'Mental Model'* of Schooling beyond the School

The 'mental model' referred to in this chapter is pervasive throughout educational institutions and is embedded in how we think about and do school. WSECP teachers, students, and the principal, like in most schools, did not exist in isolation. The Charter of the school established a Board of Directors who oversaw the school's policies and administration. The Board consisted of administrators from the partner organizations that were supporting the school—the Dean of the School of Education at UC Davis, the Sacramento City College President, the Superintendent of the Washington Unified School district and several community members from West Sacramento. In addition, educators from the Department of Education and its Assessment Division advised the Board and the Western Association of Schools and Colleges (WASC), pre-collegiate division, served as the school's accrediting agency. Throughout the Indigenous Invention process, the principal and teachers periodically informed the Board of Directors of the process and progress of the inventions made. The Board members frequently participated in school events, including public assessments, and interacted with the families of the students. They were aware of the students' accomplishments. Some periodically, though not often, visited the school, during the eight year time period described in this chapter.

For Board members the main indicator of success was the school's scores on the Academic Performance Index (API).[5] Unfortunately, after an initial gain of 85 points on the API in 2011, the school showed a decline in the two subsequent years (with losses of 5 and 35 points in 2012 and 2013, respectively).

Ironically, those who had a supervisory role vis-à-vis the school felt that the only way to improve these scores was to strongly encourage teachers to teach to the test standards and to the test items. It did not matter that this suggestion contradicted the recommendations from the National Research Council which had raised concerns about such an approach, and had stated that: "It is important to remember that the scores on the tests used with incentives may give an inflated picture of learning with respect to the full range of the content standards" (National Research Council, 2011, p. 3; Glaser & Silver, 1994).

The focus on test scores was not aligned with my own commitment and that of the teachers, and principal to educational attainment. We had spent a lot of time figuring out what particular conditions needed to be met to enable students to undertake meaningful investigations. We had spent far less time discussing and tackling test score decreases and increases. Our focus was on optimizing the conditions for learning by enhancing the behavioral, social, and emotional engagement of the students (Mahatmya, Lohman, Matjasko, & Farb, 2012). Our focus was also on a higher goal of ensuring that the students at WSECP would eventually graduate from high school and from post-secondary educational settings and be able to function in and contribute to society in a meaningful manner.

The mental model of schooling, however, directed the attention of the Board to test scores, not to how students were being engaged and to their investigations, and even less to what would become of the students later in life. As the school prepared for its WASC review, the Board wanted the school to look like a school, with its usual regularities of schooling.

The principal of the school, who had very actively encouraged the inventions and accomplishments, found herself being investigated by the Board for what has been described as 'not following usual administrative practices.' This happened during a period when she had surgery and had been on a medical leave for over eight weeks. Before the Board could complete its investigation into the charges and present its findings, she resigned as principal. A district level position in another district had become available. She applied and accepted the offer to take the position. Some two months after her resignation, the Board sent her a letter noting that they had concluded their investigations and that they had found no evidence of any wrongdoing.

In the months immediately following her departure, one of the teachers served as the interim administrator of WSECP. It was a time of uncertainty for the teachers, students and me. Yet, at the same time, everyone continued to focus on the Indigenous Invention processes and on further improving the new structures and ways of learning, which they themselves had invented and contributed to.

The Board began looking for a new principal. The teacher who served as the interim administrator applied for the position. However, someone from outside the school, community, and region was hired. The Board was not interested in preserving the alternative structures and practices that the school had created.

During this time, the teachers also prepared for an accreditation visit from WASC. As part of the accreditation process the teachers and the interim principal were required to submit a Self-Study Report with responses to the WASC's accreditation questions.

The WSECP Self Study Report reflected teachers' views and commitment to the alternative ideas and practices that they had developed. For example, a question from WASC in the Self Study Report asked for "clarification and measurement of what all students should know, understand, and be able to do through learner outcomes and academic standards."

The response by the teachers provided a summary of the results of several years of dialogue and discussion about the "Habits of Mind necessary for post-secondary success." The teachers listed the following habits

- Thinking
- Action and doing;
- Producing and innovating; and
- Application.

Under each of these headings, teachers developed questions that guided their work and students' work. Brief examples of what the WSECP teachers developed follow:

- Thinking—What are the questions you explored in your studies? Why are these questions important?
- Action and doing—How did you demonstrate sound reasoning and an inquiry stance in your investigations and conclusions?
- Producing and innovating/application—How did your study and findings contribute to an improved community and society?

The WASC reviewers received this report prior to their school visit, which took place during the new principal's first year. Unfortunately, the feedback provided by the WASC reviewers reflected that they wanted the school to fit the traditional mental model of schooling. WASC's response to the material provided in the Self Study documents was largely negative (West Sacramento Early College Prep Charter, 2014) and the WASC team determined that the school's program did not meet their criteria and would not receive full accreditation until these expectations were met. Clearly, the WASC reviewers expected that the Self Study Report would respond to the standards of a typical school, and that its program would align to those standards. Most critically, they had an expectation for grade level content standards. That meant there had to be grade level content criteria and grade level courses and classes—a regularity that had been altered at the school as previously discussed. Obviously, the school's efforts did not match these outside expectations. The response from WASC affirmed the concerns held by the Board.

By the end of the first year of the new principal's tenure, almost all the core teachers—many of whom had been at the school since its inception—left. Some chose to depart for teaching positions at other schools. Others were asked by the new principal to leave. They, too, found other teaching positions.

Reflecting on this situation and examining the shift back to the regularities that we had worked so hard to challenge during those eight years has been difficult. It has, however, offered additional insight regarding the reform process and the focus of Indigenous Invention. Our inventions did not match the mental models of schooling of those who were not a part of the direct inventive process. During the new principal's tenure, she returned the school to a school matching the regularities of schooling and the traditional mental models of schooling that the indigenous process had challenged and altered. No doubt, this also matched her mental models of schooling.

It is clear that the Indigenous Invention process had a beneficial influence on those directly involved during the eight years. It altered their mental models of schooling, but the mental models of schooling held by those distant from the ongoing invention process at the school were less influenced. Unfortunately, this directly impacted the school's ability to sustain the invented practices and structures of the school. In retrospect, it might have been beneficial had more attention been paid to allowing those affiliated with the school, such as the Board members and other policy makers, like WASC reviewers, to change *their* mental models of schooling.

In many ways, this experience echoes what Tyack and Tobin concluded about the challenges of sustaining educational reform:

> Cultural constructions of schooling have changed over time and can change again. To do this deliberately would require intense and continual public dialogue about the ends and means of schooling, including reexamination of cultural assumptions about what a 'real school' is and what sort of improved schooling could realize new aspirations. Shared beliefs could energize a broad social movement to remake the schools. To do so would require reaching beyond a cadre of committed reformers to involve the public in a broad commitment to change. (Tyack & Tobin, 1994, p. 478)

A next step for Indigenous Invention will seek to create opportunities for further dialogues beyond the school walls. Figuring out ways to extend Indigenous Invention and alter mental models of schooling to the larger community of policy makers may provide means to maintain and enhance the educational renewal and sustain the inventions created at a school like WSECP.[6]

Notes

1 This is federal designation for second language learners of English and refers to children whose native language at home is a language other than English. In the State of California this is mostly Spanish, but can also include many other languages.

2 See chapter 6, Programmatic and Behavioral Regularities, in Sarason (1996, pp. 95–118).

3 Then, each subsequent year, the school would add a new group of 6th graders and, where feasible, also add older grade students to the school community.

4 The statements quoted in this paragraph came from documents developed by the teachers and the author. They once had a place on the WSECP web site, which no longer exists.

5 The API rankings were based on student performance on the STAR (California's Standardized Testing and Reporting program) and on the California Standards Test (CST).

6 On February 16, 2017, the Board of Education of the Washington Unified School District approved the "renewal and material revision" of the West Sacramento Early College Prep Charter. These approved changes renamed the school, Washington Middle College High School. It would serve only grades 9–12, changing the age focus of the school. In this revision, the WSECP Board was dissolved and the governance of the school shifted to the school district's governing board. In addition, the partnership that established WSECP was dissolved with Sacramento City College remaining as a partner to assist students in earning an associate degree during their high school years.

References

Berman, P., & McLaughlin, M. W. (1978). *Implementing and sustaining innovations: Federal programs supporting educational change* (Vol. 8) (Prepared for the US Office of Education, DEH). Santa Monica, CA: Rand Corporation.

Bruner, J. S. (1996). *The culture of education.* Cambridge, MA: Harvard University Press.

Buchmann, M. (1987). Reporting and using educational research: Conviction or persuasion. In J. I. Goodlad (Ed.), *The ecology of school renewal, eighty-sixth yearbook of the National Society for the study of education* (pp. 170–191). Chicago, IL: The University of Chicago Press.

Churchland, P. M. (1996). *The engine of reason, the seat of the soul: A philosophical journey into the brain.* Cambridge, MA: MIT Press.

Cohen, D. K., & Mehta, J. D. (2017). Why reform sometimes succeeds: Understanding the conditions that produce reforms that last. *American Educational Research Journal, 54*(4), 644–690.

Coleman, J. S. (1966). *Equality and educational opportunity.* Ann Arbor, MI: Inter-University Consortium for Political and Social Research.

Deschenes, S., Cuban, L., & Tyack, D. (2001). Mismatch: Historical perspectives on schools and students who don't fit them. *Teachers College Record, 103*(4), 525–547.

Dewey, J. (2007). *Experience and education.* New York, NY: Simon and Schuster.

Duguid, P. (2005). "The art of knowing": Social and tacit dimensions of knowledge and the limits of the community of practice. *The Information Society, 21*(2), 109–118.

Frawley, W. J., Piatetsky-Shapiro, G., & Matheus, C. J. (1992). Knowledge discovery in databases: An overview. *AI Magazine, 13*(3), 57.

Friedman, T. L. (2005). It's a flat world, after all. *The New York Times, 3*, 33–37.

Garcia, S. B., & Guerra, P. L. (2004). Deconstructing deficit thinking: Working with educators to create more equitable learning environments. *Education and Urban Society, 36*(2), 150–168.

Gelfand, M. J., Lim, B. C., & Raver, J. L. (2004). Culture and accountability in organizations: Variations in forms of social control across cultures. *Human Resource Management Review, 14*(1), 135–160.

Glaser, R., & Silver, E. (1994). Assessment, testing, and instruction: Retrospect and prospect. *Review of Research in Education, 20*(1), 393–419.

Gleick, J. (2011). *The information: A history, a theory, flood.* New York, NY: Pantheon Books.

Gonzalez, N., Moll, L. C., Tenery, M. F., Rivera, A., Rendon, P., Gonzales, R., & Amanti, C. (1995). Funds of knowledge for teaching in Latino households. *Urban Education, 29*(4), 443–470.

Goodlad, J. I. (1954). Some effects of promotion and nonpromotion upon the social and personal adjustment of children. *The Journal of Experimental Education, 22*(4), 301–328.

Goodlad, J. I. (1975). *The dynamics of educational change: Toward responsive schools.* New York, NY: McGraw-Hill.

Goodlad, J. I. (1984). *A place called school. Prospects for the future.* New York, NY: McGraw-Hill Book Company.

Goodlad, J. I., & Anderson, R. H. (1963). *The nongraded elementary school* (Revised). New York, NY: Harcourt Brace & World.

Heath, S. B. (2014). Creative "Garages" for community and economic youth development. *Teachers College Record, 116*(14), 493–514. Retrieved October 26, 2016 from http://www.tcrecord.org ID Number: 18303.

Heckman, P. E., & Montera, V. L. (2009). School reform: The flatworm in a flat world– from entropy to renewal through indigenous invention. *Teachers College Record, 111*(5), 1328–1351.

Hong, G., & Raudenbush, S. W. (2005). Effects of kindergarten retention policy on children's cognitive growth in reading and mathematics. *Educational Evaluation and Policy Analysis, 27*(3), 205–224.

Howard, T. C. (2013). How does it feel to be a problem? Black male students, schools, and learning in enhancing the knowledge base to disrupt deficit frameworks. *Review of Research in Education, 37*(1), 54–86.

Jackson, P. W. (1973). *Life in classrooms.* New York, NY: Holt, Rinehart and Winston.

Jimerson, S. R. (1999). On the failure of failure: Examining the association between early grade retention and education and employment outcomes during late adolescence. *Journal of School Psychology, 37*(3), 243–272.

knowledge and the limits of the community of practice. *The Information Society, 21*(2), 109–118.

Kretzmann, J. P., & McKnight, J. L. (1993). *Building communities from the inside out: a path toward finding and mobilizing a community's assets.* Evanston, IL: Center for Urban Affairs and Policy Research, Northwestern University.

Kretzmann, J. P., & McKnight, J. L. (1996). Assets based community development. *National Civic Review, 85*(4), 23–29.

Lortie, D. (1975). *School teacher: A sociological perspective.* Chicago, IL: University of Chicago Press.

Mahatmya, D., Lohman, B. J., Matjasko, J. L., & Farb, A. F. (2012). Engagement across developmental periods. In L. Christenson, A. L. Reschly, & C. Wiley (Eds.), *Handbook of research on student engagement* (pp. 45–63). New York, NY: Springer.

March, J. G. (1972). Model bias in social action. *Review of Educational Research, 42*(4), 413–429.

McClusky, F. (1920). Introduction of grading into the public schools of New England, Part I. *The Elementary School Journal, 21*(1), 34–46. Retrieved from http://www.jstor.org/stable/993768

Meltzoff, A. N., Kuhl, P. K., Movellan, J., & Sejnowski, T. J. (2009). Foundations for a new science of learning. *Science, 325*(5938), 284–288.

National Research Council. (2011). *Incentives and test-based accountability in education.* Washington, DC: National Academies Press.

Olson, D. R., & Bruner, J. S. (1996). Folk psychology and folk pedagogy. In D. R. Olson & N. Torrance (Eds.) *The handbook of education and human development: New models of learning, teaching and schooling* (pp. 9-27). Malden, MA: Blackwell Publishing.

Payne, C. (2008). *So much reform, so little change: The persistence of failure in urban schools.* Cambridge, MA: Harvard University Press.

Pink, D. H. (2006). *A whole new mind: Why right-brainers will rule the future.* New York, NY: Penguin.

Pink, D. H. (2009). *Drive: The surprising truth about what motivates us.* New York, NY: Penguin.

Sarason, S. B. (1972). *The creation of settings and the future societies.* Cambridge, MA: Brookline Books.

Sarason, S. B. (1982). *The culture of the school and the problem of change.* Boston, MA: Allyn and Bacon.

Sarason, S. B. (1996). *Revisiting "The culture of the school and the problem of change."* New York, NY: Teachers College Press.

Scardamalia, M., Bransford, J., Kozma, B., & Quellmalz, E. (2012). New assessments and environments for knowledge building. In P. Griffin, B. McGaw, & E. Care (Eds.), *Assessment and teaching of 21st century skills* (pp. 231–300). Dordrecht: Springer.

Schön, D. A. (1992). The theory of inquiry: Dewey's legacy to education. *Curriculum Inquiry, 22*(2), 119–139.

Seely Brown, J., & Adler, R. P. (2008). Open education, the long tail, and learning 2.0. *Educause Review, 43*(1), 16–20.

Shepard, L. A., & Smith, M. L. (1990). Synthesis of research on grade retention. *Educational Leadership, 47*(8), 84–88.

Sirotnik, K. (1983). What you see is what you get-consistency, persistency, and mediocrity in classrooms. *Harvard Educational Review, 53*(1), 16–31.

Spillane, J. P., Reiser, B. J., & Reimer, T. (2002). Policy implementation and cognition: Reframing and refocusing implementation research. *Review of Educational Research, 72*(3), 387–431.

Strobel, K., Kirshner, B., Donoghue, J., & McLaughlin, M. (2008). Qualities that attract urban youth to after-school settings and promote continued participation. *Teachers College Record, 110*(8), 1677–1705.

Tyack, D., & Cuban, L. (1995). *Tinkering toward utopia: A century of public school reform.* Cambridge, MA: Harvard University Press.

Tyack, D., & Tobin, W. (1994). The "grammar" of schooling: Why has it been so hard to change? *American Educational Research Journal, 31*(3), 453–479.

West Sacrameno Early College Prep Charter. (2011). *WSECP assessment criteria* (Unpublished document).

West Sacramento Early College Prep Charter. (2014, November 17–19). *West Sac Early College Prep Charter Self Study Report* (WASC/CDE Focus on Learning Accreditation Manual, March 2013.)

INTERMEZZO 5

When the Sky Is Not the Limit, It Could Be the Beginning

Jan Visser

Reflections inspired by George Miley, Carolina Ödman and Pedro Russo: 'The inspiring universe'

For some of us the sky is the beginning of the lifelong process of asking questions, so important for building the scientific mind. Watching the spectacular and enigmatic beauty of the light show in the heavens above us is what links us perhaps most forcefully to our most remote ancestors. I remember well walking, as a kid, the streets of Rotterdam, site of a major mass destruction at the start of World War II, which had left the city with large open spaces, only sparsely illuminated at night, if at all. What I saw inspired me to draw maps of the sky and the surface of the moon and explore an atlas at home with information about the solar system. It instilled in me the desire, when I grew a little older, to travel to Mars and the moon, even though space technology still had to evolve out of military technology, Werner von Braun being the bridge. Ironically, one of Werner's armed rockets, a V1, on its way to London, had accidentally hit, less than two months before the end of the war, a block of houses not far from where we lived. The space art by Chesley Bonestell[1] provided further inspiration. Even now, when I watch the sky as I saw it then—which I can only still do at such places as rural Namibia and Botswana, or, most recently, the Indonesian island of Timor—I feel that something opens up in me, awakening the irrepressible urge to deepen my understanding.

I was reminded of my childhood experience when, in 2013, I visited WestSac, the school mentioned in Intermezzo 4, and a boy told me he was studying black holes. WestSac was not an ordinary school. Surely, mine had been an out-of-school experience. What occurred to this boy happened in a school environment that recognized the importance of autonomously developed interests in the attending students without forcing them through a set curriculum.

Ideally, attending to and nurturing such interests should neither depend solely on the informal environment, nor on formal settings alone. Both environments should be considered integral components of a comprehensive learning landscape. They should be aware of each other and be interoperable.

 | DOI: 10.1163/9789004416802_010

That's where Universe Awareness (UNAWE)[2] comes in. George Miley and two of his colleagues discuss this initiative, which started in 2005, in their chapter on 'The inspiring universe.' The opportunities offered by UNAWE for building, and continually nurturing, the scientific mind work across platforms, both formal and informal.

Creating awareness of the universe works obviously best if the wonders of the sky are always visible. It should not depend on destroying a city to make that possible. Combatting light pollution may be as essential to the survival of humankind as clean air and the tranquility of spaces where humans do not break the silence.

Notes

1 See https://en.wikipedia.org/wiki/Chesley_Bonestell

2 See https://www.unawe.org/

CHAPTER 6

The Inspiring Universe

George Miley, Carolina Ödman and Pedro Russo

1 Introduction

1.1 *Preamble*

Humans like to find things out. They naturally ask questions and explore the world around them. One observes such behavior already in neonatal infants. Inspiration to further develop such exploratory behavior, the quest to understand and do so deeply, is stimulated by different things, depending on the environment in which one is born and grows up. We thus do not all have the same chances to be inspired. However, there is one exception. Throughout human history humans around the world have stood in awe of the wonders of the sky, the universe. This is a source of inspiration that unites us all. The above considerations and possibly other ones have led to the Universe Awareness (UNAWE) project, which focuses specifically on underprivileged children between the ages of four and ten. Universe Awareness (UNAWE) uses the beauty and grandeur of the Universe to encourage young children, particularly those from an underprivileged background, to have an interest in science and technology and foster their sense of global citizenship from the earliest age.

Key precepts of the *Building the Scientific Mind* approach to education are (i) to cultivate habits of rational thought and (ii) to stimulate the acquisition of ethical values. These principles are also at the heart of the Universe Awareness philosophy. In this chapter we shall outline the rationale on which UNAWE is based and describe the various elements and accomplishments of UNAWE.

The famous adage "Give me a child until he is 7 and I will give you the man" is attributed to Francis Xavier, 16th century co-founder of the Jesuit Order. Except for the implicit sexism of the time, this quotation is still very relevant for education. To influence the developments of a person's character and values, it is important to begin at an early age. As argued e.g. by Gopnik, Meltzoff, and Kuhl (1999), babies and young children are curious and this natural curiosity should be stimulated. By the age of 7, many seeds of the basic attitudes and values that will govern the behavior of the future adult will have been sown. The nature of these values can determine whether a society is tolerant or intolerant, exclusive or inclusive. It is not coincidental that many world religions regard the control of schools for young children as crucial to their existence.

 | DOI: 10.1163/9789004416802_011

The philosophy that governs the education of young children can strongly influence the value systems and ethical standards of society.

2 Astronomy and Space as Educational Tools

Astronomy is an effective holistic educational tool because it embodies a unique combination of science, technology and culture (Figure 6.1). The cultural roots of astronomy satisfy the deepest philosophical yearnings by observing far into our past universal history. Unraveling the history of the Universe has been one of the great human achievements of the last half-century.

In the 21st century, astronomy has become a gateway to physics, chemistry, biology and mathematics and a driver of advanced developments in electronics, optics and information technology. Ground-based and space observatories have opened up a golden age of astronomy. Many profound discoveries have been made that have changed our view of the Universe. We have learned that some objects that inhabit our Universe are so strange that their existence could not have been predicted. Others are so spectacular that they stretch the boundaries of human wonder. The beautiful and intriguing images produced by high-tech telescopes and satellites can capture the imagination of children

FIGURE 6.1 The wide applicability of astronomy and space sciences as an educational tool and a gateway to science, technology and culture (International Astronomical Union Strategic Plan, 2011; Miley, 2012)

of all ages. Astronomers continue to push the frontiers of knowledge with global scientific collaborations. During the past decade the discovery of planets outside our Solar System and the realization that many billions of planets must inhabit our Galaxy has profound implications for human civilization.

2 The Universe Awareness Programme

2.1 *Astronomy as a Motivator of Young Children*

The excitement of astronomy and the adventure of space are particularly important vehicles for capturing the imagination of young children, introducing them to science and engineering and helping to stimulate their ethical value systems. There are several reasons for this.

- Consideration of the vastness and beauty of the Universe and of our place within it provides a special perspective that can help broaden the mind and stimulate a sense of global citizenship.
- The ability of space and astronomy to interest and *motivate* young children can be used to stimulate education in a broad holistic sense. Space-based themes can be encapsulated in stories to further language skills and to pose and solve problems that develop numerical skills.
- Astronomy provides a seductive introduction to science and technology. Many scientists and engineers trace their first interest in science to exposure as very young children to the fascinating Universe. Although the main goals of Universe Awareness are not directed at producing future scientists, exposure to intriguing aspects of space can 'light a spark' in a young child that several years later causes her to embark on a scientific or technical career.

2.2 *Genesis of Universe Awareness*

Based on the rationale presented above, UNAWE was initiated in 2005 to exploit scientific, educational and social dimensions of astronomy for the education of young children (Miley, Madsen, & Scorza, 2005). Directed at children between 4 and 10 years, UNAWE aims to broaden children's minds, awaken their imagination and curiosity in science and encourage respect, tolerance and global citizenship. A major goal is to stimulate children to develop into curious, tolerant and internationally oriented adults (Ödman, Scorza, Miley, & Madsen, 2006; Ödman, 2007, 2011).

The main ingredients of UNAWE are:

- Provision of an international network for astronomy as an educational tool for motivating very young children.
- Development of country-specific educational resources,

– Organization and stimulation of training for teachers,
– Development and implementation of a consistent evaluation framework.

UNAWE is based on the philosophy that the early formative years are crucial in the development of the human value system. From the age of 4 children can readily appreciate and enjoy the beauty of astronomical objects and can learn to develop a 'feeling' for the vastness of the Universe. On the one hand, the discovery of the universe is an excellent and exciting introduction to the scientific method and the concept that nature can be interrogated by rational means. On the other hand exposure to inspirational astronomical observations and space data can help broaden the minds and stimulate a world-view. The enormity of the Universe is also a subject for deep wonder. In particular, the Universe contains countless exotic objects that are ideal for feeding a child's imagination. To quote the European biologist J. B. S. Haldane, "The Universe is not only queerer than we suppose, but queerer than we can suppose" (Haldane, 1927, p. 286). The strange conditions that exist elsewhere in the Universe and beautiful pictures from modern telescopes are an ideal backdrop for UNAWE activities.

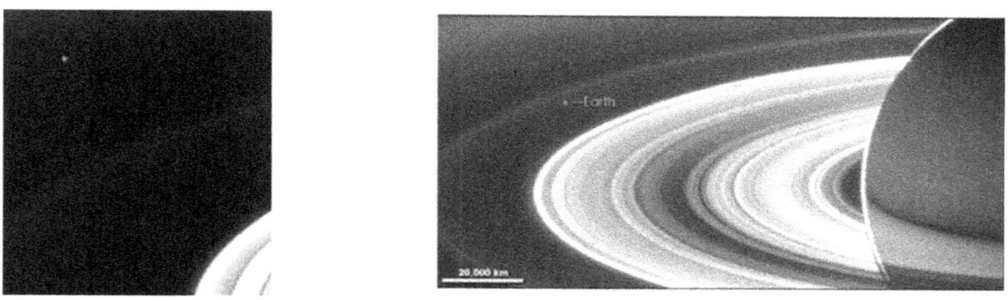

FIGURE 6.2 An iconic picture of the Earth as Sagan's "faint blue speck" (blow-up on left). The ESA-NASA Cassini spacecraft took this photo from a location close to Saturn, whose rings are shown. The blow-up is part of the larger Cassini picture of Saturn (right). (Courtesy NASA)

Studies of other planets in our solar system provide important lessons relevant to protecting the environment of the earth. To quote Carl Sagan (1980 [1990 update]):

> The discovery that such a thing as nuclear winter was really possible evolved out of the studies of Martian dust storms. The surface of Mars, fried by ultraviolet light, is also a reminder of why it's important to keep our ozone layer intact. The runaway greenhouse effect on Venus is a valuable reminder that we must take the increasing greenhouse effect on Earth seriously.

2.3 *Why Does* UNAWE *Emphasize Underprivileged Communities?*

Although the resources of UNAWE are open to all, UNAWE concentrates on children from relatively underprivileged communities. The rationale for this is threefold:

1. Early educational interventions during the young formative years are crucial in stimulating child development (OECD, 2006; Worth & Grollman, 2003; Rocard, 2007).
2. There is an enormous untapped potential and talent in underprivileged communities. Such communities are frequently alienated from society at large and this alienation begins at childhood. According to a UNICEF report (UNICEF, 2006), "The opportunity to help disadvantaged children have a more equal start in schooling is in the earliest years when the basis for their cognitive, social and emotional development is being formed" (p. 1).
3. Educating very young children from underprivileged communities is highly cost effective (Schweinhart, Mont, Xiang, Barnett, Belfield, & Nores, 2005), but needs a special approach (Early Years Learning Framework for Australia, Commonwealth of Australia, 2009; Arnold & Doctoroff, 2003). A series of important studies by James Heckman, recipient of the Nobel Memorial Prize in Economic Sciences, and his group emphasizes the cost-effectiveness of early educational intervention in underprivileged communities with pre-school activities that motivate very young children (Cunha, Heckman, Lochner, & Masterov, 2006; Heckman & Masterov, 2007; Heckman, 2000, 2008) (see Figure 6.3). However, it is clear that further such studies are needed in a range of diverse environments and cultural settings.

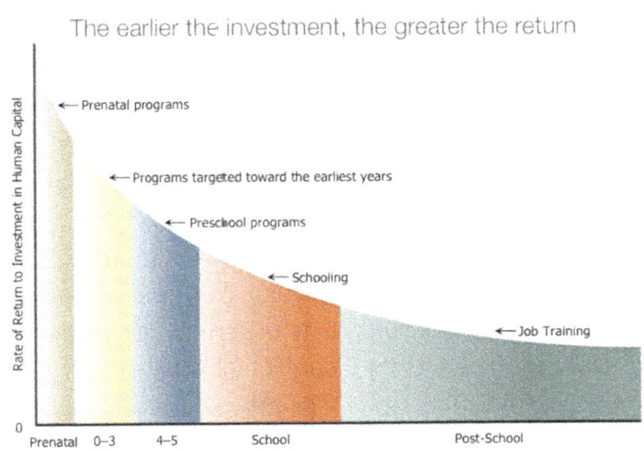

FIGURE 6.3 This graph demonstrates that the rate of return on human capital investment decreases with age, with the highest return on investments at preschool age (from Heckman, 2008)

2.4 *Why Does UNAWE Target Teachers?*

One of the most important components of UNAWE programs is teacher training. There are several reasons for this.

- The first is to increase the reach of the program. The effectiveness of reaching children is multiplied by a large factor through targeting their teachers. Enthusiastic teachers can 'infect' thousands of children with their enthusiasm.
- The second reason is that, while elementary school teachers have a substantial influence on the first school years of young children, they often feel inadequate to teach STEM (science, technology, engineering and mathematics) issues (e.g. Peacock, 1991; Beilock, Gunderson, Ramirez, & Levine, 2009). Universe Awareness training is designed to build the confidence of teachers in teaching these topics. We demonstrate to teachers that space can be an important tool for motivating young children and introducing them to the excitement of science and technology. After they are motivated, this excitement can also be exploited e.g. to teach children language skills through reading space stories or math skills through solar system problems.
- An indirect factor in reaching out to teachers is that elementary school teachers are highly undervalued in many countries. Membership of the international (multidisciplinary) UNAWE network can help enhance the prestige of school teachers, particularly in some developing countries.

2.5 *Furthering Gender Equality*

In many parts of the world, women show less interest in science and engineering than men. Several reports and studies have identified reasons why girls are significantly more negative about certain science subjects and careers than boys and traced the views and attitudes of boys and girls about science to influence from an early age (e.g. Osborne & Collins, 2000). Gender imbalance needs to be addressed at all levels of education and all career stages. It is crucial to dispel the myth that girls can't do science and that technical subjects are better left to the boys.

Since its inception UNAWE has had furthering gender equality as an important goal. The education and upbringing of very young children can play a substantial role in helping to tackle gender imbalance. During ages from 4 to 10 years the value systems of children form and preconceptions about gender roles begin to take shape. Furthermore, in most countries throughout the world elementary school teachers tend to be women. By building up the confidence of these teachers in science-related topics, the proposed teacher training courses will contribute to projecting an image of women who are confident in science and engineering.

3 Growth and Accomplishments of UNAWE

3.1 *UNAWE—A Global Activity*

Universe Awareness was developed as an international partnership with a bottom-up structure. The program is unique in that it is the only global project that links top international researchers to the education of young children and the training of their teachers. Local experts drive the implementation of the program in each partner country and give special attention to local cultural needs and histories. Since its initiation in 2005 UNAWE has grown into a multidisciplinary network with more than 500 participants and is active in more than 57 countries worldwide (see Figure 6.4), with more than 22 different languages.

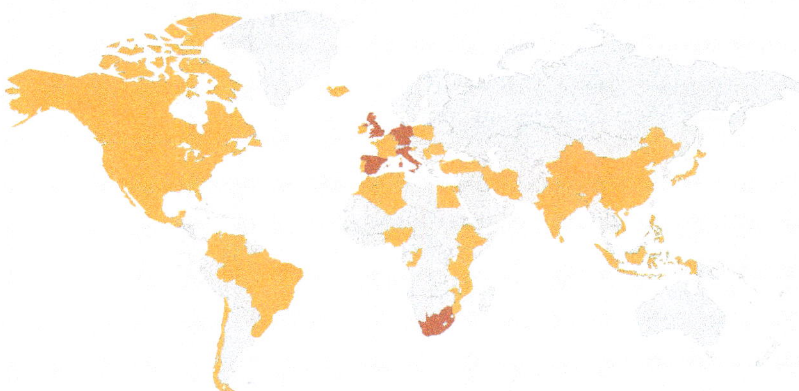

FIGURE 6.4 The 57 UNAWE member countries (in brown), and the 6 countries participating in the FP7 DG Enterprise and Industry EU-UNAWE project in red

UNAWE was adopted as a cornerstone program of the UN-ratified International Year of Astronomy 2009 and an official activity of the International Astronomical Union. At its inception UNAWE was endorsed by several Nobel Prize winners (for example Professor Riccardo Giacconi, 2002 Nobel Prize winner; Professor Joseph Taylor: 1993 Nobel Prize winner, Professor Charles Townes: 1964 Nobel Prize winner). UNAWE has also received considerable attention internationally, including a feature article in *Physics Today*, the professional magazine of the American Institute of Physics (Feder, 2007). A highlight was the award to UNAWE of the 2011 Science Magazine Science Prize for Online Resources in Education (SPORE) (Ödman-Govender & Kelleghan, 2011). Because of its success, UNAWE was mandated as expert for e.g. the UNESCO World Report: "Investing in Cultural Diversity and Intercultural

Dialogue" and invited to participate in the setting up of an Outer Space Committee of the Conference of NGOs in consultative relationship with the United Nations (CONGO).

In 2010, UNAWE received a considerable boost with the award of a 2 million Euro 3-year project by the space program of the European Commission. "EU Universe Awareness (EU-UNAWE) is implementing substantial UNAWE programs in five European countries (Germany, Italy, the Netherlands, Spain, and the UK) and South Africa (e.g. Russo et al., 2011; Schrier, Nijman, & Russo, 2012). EU-UNAWE exceeded its original goals by a wide margin. EU-UNAWE trained almost 1100 teachers and reached more than 60,000 pupils through direct and indirect activities and was evaluated highly by independent reviewers appointed by the European Commission.

During the course of EU-UNAWE an extensive assessment was conducted among teachers and educators in all six participating countries to study the effectiveness of the teacher training and the resources. It was concluded that the project was highly effective and efficient in developing motivation, critical thinking, knowledge and interest in space, and intercultural attitudes.

3.2 *Universe Awareness Resources and Their Development*

The development of suitable resources for teachers is an essential element of UNAWE. These resources attempt to assist teachers to deal with important concepts while at the same time emphasizing the wonder and excitement of the Universe.

A major achievement of the EU-UNAWE project has been to create a repository of open-source age-related educational materials in several languages for teaching children ages 4 to 10 and to make them available freely to teachers and parents via various national websites as well as the UNAWE international website. UNAWE resources undergo stringent peer review to ensure their pedagogical quality and accurate scientific content. This review process developed in the context of the EU-UNAWE project was adopted by the "Task Force for Schools and Children" of the International Astronomical Union Office of Astronomy for Development and is embodied in the IAU-endorsed *astroEDU* platform for educational resources[1] (Russo et al., 2015).

Almost 200 resources have been produced to date. UNAWE 'flagship' resources are the popular *UNAWE Earthball* and *Universe in a Box*. The *UNAWE Earthball* is a simple inflatable globe without borders. It shows the earth from the perspective of Space and illustrates important aspects of our planet such as its shape, its axial tilt and its surface composition. The accompanying activity booklet contains 50 interactive, hands-on activities for children aged between

6 and 12 years that can be put into practice in schools, playgrounds or at home. They cover a broad range of topics, including astronomy, geography, environmental awareness, global citizenship and even biology.

Universe in a Box is an educational kit that provides teachers and educators with more than 40 practical activities as well as the materials and models required to carry them out. It has a modular design with five modules: Our Fascinating Moon; The Earth—Our Home Planet; The Sun—Our Home Star; Our Solar System; and The World of Constellations. In accordance with the UNAWE philosophy these were designed to

- Encourage inquiry-based learning among children, by involving observing, discussing, drawing conclusions and presenting.
- Link astronomical topics with other subjects such as mathematics, art, and philosophy to support interdisciplinary learning and present a more holistic view of our universe.
- Make children aware that they are inhabitants of a small fragile planet.
- Foster respect for other cultures.

Another accomplishment of EU-UNAWE was the development of 'Space Scoop.' Space Scoop is an astronomy news service for children aged 8 and above. The latest press releases from leading astronomical ground and space observatories are rewritten, so that they are understandable for children.[2] The Space Scoop case study is described in more detail in the next section.

4 Universe Awareness Case Studies in Innovation

4.1 *Case Study 1: Teacher Training—Duo Internships*

During 2009–2010 and in 2014, the Universe Awareness project organized and coordinated a pilot project in teacher training in the Netherlands involving 'duo-internships.' During these internships, pairs of student primary school teachers and astronomy university students were coupled and charged with developing astronomy lessons and implementing these lessons at primary schools. Nineteen such 'duos' participated in activities at primary schools in Groningen, Arnhem, The Hague, Rotterdam and Utrecht. The combination of an astronomy student and a student teacher ensured that lessons were both scientifically correct and pedagogically sound. Moreover, the students learned from each other. The astronomy students learned how to explain science in a way that the children could understand and the aspiring teachers learned to master some simple inspirational scientific concepts. These duo-internships were experienced by all concerned as highly successful.

FIGURE 6.5 Girls taking part in UNAWE activities in Iran

4.2 *Case Study 2: Educational Resources—Space Scoop News Service*

An innovative resource developed in the course of implementing the EU Universe Awareness project was 'Space Scoop,' a news service for older children.[3]

Science is often perceived by children as a dull subject (Allodi, 2002) and an aim of Space Scoop is to change this. By sharing exciting topical astronomical discoveries with them, Space Scoop attempts to inspire children and cultivate their interest in science and technology. Space Scoops are short news articles about astronomical discoveries written in a child-friendly language and accompanied by a stunning astronomical image. Space Scoop is a tool that can be used in many different settings to teach, share, and discuss the latest astronomy news.

The main educational goals of Space Scoop are the following ones:

- Share the excitement of new discoveries with children.
- Enhance understanding of the world and encourage critical thinking.
- Broaden children's minds and cultivate world citizens.

Each Space Scoop begins with a new discovery or observation in the field of astronomy and space science. Press releases of partner organizations (including ESO,[4] NAOJ,[5] ESA,[6] NASA[7]) are translated into a child-friendly language. Similarly, Space Scoop undergoes an internal educational review to assure the highest quality. The final texts are then sent to our group of volunteer translators from all over the globe, where they are translated into up to 33 languages. The final Space Scoops are released to the public at the same time as the original scientific press release.

Space Scoop is a versatile resource which can be adapted to many different formats and digested by many audiences, including children aged 8 and above, the visually-impaired, teachers and science museums. Its wide reach demonstrates that it meets a considerable demand. Among outlets for Space Scoop are several magazines for children, including *Anorak* and *Timbuktu*; the American Association for the Advancement of Science; *Eurekalert!*; a global news service, Goa's *Nahvind Times*; the Dutch science magazine, *Universum*; *Wired.com*; the popular Indonesian website, *Langitselatan*; *National Geographic Indonesia*; and the Slovenian magazine and website, *Portal v Vesolje*. A Space Scoop article, "The Universe is Big, Beautiful … and Mostly Invisible," was featured in an official South African school textbook,[8] which is used in primary schools all over the country. Space Scoop content is also used in the Cambridge University Press English Language Book, *LGCSE English as a Second Language*.[9]

Digital media is presently accessible in most parts of the world (Rosen, 2010), and a Space Scoop app has been created for Android mobile devices to help widen the audience base. Space Scoop can also be used as the basis for several educational activities. Space Scoop Storytelling uses the articles as the basis for a creative cross-curricula activity that involves astronomy, literature, and presentation skills.

Space Scoop is available in English, Arabic, Dutch, Italian, German, Spanish, Polish, Albanian, Chinese, French, Icelandic, Indonesian, Japanese, Maltese, Norwegian, Portuguese, Romanian, Russian, Sinhalese, Slovenian, Swahili, Turkish, Ukrainian, Vietnamese and Tétum. During 2011 and 2013 there have been 200 Space Scoops released and more than 2100 separate translated Space Scoops.

TABLE 6.1 Space Scoop impact numbers from March 2011 to November 2013

Number of Space Scoops published	196
Languages in which Space Scoops are available	22
Number of Space Scoops translated	2108
Number of Space Scoop Partners	10
Page views on UNAWE website	100,000 (80% unique page views)

5 Future of UNAWE

5.1 *Innovation and Research Potential*

The long-term ambition of the UNAWE team is to expand the reach of UNAWE substantially and stimulate the implementation of UNAWE activities in many more countries, particularly in Africa. To this end there is a need to develop innovative ways for enlarging the reach of the program and new innovative scalable modes for delivery.

One way of expanding the reach of the program is to make courses and demonstration videos for teachers (mini-MOOCs [Massive Open Online Courses]) matched to the language and culture of the targeted country. Another interesting tool for reaching out to children in remote areas is the mobile 'Astrobus' pioneered by UNAWE-Tunisia, an initiative of La Cité des Sciences in Tunis. The bus transports a small telescope, a mini-planetarium and an exhibition throughout Tunisia. Several mobile teacher training pilot projects have been carried out in South America, Africa and India by the *Galileo Mobile* team.[10] The concept of an Astrobus could be applied in many more countries to reach out and deliver UNAWE to remote villages.

There is an additional aspect of UNAWE that is necessary for the future, namely research. A direct research goal of several of UNAWE projects (such as EU-UNAWE) is to evaluate the effectiveness in reaching the UNAWE goals. Evaluation of short-term impact on aspects such as comprehension is relatively simple. However, the long-term effect of such programs in influencing the development of values and attitudes is extremely difficult to evaluate. There are still very few studies that attempt to investigate the long-term effect of education of children on their performance as adults in society. A well-known example is the High/Scope Perry Preschool study (Schweinhart et al., 2005) that follows the history of two groups of American children from an inner city community through age 40, one of whom has received preschool education. This demonstrates that preschool education has a dramatically positive effect on success. Additional long-term studies in various environments are needed. The global reach of UNAWE, the unique nature of astronomy as a motivational tool, and the educational and social goals of UNAWE make it an excellent test bed for continuing long-term latitudinal studies to investigating the lasting effectiveness of programs that foster the motivation of young children. In general, UNAWE can serve as a unique test bed for interdisciplinary research programs in multiple domains. These include projects to investigate inter-cultural aspects of early child development and the economic effectiveness of motivational educational programs directed at young children in a range of environments and cultures.

5.2 *Sustainability of UNAWE*

Future ambitions of UNAWE are of course contingent on the availability of sufficient funds. On the long term there are several possibilities for obtaining financial support.

1. UNAWE is an integral part of the International Astronomical Union Strategic Plan 'Astronomy for Development' 2010–2020, whose implementation is being coordinated by the IAU Office of Astronomy for Development (OAD) in Cape Town. The OAD was inaugurated in 2011 as a joint partnership between the IAU and the South African National Research Foundation. The OAD, with strong backing from the South African authorities, is now well positioned to raise substantial funds. During the next 3 years the OAD will conduct an intensive and widespread fund raising campaign targeted at e.g. international and regional development agencies, multinational companies, philanthropic foundations and individual philanthropists. Support for school education and UNAWE will be important aspects of this campaign.
2. The power of astronomy, in particular, radio astronomy and the associated sophisticated technologies, as a tool for capacity building in Africa has been acknowledged by official resolutions of both the European Parliament and the African Union. It has been decided to site half of the Square Kilometer Array, a next-generation multi-billion dollar global radio telescope in Africa, where it will span 8 African countries. Furthermore the Chinese President Xi Jinping has praised the importance of astronomy as a tool for development. UNAWE is in an excellent position to exploit such developments because it can help inspire the next generation of scientists, engineers and the public with astronomy and thereby help build human capacity globally.
3. UNAWE has substantial support within the European Parliament and the European Commission.
4. Because of its inspirational goals, global reach and demonstrated success, UNAWE is an attractive high-profile program for support by philanthropic foundations and individuals.

7 UNAWE and the "Scientific Mind"

As mentioned at the outset, the philosophy of BtSM fits well with the precepts of the *Building the Scientific Mind* (BtSM) approach to education. However, it may be questioned whether the term "scientific mind" is a misnomer and could be counterproductive in achieving the BtSM goal of reforming the education system. Although derived from the Latin *scientia* or 'knowledge,' in modern English science is usually taken as referring to disciplines that seek to explain the phenomena of the material universe and in this it is distinguished from the humanities.

In 1959, C.P. Snow delivered a famous set of lectures that drew attention to the two cultures (Snow & Collini, 2012). He spoke of scientists who could scarcely struggle through a novel by Dickens, but more importantly of humanities professors who were ignorant of the Second Law of Thermodynamics, who sneered at science as an inferior branch of learning that no really cultured person needed to trouble with. "If the scientists have the future in their bones," he claimed, "then the traditional culture responds by wishing the future did not exist" (p. 11). "There is only one way out of all this," Snow argued. "It is, of course, by rethinking our education" (p. 18).

BtSM rightly advocates the mobilization of children's curiosity from the earliest age on so as to stimulate their ability to think rationally and to develop their sense of ethical values. Surely, cultivating BtSM values and skills is equally important for the appreciation and pursuit of art, music and literature and for the correct interpretation of history as it is for doing science. BtSM is a holistic educational concept designed to stimulate the human mind as a whole and stimulate rationality and creativity and not just the ability to do science.

In recent years the need for radical changes in approaches to education has received a new impetus from several directions. Search engines and the web have resulted in a revolution in our access to knowledge, so that teaching facts is regarded as not nearly as important as was previously the case (Wagner, 2008). The skill sets that are needed to function optimally in the modern global economy and realize a person's full potential are creative ones which stimulate the recognition of links between seemingly unrelated facts and help one to build innovatively from existing facts. Above all, motivating children to learn is crucial and it can be argued that the development of an ethical value system is critical to the survival of human civilization. We have attempted to demonstrate in this chapter that the use of astronomy, space and the philosophy of Universe Awareness can contribute substantially to the needed transformation of education globally and to the BtSM approach to education.

Notes

1 http://www.iau.org/astroEDU

2 http://www.unawe.org/resources/spacescoop/

3 http://www.unawe.org/resources/spacescoop/

4 European Southern Observatory.

5 National Astronomical Observatory of Japan.

6 European Space Agency.

7 National Aeronautics and Space Administration.

8 Natural Sciences Grade 8 Textbook, Heinemann Educational Publishers (South Africa).

9 http://www.amazon.com/Cambridge-IGCSE-English-Second-Language/dp/1444191624
10 http://galileo-mobile.org/node/35

References

Allodi, M. W. (2002). Children's experiences of school: Narratives of Swedish children with and without learning difficulties. *Scandinavian Journal of Educational Research, 46*(2), 181–205.

Arnold, D. H., & Doctoroff, G. L. (2003). The early education of socioeconomically disadvantaged children. *Annual Review of Psychology, 54*, 517–545.

Beilock, S. L., Gunderson, E. A., Ramirez, G., & Levine, S. C. (2009). Female teachers' math anxiety affects girls' math achievement. *Proceedings of the Academy of Sciences, 107*, 1860–1863.

Commonwealth of Australia. (2009). *Belonging, being and becoming: The early years learning framework for Australia.*

Cunha, F., Heckman, J. J., Lochner, L. J., & Masterov, D. V. (2006). Interpreting the evidence on life cycle skill formation. In E. A. Hanushek & F. Welch (Eds.), *Handbook of the economics of education* (Chapter 12, pp. 697–812). Amsterdam: North-Holland.

Feder, T. (2007). Astronomy outreach used for social good. *Physics Today, 60*, 30–31.

Gopnik, A., Meltzoff, A. N., & Kuhl, P. K. (1999). *The scientist in the crib: Minds, brains, and how children learn.* New York, NY: William Morrow.

Haldane, J. B. S. (1927). *Possible worlds and other papers.* London: Chatto and Windus.

Heckman, J. J. (2000). Policies to foster human capital. *Research in Economics, 54*(1), 3–56.

Heckman, J. J. (2008). Schools, skills and synapses. *Economic Inquiry, 46*(3), 289–324.

Heckman, J. J., & Masterov, D. V. (2007). The productivity argument for investing in young children. *Review of Agricultural Economics, 29*(3), 446–493.

IAU Strategic Plan 2010–2020. (2010). *Astronomy for development* (Rev. ed., 2012). International Astronomical Union. Retrieved from http://iau.org/static/education/strategicplan_2010-2020.pdf

Institute of Physics Report (IOP). (2006). *Girls in the physics classroom; A teachers guide for action.*

Miley, G. (2011). "Astronomy for international development", The role of astronomy in society and culture. *Proceedings of the International Astronomical Union, IAU Symposium, S260*, S539–S546.

Miley, G. (2012). *The IAU astronomy for development programme, organizations, people and strategies in astronomy* (Vol. 1, pp. 93–111, A. Heck, Ed.). Venngeist: Duttlenheim.

Miley, G., Carignan, C., & Govender, K. (2012). Building on IYA2009: IAU strategic plan "Astronomy for the developing world" tracing the ancestry of galaxies (on the land of our ancestors). *Proceedings of the International Astronomical Union, IAU Symposium, 277*, 203–210.

Miley, G., Madsen, C., &Scorza de Appl, C. (2005). Universe awareness for young children. *ESO Messenger (European Southern Observatory), 121*, 66.

OECD Report. (2006). *Evolution of student interest in science & technology*. OECD Global Science Forum. Retrieved from http://tinyurl.com/oecd2006report

Ödman, C. J. (2007). Education & public outreach: Odman: Universe awareness: Making young children aware of the universe. *Astronomy & Geophysics, 48*(4), 4.20–4.21.

Ödman, C. J. (2011). "Universe awareness: Inspiring young children around the world", Role of astronomy in society and culture. *Proceedings of the International Astronomical Union, IAU Symposium, 260*, 650–659.

Ödman, C. J., Scorza, C., Miley, G. K., & Madsen, C. (2006). *Universe Awareness. An inspirational programme for economically disadvantaged young children*. Paper presented at Innovation in Teaching/Learning Astronomy Methods, 26th meeting of the IAU, Special Session 2, 17–18 August, 2006, Prague, International Astronomical Union, SPS2, #39.

Ödman-Govender, C. J., & Kelleghan, D. (2011). Astronomical perspectives for young children. *Science Prize Essay, Science, 333*, 1106.

Peacock, A. (1991). *Science in primary schools, the multicultural dimension*. London: Routledge.

Rocard, M. et al. (2007). *Science education NOW*. European Commission, Directorate for Research. Retrieved from http://tinyurl.com/rocard-report

Rosen, L. (2010). "Welcome to the iGeneration!". *Education Digest: Essential Readings Condensed for Quick Review, 75*(8), 8–12.

Russo, P. (2015). *Design, implementation and evaluation of transnational collaborative programmes in astronomy education and public outreach* (PhD Thesis). Leiden University, Leiden.

Russo, P., Gomez, E., Heenatigala, T., & Strubbe, L. (2015). Peer-review platform for astronomy education activities. *eLearning Journal*. (Paper #40)

Russo, P., Miley, G., Westra van Holthe, F., Schrier, W., & Reed, S. (2011). European universse awareness. In *Proceedings EPSC-DPS Joint Meeting* (p. 1759). Nantes, France. Retrieved from http://meetings.copernicus.org/epsc-dps2011

Sagan, C. (1980 [1990 update]). *Cosmos: A personal voyage—Episode 13: Who speaks for Earth*. U.S. Public Broadcasting Service. Retrieved from https://en.wikiquote.org/wiki/Carl_Sagan

Schrier, W., Nijman, I., & Russo, P. (2013). Universe awareness: Inspiring every child with our wonderful cosmos. In J. Barnes, C. Shupla, J. G. Manning, & M. G. Gibbs (Eds.), *Communicating science: A national conference on science education and public outreach* (p. 141). San Francisco, CA: Astronomical Society of the Pacific.

Schweinhart, L. J., Monti, J., Xiang, Z., Barnett, W. S., Belfield, C. R., & Nores, M. (2005). *High/scope perry preschool study through age 40*. Ypsilanti, MI: High/Scope Press.

Snow, C. P., & Collini, S. (2012). *The two cultures* (1959 Rede Lecture). Cambridge: Cambridge University Press.

UNICEF. (2006). *Report on programming experiences in early child development*. New York, NY: United Nations Children's Fund. Retrieved from http://tinyurl.com/unicef-childhood-report

Wagner, T. (2008). *The global achievement gap: Why even our best schools don't teach the new survival skills our children need – and what we can do about it*. New York, NY: Basic Books.

Worth, K., & Grollman, S. (2003). *Worms, shadows, and whirlpools: Science in the early childhood classroom*. Portsmouth, NH: Heinemann.

INTERMEZZO 6

Never Ever without Passion

Jan Visser

Reflections inspired by Jos van den Broek: 'Playing the role of facilitator: Questioning the curious mind'

Children are full of questions. How should we address them? What role can having questions play in the building of the scientific mind?

There is a huge difference between having been told or demonstrated something in response to a question you had and having discovered what you tried to find out yourself. In the former instance you will only have been mentally active and will have focused on memorization. In the latter instance you will often have been involved, body and mind, intellectually as well as emotionally. The experience will have been more profound and it will have changed you. You have not only learned something new but will also have added to the set of mental tools you develop along the lifespan to discover and understand ever more.

Harry Kroto sees as the main task of a teacher "to unlock the creative potential of each and every young person ... [so that students] become responsible citizens; citizens who can make wise decisions without being told by others what they should think and do." I saw Harry in action doing exactly that at a workshop session with school kids to which he had invited me. He wanted the students to construct so-called 'buckyballs,' starting from scratch. Buckyball is the popular name for the structure of C_{60} molecules, the 60 carbon atoms of which are positioned on the vertices of a ball like structure made of 20 hexagons and 12 pentagons. Kroto, a chemist of world renown, had been a co-discoverer of the structure of C_{60} (for which he shared the 1996 Nobel Prize in Chemistry with Curl and Smalley). Now he wanted the students to go through a similar experience. There was no explaining on his part of how to do it. Being confronted with the problem and motivated by their facilitator, the students passionately started working to meet the challenge. And they did. After the session was over, Harry continued the conversation with the students. "Have you ever worked on something you were not passionate about?" he asked. Many students raised their hand. "Never ever work on something you are not passionate about," he reacted, and then continued sharing his passion

 | DOI: 10.1163/9789004416802_012

Buckyballs in the hands of students constructing them

as a sportsman, scientist, and graphic designer (he had won a graphic design award that meant more to him than his Nobel Prize).

This brief intermezzo is written in memory of my friend Harry Kroto, who died in 2016 from complications of amyotrophic lateral sclerosis (ALS). Harry was therefore not able to contribute to this book, as he had intended. The essay is dedicated to my friend Jos van den Broek, author of the chapter on 'Playing the role of facilitator: Questioning the curious mind.'

CHAPTER 7

Playing the Role of Facilitator: Questioning the Curious Mind

Jos van den Broek

1 Introduction

Gavage—from the French word for 'to stuff'—refers to force-feeding of ducks or geese in order to fatten their livers in the production of *foie gras*. School gave me a fat liver. I hated school. I hated languages and always had low grades for English, German as well as French. History was not my thing. Writing didn't interest me. Cosmology? Boring! Physics? Mwah!

At the same time—I know now for sure—I love to learn. It's even my favorite pastime. But, I preferably learn in my own time, at my own pace, and only the subjects I like. Fortunately there are numerous subjects I like, I found out only *after* formal training. I now feel at ease in three foreign languages and have some basic knowledge of a few more, a great advantage when going abroad. I was a journalist for years, I wrote quite a number of books, interestingly even one on history and one on cosmology, I gave numerous talks about these and other subjects, and made several popular-science posters about the latest Nobel Prizes in Physics and Chemistry. Things can change. I bet that none of my former teachers would believe the progress I made during the almost fifty years after leaving school. I don't blame them: most of them were force-feeders by training. They didn't have a keen eye for—intrinsically—curious learners like me. (With a few exceptions: two or three of them were great scholars who were able to trigger my curiosity. They were—and still are—my heroes.)

When starting to read *Learning to Make a Difference: School as a Creative Community* (Leadbeater, 2014) I stumbled upon an incredibly important issue: "Imagine for a moment that this is the beginning of 1914. You have been charged with coming up with a list of 20th-century skills that learners will need to equip them for the century to come" (p. 8). This is what it is all about! My teachers were in fact not equipped to equip me for my present duties, most probably forced by the curriculum and the prevailing consensus among the majority of people that schools should work the way they always worked. Not much seems to have changed in this regard in school systems around the world.

 | DOI: 10.1163/9789004416802_013

2 Curiosity-Driven Educational System

Over a century ago the legendary educational innovator Maria Montessori already invented a curiosity-driven educational system with a student-driven choice of activities from within a prescribed range of options, with uninterrupted blocks of work time, ideally three hours, and—last but not least—a constructivist or 'discovery' model, where students learn concepts from working with materials, rather than by direct instruction. Not by *gavaging* but by making them intrinsically curious and well equipped to serve society in the best possible way. Montessori demonstrated her work in 1915 at the Panama–Pacific International Exposition in San Francisco, California. A glass-walled classroom was put up at the exposition, and thousands of observers came to see a class of 21 students (Kramer, 1976).

Did anything change positively during the past 100 years, one might ask? The wish to change must be there, without doubt. In fact, a world where all information can be 'googled' within seconds—at any place and any moment—does not need the 20th century system of force-feeding education. But when I, 100 years after Maria Montessori's precious performance at the exposition in San Francisco, had to prepare a lecture on eugenics for a MA course on human genetics, I was asked by the instructor to prepare a few multiple-choice questions for an examination on the subject, just for the sake of efficiency. Multiple-choice questions? In an era with all possible information at one's fingertips? I declined the request and instead asked the students to write down their thoughts about a difficult ethical genetics-related problem. They all enjoyed this way of testing.

Without any doubt most of these students will now remember the names of Plato and Seneca, Thomas Malthus and Herbert Spencer, Francis Galton and Charles Darwin. They now have heard about the comparison between eugenics during antiquity and that of the 19th and 20th century. If they need any further information on these subjects then they can easily look it up online. But most of all I hope that I was able to give them insight in the connection between 'positive eugenics' (encouraging reproduction by persons presumed to have inheritable desirable traits), 'negative eugenics' (discouraging reproduction by persons having genetic defects or presumed to have inheritable undesirable traits), and 21th-century human enhancement with its innumerable possibilities and the innumerable ethical issues that will go hand in hand with the introduction of the converging technologies that are expected to enhance our physical and mental capabilities. We can't teach our students what these technologies will look like, because we only have a vague idea. But the same students will be the inventors, care-takers and watchdogs of these future technologies. That's what teaching is all about: equipping our students for their future duties. Preparing them for an unknown future. Making them eager for the new. But multiple-choice questions just for the sake of efficiency? No way!

3 Eager for the New

The Dutch word for 'curious' is 'nieuwsgierig', with equivalent words in the Germanic languages Norwegian ('nysgjerrig'), Danish ('nysgerrig'), German ('neugierig'), Frisian ('nijsgjirrich') and Afrikaans ('nuuskierig'). 'Nieuwsgierig' literally means "eager for the new." In that sense its origin is quite different from that of the English 'curious', which is, in fact one of the many words in the English language with Latin roots. It originates—via the French 'curieux'—from the Latin word *curiosus,* meaning "careful, diligent, thoughtful, devoted; elaborate, complicated," or in the bad sense "meddlesome," akin to *cura* "care." In all Romance languages, plus the Semitic language Maltese ('kurjużi'), the word for 'curious' stems from the Latin stem *cura.*

A child's mind is far more 'eager for the new' than 'careful or diligent', and personally I feel more affection for the Germanic versions—with the exception of the English one—than for the Romance versions of 'curious'. In their early years those 'eager for the new' have one thing in common: so-called "Why?" and "How?" questions are strongly in their favor. Kids have an inquisitive mind.

I myself was no exception: I was 'eager for the new'. And fortunately, history works here like falling dominoes: about sixty years after I first started to ask my parents an avalanche of questions triggered by my own eagerness for the new, I have the great privilege of receiving scores of children's questions. These questions—largely why-like—mostly relate to natural events, like:

- Why are roses red whereas the leaves of trees are green?
- Why does my cat have white feet?
- Does a ladybug earn another spot on its birthday?
- Why do apples turn brown when they fall?
- Clouds are white, but why are they grey when it rains?
- Why does the Moon travel with us when we're sitting in a car?

Quite often—as was the case with the red roses and the green leaves—I have to tell that the real answer to a "Why?" question often doesn't exist: in my opinion Mother Nature has neither reasons nor purposes. She is a tinkerer, and if a solution fits under certain circumstances it works. Natural selection does not take place according to the neo-Darwinian "survival of the fittest" but following the Darwinian "survival of the fit enough."

Some of the children's questions bear testimony to the presence of an inquisitive mind and a remarkable observational power. It's for instance far from obvious for most of us that black cats often have white paws but that white cats with black paws are like needles in a haystack! It's where the scientific mind starts to operate. It was Martha (7) who came up with this beautiful question—and with many more. Her avalanche of curiosity led to a question and answer children's book on nature awareness (Van den Broek, 2012). This gave rise to many more questions from children with ages ranging from 4 to 12; to a new book on universe awareness (Van den Broek, 2013); and a question and answer column for children in a number of Dutch regional newspapers, leading to a third book (Van den Broek, 2015).

4 Building the Scientific Mind Is Breaching the "Walls of in-Principle-ity"

But what in fact is for me 'the scientific mind?' The *Encyclopædia Britannica*—quite narrowly—defines 'science' as follows: "Any system of knowledge that is concerned with the physical world and its phenomena and that entails unbiased observations and systematic experimentation. In general, a science involves a pursuit of knowledge covering general truths or the operations of fundamental laws" (Encyclopædia Britannica online, 2014). When we talk about 'scientists', we generally mean those men or women who know about or study the natural world based on facts learned through experiments and observation in a particular area of scientific study such as biology, physics, astronomy, medicine or chemistry.

Within this definition of 'science' or 'the sciences' we run the risk of falling short, because the humanities as well as the social sciences should definitely be included when talking about 'the scientific mind'. In my definition 'the scientific mind' therefore is an inquisitive, curious and creative mind with interests that stretch far beyond the borders of 'the sciences;' it is a socially, culturally and philosophically aware mind that is able to break barriers and cross boundaries during its exploratory adventures. My then seven-year-old friend Martha made me aware of that with her question about the—in her mind—illogical spelling difference for the diphthong [ɛi] in Dutch. In standard Dutch, and most Dutch dialects there are two possible spellings for this diphthong *ij* and *ei*. This causes confusion for schoolchildren, who need to learn which words to write with *ei* and which with *ij* although they sound exactly the same.

It is the task of a facilitator, teacher or mentor to stimulate this unbiased inquisitive attitude. In the curriculum of the Department of Science Communication and Society at Leiden University where I work—offering a one-year

Master's specialization in Science Communication—cultural awareness is one of the key issues of our educational program. Almost without restrictions our students are allowed to make dance, film, comics or theatre productions to show what they have learned. "I haven't been creative since kindergarten!" we sometimes hear. This sad remark tells something about the present, failing educational system, but at the same time this fills us with great pride. Indeed, getting students to go this route often asks for some persuasion. Bureaucratic structures are very good at building walls around disciplines, subjects, faculties or institutions. If we hear that 'in principle' something is not possible or allowed, we generally are able to breach these—in most cases quite weak—walls of 'in-principle-ity': when 'in principle' it's not possible, there's a fair chance that it will ultimately be possible.

5 Facilitating the Curious and Creative Scientific Mind

What is the role of the facilitator in creating the—curious and creative—scientific mind (see Figure 7.1)? Firstly by creating a Montessori-like environment where the learner can flourish, an environment without bureaucratic restrictions. Here I limit myself to facilitating learning by experimentation, a world that I know best, as a trained biochemist and pharmacologist, the

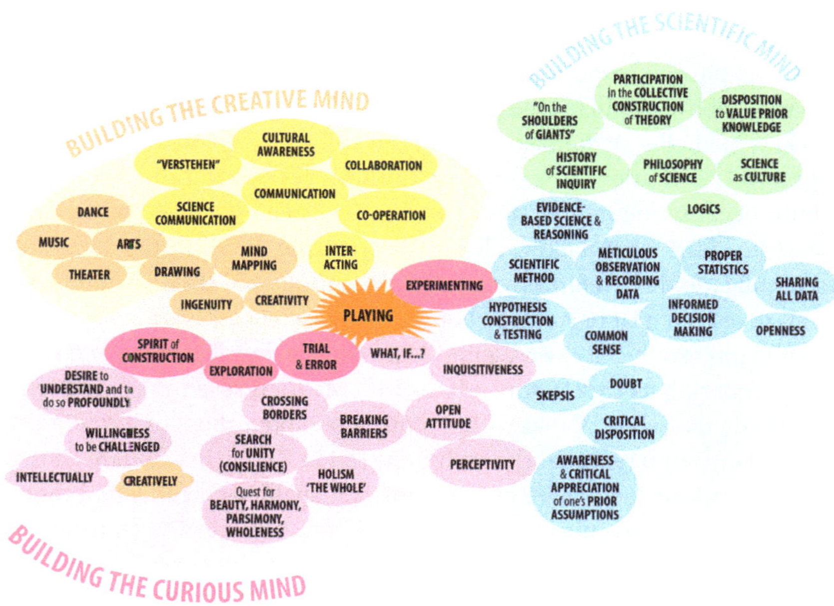

FIGURE 7.1 Multidimensional model of the curious and creative scientific mind[1] (CC-BY-SA, Van den Broek, 2016)

designer of numerous 'kitchen experiments' and the (co-)author of several books about these experiments (Van den Broek & Schulp, 2004; Schulp & Van den Broek, 2005; Van den Broek, 2013).

6 Facilitating the Inquisitive Mind—without a Bodice

Jan Visser taught me[2] that for children to become real scientists—to help them build their scientific mind—you have to make them *work* as well as *think*. In his thoughtful essay *Constructive Interaction with Change: Implications for Learners and the Environment in Which They Learn* Visser pleads for the role of mentoring, which he considers "one of the best sides of human nature," with—as he calls it—"the disposition to dedicate oneself to the well-being of another person" (Visser, 2008, p. 29). A facilitator can be seen as an 'incidental mentor' in the process of incidental or informal learning.

Learners construct new knowledge from existing knowledge. According to Bransford, Brown and Cocking (2000) a logical extension of this view is "that teachers need to pay attention to the incomplete understandings, the false beliefs, and the naive renditions of concepts that learners bring with them to a given subject" (p. 10). Teachers then need to build on these ideas in ways that help each student achieve a more mature understanding. In my view, a facilitator and the 'incidental mentor' play similar roles in this process.

But such an inspirational person has more to add. A facilitator does not bring the truth, but helps with finding a scientifically sound method—helps with 'building the scientific mind'—by teaching how to add changes in a structured way, in Visser's (2008) dialectic process of 'constructive interaction with the changing environment.'

7 Learning by Experimenting

In what I call the 'Learning by Experimenting Model' (Figure 7.2), I added Bransford's earlier mentioned role of a teacher (or mentor) to pay attention to the learner's incomplete understandings, false beliefs, and naive renditions of concepts. The inner circle of the Learning by Experimenting model—inspired by Kolb's learning styles (Kolb, 1984; Kolb & Kolb, 2005, 2008)—shows the constantly changing roles of the learner when performing a series of experiments and changing the experimental conditions.

'Building the scientific mind' means 'facilitating the inquisitive mind.' In the ExperiMental Cycle depicted in Figure 7.2 the role of the facilitator (the

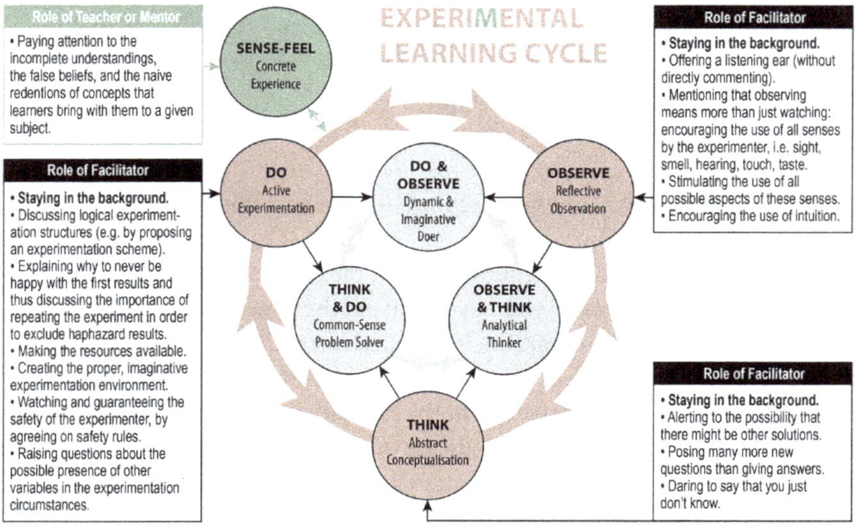

FIGURE 7.2 Learning-by-experimenting model. Outer circle: learning by experimenting. Inner circle: learning styles. The roles of facilitators, teachers and mentors in the learning process are given in the boxes. The circular arrows are not unidirectional, as is the case in many circular models, but bidirectional. Inquisitive and playful learning does not require a certain direction (CC-BY-SA, Van den Broek, 2016)

'incidental mentor') becomes manifest in a variety of ways, in all stages of the learning process mentioned below:

- In the step of '*doing*': e.g. discussing a logical experimentation structure (see for instance Table 7.1 in Exhibit 1); explaining why to never be happy with the first results and thus discussing the importance of repeating the experiment in order to exclude haphazard results; making the resources available; and keeping an eye on the safety conditions, like proper ventilation and the use of safety goggles.
- In the step of '*observing*': e.g. mentioning that there is more than mere watching, but that all senses can be used, and that even when using all the senses the observation of a number of variables is easily forgotten. We can see the colour of an object change, but when we forget to observe that the surface changes from shiny to dull we might miss an important experimental outcome. The same holds true when we don't observe temperature or structural changes (e.g. stickiness), remarkable sound effects, smells or—tricky!—even changes in taste.
- In the step of '*thinking*': e.g. suggesting the existence of alternative explanations and by suggesting alternative lines of questioning. In the case of Sacha's balloons (see Exhibit 1) it could have been the age of the balloons, or the influence of light or heat.

But in all steps the 'task' of the facilitator in enabling the building of the scientific mind is first and foremost to stay in the background as much as possible. Secondly, to ask new questions without giving answers. And lastly, to dare to say: "I just don't know! Let's give it another try."

8 The Central Role of Playing

The scientific mind can only function in an environment where curiosity and creativity are prerequisites, where there is time and space for exploration and—not the least—for playing. 'What if?' questions lead to playing, and vice versa. Even the [ɛi] diphthong problem can lead to playing: to playing with words, with poems, et cetera. A playful mind explores, crosses boundaries, breaks down and builds new constructions from the building blocks and—often literally—from waste. It learns from its own errors.

During my life I made many errors, and every once in a while these have led me to a success. For instance, almost fifty years ago I converted a brand-new Philips mono gramophone into a stereo version, with the use of an old radio, bringing my father to great rage. It was my strong opinion that the Dutch electronics company had fooled me by soldering the wires of the left and right channels together. By mere tinkering I learned how a stereo gramophone functions, in my eyes a rebellious act of *con*struction instead of *de*struction. And at the same time it probably gave me for the very first time insight in how the scientific mind works: it is—and should be—unstoppable.

9 Exhibit 1: Sacha's Balloons

Through the Facebook page of his father Stefan I received a quite intriguing question from Sacha (8). A photo of two balloons sticking together (Figure 7.3) accompanied Sacha's message. One balloon was marked "water"; the other "koolzuur"—literally 'carbon acid', the Dutch word for carbon dioxide. Sacha asked: "Why does a balloon with water stick to a balloon with carbon dioxide?" To be honest: I didn't have the slightest idea. This turned out to be a great advantage: the difference between 'master' and 'apprentice' disappeared. The only role I played was that of the learned facilitator who offered tools for a scientific approach.

Children are scientists to be. Sacha's question was a real question of a scientist to be. How I knew? Because I could see from the photo that Sacha started experimenting himself. As a real scientist he carefully marked the contents of his balloons. And he documented the outcome.

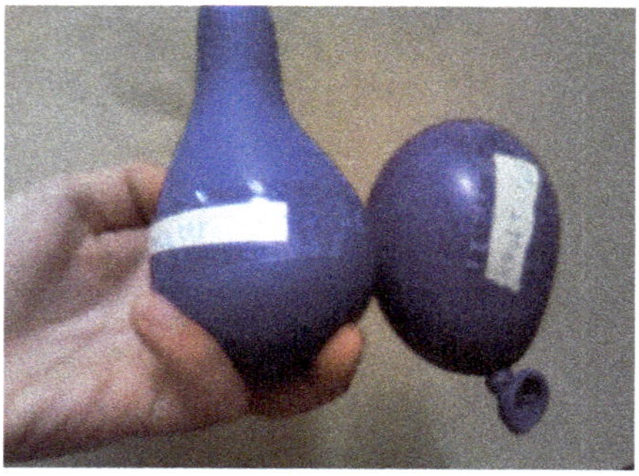

FIGURE 7.3
A balloon filled with carbon dioxide sticking to a balloon filled with water. (Photograph: Sacha Tax)

I didn't send an answer to Sacha's difficult question—not even suggestions—but instead replied with a number of questions:

> 1. How old are you?
> 2. How did you get the carbon dioxide into the balloon?
> 3. Does a balloon you inflated with your mouth also stick?
> 4. And how about a balloon you inflated with a bicycle pump?
> 5. And what happens when you put carbonated water into the balloon instead of tap water?
>
> I wish you luck with your experiments. I'm curious!

Sacha facebooked in return:

> Dear Jos,
> 1. 8;
> 2. I put the balloon on top of a bottle of carbonated water;
> 3. No!;
> 4. That doesn't stick either.
> 5. That's what I'm going to do tomorrow because we're out of carbonated water.
>
> Kind regards, Sacha.

That asked for even more experiments—Sacha definitely appeared to enjoy it. I sent him the following suggestions:

- Experiment 1. Put a little bit of soda (or baking powder) into a glass bottle. Add some vinegar and immediately put the balloon on top of the bottle. Soda (or baking powder) + vinegar gives carbon dioxide!

– Experiment 2. Inflate with a bicycle pump a balloon until it's exactly as big as the one with carbon dioxide. (Because you exhale carbon dioxide, and I want air with as little carbon dioxide as possible.) Which one of the two balloons deflates fastest: the air-filled balloon or the one filled with carbon dioxide? Any idea why there's a difference if there is any?
– Experiment 3. Fill a balloon with some sunflower oil without making the outside slippery. A job to be performed with care. Does the carbon dioxide-filled balloon stick to it?

When you have done these experiments, we continue our conversations.

It soon became clear that Sacha—as a well-documented real scientist—had picked up the idea of soda and vinegar from one of my Q&A books.

Later I invited Sacha to do a series of experiments in our kitchen, a proposal he and his father Stefan accepted with great enthusiasm. On a Saturday morning we transformed our kitchen sink into a lab table, filled with glassware and chemicals from the kitchen cupboard. Lab coats, gloves and real safety goggles added to the serious fun. And, not to forget, a notebook and a camera.

How to approach Sacha's question? Firstly by investigating what sticks to what. According to Jan Visser (2008), learning comes down to constructive interaction with the changing environment. This asks for a systematic

TABLE 7.1 Systematically approaching Sacha's question: What sticks to what?

Content	Mouth-inflated	Pump-inflated	Carbon dioxide	Carbonated water	Water	Sunflower oil
Mouth-inflated						
Pump-inflated						
Carbon dioxide						
Carbonated water						
Water						
Sunflower oil						

approach. One of the most important tasks of the learned facilitator is to offer tools for such an approach. In this case, Sacha and I prepared a fill-in table. With the altogether twelve balloons we step-by-step investigated the stickiness of pairs of balloons.

Did we find a solution to Sacha's question? To be honest: not yet! But the lack of a firm answer didn't minimize our enthusiasm. We were researchers doing a proper job. For a moment the difference of 55 years between 'master' and 'apprentice' disappeared.

10 Exhibit 2: Learn Cleverly, Eat an Apple

When writing books in which you answer children's questions with a scientific twist you can expect loads of new requests for answers. Felien (9), for instance, asked me why bruised apples or pears get brown and why they rot. At first sight quite a simple question, but at the same time a question that gives an enormous amount of opportunities to do—systematic—experimentation. 'Building the scientific mind.' Following is an exchange of notes about apples and science.

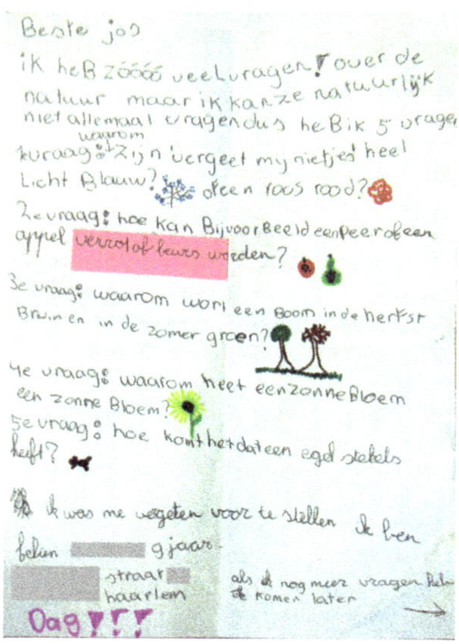
Beste jos
ik heB zóóóó veelvragen! over de natuur maar ik kan ze natuurlijk niet allemaal vragen dus heBik 5 vragen
1e vraag: waarom zijn vergeet my nietjes heel Licht Blauw? of een roos rood?
2e vraag: hoe kan Bijvoorbeeld een Peer of een appel verrot of beurs worden?
3e vraag: waarom wort een Boom in de herfst Bruin en in de zomer groen?
4e vraag: waarom heet een zonneBloem een zonne Bloem?
5e vraag: hoe komt het dat een egel stekels heeft?
ik was me vergeten voor te stellen ik ben felien 9 jaar.
straat
haarlem
als ik nog meer vragen heb komen later
Dag!!!

Dear jos
i haVe sóóóó many questions! about nature but of course I can't ask them all so I haVe 5 questions
1st question: why are 'forget-me-nots' very light Blue [drawing] oR a rose red? [drawing]
2nd question: how can For Example a Pear or an apple bruise of rot? [drawing]
3rd question: why dos a Tree get Brown in the autumn and green in the summer? [drawing]
4th question: why do we call a sunFlower a sun Flower? [drawing]
5th question: how come that a hedgehog has spines [drawing]
~~f~~i have forgotten to introduce myself i am felien [...] 9 years. [...] street [...] haarlem
if i have more questions ~~f~~come later
Bye!!!

FIGURE 7.4 Felien's letter, containing lots of 'why-questions'

10.1 *My First Fictitious 'Note to Felien', as Published in Our Local Newspaper*

Dear Felien,

I wonder, when hearing such a question, what in fact is 'bruised' or 'rotten.' Is bruised the same as rotten? No. The bruised brown apple is in a way comparable with a bruised, blue shinbone. An apple gets bruised if you break the cell walls of the fruit by a strong force. For instance by dropping it. If the cell walls are broken, the oxygen in the air is able to break down the contents of the cells. This is called 'oxidation': the fruit pulp browns. An apple starts to rot when microorganisms like bacteria and molds start devouring the bruised cells. However, that's the answer of someone who doesn't make other people think for themselves. This can be done differently. I thus thought out a few experiments to figure out how the bruising and rotting affects an apple:

- What happens to an apple when you let it bounce a few times, compared with an apple you treat as carefully as possible?
- Does the browning go faster or slower in the light or in the dark, in the fridge or at room temperature?
- What happens if you make little holes in the apple with a pin? Does that stimulate the browning process?

I'm quite curious about your results.
Please keep me informed!
Cheers,

Jos

10.2 *My Second Fictitious 'Note to Felien', Published One Week after the First*

Dear Felien,

I hope you enjoyed the experiments.
I have a new experiment for you. Please cut an apple in eight similar parts. Does the browning process go faster or slower if you put a part in boiling water for a minute? Why?

- Does it go faster or slower if you keep a part under water?

- Does it go faster or slower if you mash a part with a fork? Why?
- Does it go faster or slower if you rub in a part with lemon juice? Why?
- Does it go faster or slower in a plastic bag you sucked vacuum?
- Put a part into an egg cup. Put that egg cup into a marmalade jar with some soda. Carefully add some vinegar to the soda. Cover the jar when the reaction has almost come to an end. What happens to the part of the apple? Why?
- Does an apple part rot faster if you cover it with fungal spores?

Good luck!
Cheers,

Jos

10.3 *My Third Fictitious 'Note to Felien', Published Two Weeks after the First*

Dear Felien,

"Yes, Jos," I hear you mumble after having performed my/your apple experiments, "how certain are you that bruised apples turn brown due to oxidation?" Good question!

This is why I challenge you to repeat the experiment with the most 'oxygenic' substance in your mom's kitchen cupboard: the cleaning powder beyond praise *Vanish Oxi Action*. "Contains active oxygen," the container says. I bet that the active oxygen will do its utmost best when you put some concentrated *Vanish* solution on a bruised and browned apple cut in halves. (Please ask your mom or dad to help.) Convincing, isn't it?!

The most beautiful experiment follows next. First get some pure vitamin C from the pharmacist. Then cut the half bruised and browned apple that you treated with *Vanish Oxi Action* in two equal parts. Keep one quarter apart. The other quarter you rub in with a concentrated solution of vitamin C. The result made me quite happy! How about you? Can the effect of oxygen and vitamin C be proven more clearly? Long live science!

Warmly,
Jos

Acknowledgements

The author wishes to thank Kim Elberse, MA student in Biomedical Sciences and Science Communication at Leiden University, for her help with the preparation of this chapter by interviewing Professor George Miley (UNAWE, Department of Astronomy, Leiden University), Professor Jan Visser (Learning Development Institute), Dr Anne Land (Department of Science Communication & Society, Leiden University) and Jeroen van der Brugge, M.Sc. (Department of Education, Netherlands Biodiversity Center Naturalis). The discussions with and the materials offered by Jan Visser and Lya Visser (Learning Development Institute) and George Miley and Pedro Russo (UNAWE) were indispensable.

Notes

1 This is an elaboration of how the multidimensionality of the curious and creative scientific mind is depicted in Chapter 1.

2 Personal communication.

References

Bransford, J. D., Brown, A. L., & Cocking, R. R. (Eds.). (2000). *How people learn: Brain, mind, experience, and school: Expanded edition*. Washington, DC: National Academy of Sciences. Retrieved from http://www.nap.edu/catalog/9853.html

Encyclopædia Britannica online. (2014). Retrieved September 22, 2014, from http://www.britannica.com/EBchecked/topic/528756/science

Kolb, A. Y., & Kolb, D. A. (2005). Learning styles and learning spaces: Enhancing experiential learning in higher education. *Academy of Management Learning & Education, 4*(2), 193–212.

Kolb, A. Y., & Kolb, D. A. (2008). Experiential learning theory: A dynamic, holistic approach to management learning, education and development. In S. J. Armstrong & C. Fukami (Eds.), *Handbook of management learning, education and development*. London: Sage Publications.

Kolb, D. A. (1984). *Experiential learning: Experience as the source of learning and development*. Englewood Cliffs, NJ: Prentice-Hall. Retrieved from http://academic.regis.edu/ed205/kolb.pdf

Kramer, R. (1976). *Maria Montessori* (pp. 212–215). Chicago, IL: University of Chicago Press.

Leadbeater, C. H. (2014). *Learning to make a difference: School as a creative community. WISE Matters 01.* Qatar: Qatar Foundation. Retrieved from http://www.wise-qatar.org/wise-matters

Schulp, A., & Van den Broek, J. (2005). *25 Nieuwe wetenschappelijke experimenten voor huis, tuin en keuken* [25 New scientific experiments for the home, garden and kitchen]. Rijswijk: Veen Magazines.

Van den Broek, J. (2012). *Beste Martha. Vragen, antwoorden en nog meer vragen over huisjes, boompjes en beestjes* [Dear Martha. Questions, answers and still more questions round and about the house, the yard and the farm]. Leiden, the Netherlands: Uitgeverij Micromys.

Van den Broek, J. (2013). *Wie is de baas van de zon? Vragen, antwoorden en nog meer vragen + een heleboel proefjes over Aarde, Zon en het Heelal* [Who is boss of the sun? Questions, answers and still more questions + lots of experiments about the earth, the sun and the universe]. Leiden: Uitgeverij Micromys.

Van den Broek, J. (2015). *Daarom! Vragen, antwoorden en nog meer vragen over van alles en nog wat* [Because! Questions, answers and still more questions about anything and everything]. Leiden: Uitgeverij Micromys.

Van den Broek, J., & Schulp, A. (2004). *Wetenschappelijke experimenten voor huis, tuin en keuken* [Scientific experiments for the home, garden and kitchen]. Rijswijk, the Netherlands: Veen Magazines.

Visser, J. (2008). Constructive interaction with change: Implications for learners and the environment in which they learn. In J. Visser & M. Visser-Valfrey (Eds.), *Learners in a changing learning landscape. Reflections from a dialogue on new roles and expectations.* Dordrecht: Springer.

INTERMEZZO 7

On Dialogue

Jan Visser

Reflections inspired by Paul Webb: 'How you talk is how you think; how you think is how you understand'

Building the scientific mind is a lifelong process. Like any process of growth, also the growth of the mind depends on propitious conditions surrounding it. Being in environments that lead one to question the world is part of those conditions. Being surrounded by people who care and who are willing to enter into dialogue with their fellow human beings is similarly crucial.

David Bohm (1917–1992) was a renowned theoretical physicist as well as a profound thinker about almost anything beyond the realm of physics. In his 1980 major work on *Wholeness and the Implicate Order*, he explores the nature of reality and of consciousness, seeking to understand both as part of a coherent whole, suggesting therefore that thought may be understood as an integral part of reality as a whole. In another important publication, *On Dialogue*, Bohm (1996) emphasizes the cruciality of "collectively shared meaning" in the life of our species, or, as he puts it, "The collective thought is more powerful than the individual thought." And he goes on to say that "… individual thought is mostly the result of collective thought and of interaction with other people. The language is entirely collective, and most of the thoughts in it are. Everybody does his own thing to those thoughts …" (p. 15). It is thanks to the evolution of collectively shared meaning that we are able to evolve culturally at a much faster pace than would be possible through biological evolution alone. This is what makes human learning so different from learning in other animals and from machine learning in its current incarnation. The Data character in *StarTrek The Next Generation* tries hard to change this, but success remains modest despite the android's commendable efforts.

The term 'discussion' has, etymologically speaking, a different meaning, namely 'shaking things apart.' It doesn't have its focus on contributing to collectively shared meaning. In fact, it sounds like the opposite of 'dialogue' in the Bohmian sense of the word as it aims at distinguishing one way of understanding from another one.

Paul Webb's chapter on 'How you talk is how you think; how you think is how you understand' recognizes this difference, and thus has its focus on

 | DOI: 10.1163/9789004416802_014

'*productive* discussion' (my emphasis), also known as 'collaborative reasoning.' It aims at exploring the world and doing so in a shared fashion. The chapter—which in itself owes much to the productive interactions between author and editors—is of great value for improved pedagogical practice in the classroom. It thus is also particularly pertinent to teacher preparation. Technologies such as microteaching may well, in my experience, help develop the relevant skills for this purpose.

References

Allen, D. W., & Ryan, K. (1969). *Microteaching*. Reading, MA: Addison-Wesley Publishing.

Bohm, D. (1980/2005). *Wholeness and the implicate order*. London & New York, NY: Routledge Classics.

Bohm, D. (1996/2004). *On dialogue*. London & New York, NY: Routledge Classics.

CHAPTER 8

How You Talk Is How You Think; How You Think Is How You Understand

Paul Webb

1 Introduction

The *Building the Scientific Mind* (BtSM, n.d.) program, developed and implemented by the Learning Development Institute between 2005 and 2015, characterizes a scientific mind as one that comprises attitudes and skills—cognitive and meta-cognitive ones—that disposition people to question facts and critically challenge the 'givenness' of any *a priori* knowledge and authority. The BtSM website reminds us that such a mind is not only a characteristic of scientifically literate people, but of citizens with the scientific spirit required to creatively and constructively respond to the challenges of our time.

A scientific mind is thus a mind that is consistently harmonious with deep understandings that are framed within the heritage of the millennia-long history of the human pursuit of knowledge constructed through openness to scrutiny. It comprises habits of thinking, dispositions, values, ethical concerns, aesthetic considerations and attitudes which are the products of a complex array of skills and mental capabilities in varied domains, built up over a long period of time (BtSM, n.d.; Visser, Chapter 1, this volume).

In this chapter it is argued that a critical component of the successful development of this type of mind rests on the use of language and argument which includes and respects the creative co-presence of other voices and ideas. In other words, such a mind acquires conceptual agency through participation in 'conversations that matter' (Greeno, 2006) and recognizes that thinking is something that is developed through openness to the ideas of multiple 'others.' In the epoch of the Anthropocene the development of such a mind-set has become vitally important as, for the first time in the existence of human civilization we are faced with choices that have the potential to determine our survival as a species.

 | DOI: 10.1163/9789004416802_015

2 The Notion of 'Productive Discussion'[1]

A research conference sponsored by the American Educational Research Association in 2011 called for the sharing of findings on the effects of discussion in school teaching and learning. This conference brought together leading scholars from across the world in education, learning sciences, psychology, linguistics, and computer science, and their findings were collated in a book edited by Resnick, Asterhan, and Clarke (2015). Collectively, the contributing authors revealed how teaching and learning through discussion raised student achievement and retention of knowledge in virtually every school subject. Some of the studies showed how, after learning how to participate in academic discussions, students' grades improved in mathematics, science and reading comprehension. Others looked more deeply into how this type of learning transferred to disciplines and skill sets that had not been taught directly. While the editors considered these findings to be 'startling,' they noted that the practices and achievements that had been reported were responses to a changing landscape of education practice and social theory that had been developing for more than a century.

Each chapter in Resnick, Asterhan and Clarke's book reveals evidence that knowledge develops and is constructed through social interaction, and that social context plays a central role in the process of 'making meaning.' The crucial importance of awareness of context for meaningful understanding is also argued by Greenberg in Chapter 2 (this volume). The notion of the importance of social interaction and social context for learning has been around for a long time. Vygotsky, in his treatise on *Mind in Society*, written in 1930 and published posthumously in 1978, points out that the basis of effective learning lies in the type of social interaction that takes place between two or more people with different levels of skills and knowledge. Many of the science education research articles produced over the past thirty years reflect this type of thinking which represents a shift in emphasis from a Piagetian view of an individual struggling to make sense of the world to the Vygotskyan notion of the social construction of knowledge (Hodson, 2009). This shift has largely been the result of the accumulation of evidence for the efficacy of interventions that successfully engage children in what many now call 'productive discussion'—a concept and approach for which there is a fair degree of agreement among scholars as to its meaning (Resnick et al., 2015). Typically, 'productive discussion' is structured and focused, but not dominated by a teacher. It is framed in a series of open-ended questions that generate individual and collective reasoning within a group. The participants have opportunities to engage in individual and collective reasoning about issues and to provide explanations for their claims by drawing on their experience and prior discussions. In other words, they are able to listen and react to each other's ideas, reason together and co-construct understanding (Wilkinson, Murphy, & Binici, 2015).

Different authors have used a range of names for what is ultimately 'productive discussion.' Examples are 'collaborative reasoning' (Chinn & Anderson, 1998; Reznitskaya et al., 2009); 'critical discussion' (Keefer, Zeitz, & Resnick, 2000); 'accountable talk' (Michaels & O'Connor, 2002); and 'argumentation approaches to learning science' (Osborne, 2010). Similarly, 'exploratory talk,' which is the focus of this chapter, is an approach which has been described as 'thinking aloud with others' and which 'foregrounds student reasoning' (Monaghan, 2005). As a form of 'productive discussion' exploratory talk naturally uses the open sharing of ideas, receptiveness to the ideas of others, constructive conflict and well-argued counter proposals in order to reach consensus in groups (Mercer & Littleton, 2007). However, simply getting children to talk does not automatically mean children will learn, and research has shown that for 'productive talk' to be effective and take root in educational settings it has to be taught explicitly and practiced continually.

Some of the evidence for the positive effects of talk on learning has been around for a long time. For example, Matthew Lipman, who founded the *Philosophy for Children Project* in the early 1970s, was particularly interested in developing children's reasoning skills by getting them to talk about science. He firmly believed that children possess the ability to think abstractly from an early age and that by talking about issues they could learn to think and argue logically. This belief was supported by the results of a number of research studies on his project which all showed that the participants did better than other children of the same age in terms of reading, reasoning and thinking in general after learning how to discuss issues effectively (Lipman, Sharp, & Oscanyan, 2010; Trickey & Topping, 2004).

Similarly the *Cognitive Acceleration in Science Education* (CASE) project, which was developed by Michael Shayer and Philip Adey at King's College London in 1981, used an approach which focused on group work and discussions in science classes. Teachers on the project were trained to facilitate discussions effectively. When the children were talking the teachers were only allowed to ask questions to get an idea of what and how the children were thinking. Like Lipman's project, studies on the CASE project also produced some very interesting results. For example, after three years in the program when tested in their final year of compulsory schooling (leading to the General Certificate of Secondary Education in the United Kingdom), the participants were performing at one grade level higher in science than what was expected of other children of the same age. Not only were they achieving more than expected in science, but their mathematics and English grades were better in approximately equal measure. Another important aspect of note was in terms of social mobility, namely that it was children from disadvantaged socio-economic backgrounds who benefited the most (Shayer & Adey, 2002).

3 Measuring the Effects of 'Productive Discussion'

There have been many other claims over a wide field of research studies that show that students who engage in talk in class (dialogic teaching situations) perform better on tests than those who do not. However, there has been criticism of claims that talk improves the ability to reason or think. Such criticisms include the charge that reported research only showed improvements in areas that were specifically taught, in other words what was tested was not independent of what the discussions had been about. Another criticism is that only a small number of studies that showed positive results used measures that were standardized. In other words, doubts were raised as to whether all test takers had to answer the same questions and were scored in a way which makes it possible to compare their relative performance. Put differently, the studies were criticized for not having sufficient methodological rigor (Trickey & Topping, 2004); not enough was shown to make the findings convincing.

Apart from the criticisms of the methods used, some critics questioned the statistical data that were produced in the studies where statistical analyses were done. One such question skeptics would ask is: 'To what extent can we trust the differences that were found? To what extent were the findings statistically significant, or were they perhaps merely due to chance?' Even when that question can be answered to the satisfaction of the skeptics, and they are persuaded that it would be very unlikely that the differences in results could be attributed to chance, a further question would be: 'But are the differences found also practically significant?' For example, in the case of the positive effects of discussion on reasoning, how strong is the relationship between the quantity or quality of talk in the classroom and the students' ability to reason effectively. In other words, what is the effect size? Is the size of the effect large enough to justify investing in productive discussion by training teachers to competently facilitate such discussion?

Among the studies alluded to earlier, research on Lipman's *Philosophy for Children study* is considered to be one of the most convincing. He used a standardized test (the Iowa Test of Basic Skills) which showed statistically significant results, demonstrating a reasonable effect on students' comprehension after three years of instruction (the overall Cohen's d score was 0.55, i.e. a medium effect size). Unfortunately, when these measures were used in other studies which also used multiple-group designs and standardized tests, the practical significance scores (Cohen's d) were usually only around 0.2, that is the effect size was small, indicating weak practical significance (Wilkinson et al., 2015).

Another criticism is that in studies where suggestions of 'transfer' via discussion to other disciplines or different skills such as reasoning were made, there was little to no clarification as to exactly what was transferred or how the transfer was measured. This criticism is particularly important when anyone tries to claim that 'far transfer' takes place; for example claiming that talking about subject-specific topics in science and/or mathematics can lead to seemingly completely unrelated abstract reasoning abilities in other areas.

While acknowledging the criticisms above, what I want to put forward in this chapter is that there has been a set of studies that meet the methodological requirements of the skeptics and which does provide the kind of 'robust' data needed to convincingly show that 'productive discussion' on science and mathematics topics does indeed have the ability to improve children's abstract reasoning. I will also use these findings to illustrate what is meant by near and far transfer, how these concepts can be measured, and why knowing that they are different is important.

4 The Notion of Transfer

One always expects that when people are taught something that they will learn about the specific thing that is being taught—with some learning more than others. However, something that is quite different and much more exciting is to be able to teach people to perform well in new situations, and to get them to be able to adapt what they have been taught to solve problems and do things that are new or different from what they learned. This ability, although largely aspirational, is a major aim of education. For example, the United States Common Core Standards for English Language Arts (ELA) (2010) states that "The skills and knowledge captured in the ELA/literacy standards are designed to prepare students for life outside the classroom" and "Students will learn to use cogent reasoning and evidence collection skills that are essential for success in college, career, and life" (p, 3.) Clearly, this is an expectation of transfer beyond what is being taught but, sadly, not much is said about how it should be done and measured.

Arguments around detecting and measuring transfer have been around for a long while. Fairly recently, two psychologists wrote that there is "little agreement on the nature of transfer, the extent to which it occurs, or on understanding of its underlying mechanisms." They even went as far as to say that "Despite a century's worth of research, arguments surrounding the question of whether far transfer occurs have made little progress towards resolution" (Barnett & Ceci, 2002, p. 612). Their response to this problem was to draw up a framework one can use to judge whether transfer had taken place or not and for distinguishing between 'near' and 'far' transfer (or anywhere in between). They defined near transfer as learning that is different but related to what was taught, while far transfer results in the ability to apply what was learned to new or different situations. They used two major dimensions—content and context. *Content* looks at whether transfer involved specific information or procedures on the one hand, or involved a general strategy or rule on the other. It also looks at whether what is measured is the same as or different to what was taught, and what sorts of memory demands the transfer task places on the participants. *Context* looks at whether the transfer context is in the same or a different knowledge domain (same subject or a different one), the same physical context (at school or outdoors in a forest), the amount of time between teaching and testing (long versus short), the same social context (alone or in groups, with other children or with adults), the same functional context (talking or writing, playing music or doing mathematics), and the modality of presentation (being shown how to follow a trail in the outdoors versus drawing a topographical map). These are but a few examples, but they give an idea as to how Barnett and Ceci's framework can be used. As would be expected, they point out that 'near transfer' involves tests which are close in time and similar in content and form to the training, while 'far transfer' involves tests which are far apart and different in content and form from the initial learning experience.

We can use Barnett and Ceci's framework when we look at one of the first studies linking classroom discussion and improvements in abstract reasoning abilities. Nearly two decades ago Neil Mercer, Rupert Wegerif and Lyn Dawes (1999) used a form of 'productive discussion' called 'exploratory talk' and then tested if this type of talk had any effect on the ability of eleven-year-old children to engage in abstract reasoning. They employed a widely used, well-established, reliable standardized psychological test of non-verbal abstract reasoning to measure the children's reasoning skills before and after engaging in 'exploratory talk' in their science classrooms, namely the Raven's Standard Progressive Matrices test (Raven, Raven, & Court, 2003).

The problems presented in the Raven's test are a series of diagrams or designs (abstract representations) that can be used to measure and compare a person's capacity for observation and clear thinking relative to other people. The test can be used irrespective of past experience or present ability for verbal communication, and across a range of ages. It has been widely believed that ability to solve non-verbal reasoning tests like those used in the Raven's test is an *inherited* ability and cannot be changed by social or cultural factors (in other words, one cannot learn how to do them). Mercer, Wegerif and Dawes obviously did not believe this to be the case as they used the Raven's test to measure abstract reasoning abilities that they assumed to be *dependent* on (having been caused by) what happened in the classroom, namely the children engaging in 'exploratory talk.' They found that, as the children's ability to engage in 'exploratory talk' improved, so did their scores on the Raven's test. This finding is important as solving problems based on abstract representations is an important mental ability for thinking 'scientifically,' and also because the improvements shown in this ability after engaging in 'exploratory talk' suggested that 'transfer of learning' had taken place.

Using Barnett and Ceci's taxonomy it was clear in this case that on the content dimension there was no specific information, procedure, general strategy or rule given by the teacher which would help the children solve the Raven's problems. In terms of context the children went from talking in class groups to writing an abstract reasoning test, and the time over which this all happened was relatively long (at least 12 lessons). However, the talk that took place was about how to solve the Raven's problems, and the children then solved these problems in groups. As such the transfer that took place would have to be categorised as 'near transfer.' If there was no connection between what the children spoke about, for example something like connecting light bulbs and cells in different ways in science class, and if they wrote the Raven's tests on their own, outside the classroom, then the transfer could be considered to be 'far transfer,' something that is highly valued in education and supposedly very difficult to achieve. It is exactly this sort of evidence—namely that the process of 'productive talk' can generate better abstract reasoning skills unrelated to the topics of discussion—that was sought when reviewing a number of related studies that focused on talking in science and mathematics classrooms (Webb, Whitlow, & Venter, 2016).

5 Finding Far Transfer

When trying to measure the effect of exploratory or any other talk in classrooms one must firstly be sure that the type of talk that one wants to happen has actually taken place. In 'exploratory talk' ground rules are set for talking and listening. These rules are usually developed by the teacher with the children, but always include issues of mutual respect for persons and ideas; the sharing of all relevant information, acceptance of challenges and rebuttals; and striving towards agreement on claims made. When observing discussions, researchers are able to record the degree to which the children comply and they can log the frequency of certain key words such as 'because,' 'agree' and 'I think,' as well as the context and depth of the talk that took place. This is the way that Mercer, Wegerif and Dawes (1999) determined whether 'exploratory talk' was actually taking place or not in their studies. Their method has been used in a number of other studies including seven South African 'exploratory talk' studies which took place over a period of ten years.

Almost all of these studies used the Raven's Standard Progressive Matrices Test (2003) as a measure of abstract reasoning (one study used reasoning from word to numerical problems in mathematics). The main difference however was that in these studies there were interventions where the children engaged in 'exploratory talk' that was *not related in any way* to solving the test. The children were tested before and after the intervention and their changes in scores were contrasted with comparison groups of children of similar age and ability in the same schools, but who were not included in the 'exploratory talk' process. Each exploratory talk intervention took place over a period of at least six months and the participants were between 11 to 14 years old, except for one study which worked with young adults. The first study used science practical work sessions, 'Big Book' stories and science scenario 'prompt posters' to stimulate 'exploratory talk' during science classes (Webb & Treagust, 2006). The teachers in this study were able to get their children to talk productively when doing practical work and discussing science-based shared readings from especially prepared science stories, but not when using the posters. The method that the teachers were taught when using the prompt posters was reviewed and a refined version was used in a second (larger) study where all scenarios resulted in the students engaging successfully in 'exploratory talk.' Mary-Grace Villanueva (2010) used practical science activities to stimulate 'exploratory talk,' but in her first attempt she was not able to get the teachers to facilitate it successfully and the Raven's pre-post tests showed no statically significant change in scores. She revised the teacher-training design aspects of her project and then worked with a new set of teachers. Her improved teacher-training workshops were successful and the groups of participating teachers were all able to successfully engage their students in 'exploratory talk' which resulted in an improvement in Raven's scores compared to the comparison group's scores.

Lyn Webb (2010) worked with mathematics teachers who used mathematics 'concept cartoons' to get their children to engage in 'exploratory talk' and, again, their Raven's test scores after the intervention were statistically and practically significantly better than the comparison groups scores. Percy Sepeng

(2011) used mathematics word problems to stimulate discussion and also got statistically and practically significant improvements on reasoning abilities in the experimental group. Shirley-Ann Boschmans (2013), who worked with Pharmacy students at university level, was unable to engage them in 'exploratory talk' (they talked about the content of their studies) and there were no changes in their Ravens scores after the intervention.

All of the above studies which enabled teachers to facilitate 'exploratory talk' amongst their students produced statistically significant improvement in reasoning skills with large or medium effect sizes. What is also important to note is that, when the teachers were not able to get real 'exploratory talk' going in their classrooms there were no statistically or practically significant changes in the Raven's scores. The results of these studies are presented in Table 8.1 where the numbers of children who participated are indicated as n, and the statistically significant changes are recorded as *probability scores* represented as a p value.

The reasoning scores that improved in these studies were not linked in any way to the knowledge domain (from 'exploratory talk' around science and mathematics topics to solving Raven's Standard Progressive Matrices tests) or in terms of temporal context (pre-post testing over six months to a year). Although the physical context was similar (both activities occurred at school), there was considerable distance in terms of social context and modality (from group discussion in science or mathematics to individual completion of the written Raven's tests). As such, it can fairly confidently be said that, as the improved ability to reason better on problems that were unrelated to the topics of discussion, they indicate far transfer (Barnett & Ceci, 2002). Also, as noted earlier, evidence that when no 'exploratory talk' took place no statistically significant improvement in reasoning was recorded, provides credible support to the claim that it was the 'exploratory talk' that made the difference.

6 Explaining Far Transfer

As noted earlier, 'exploratory talk' is one among a number of variants of 'productive discussion,' all of which share the same basic features. Thus any explanation about the effects of 'exploratory talk' can probably be used for all. One explanation, based on the well-established associations between working memory capacity and measures of problem solving (see, e.g., Ackerman, Beier, & Boyle, 2005; Kane, Hambrich, & Conway, 2005; Oberauer, Schulze, Wilhelm, & Suss, 2005), is that 'exploratory talk' provides practice in how to manage working memory capacity and thereby reduces cognitive load by explicitly teaching

TABLE 8.1 Results of the studies under review focusing on 'exploratory talk' and problem solving in terms of both statistical and practical (Cohen's d) significance

Study	Focus	Age	Students engage in 'exploratory talk'	Significance on Raven's SPM
Webb 2003 (first study; n=146)	'Exploratory talk' in science classes	11–12	Yes when using readings and practical work to stimulate discussion.	$p \leq 0.05$; d=0.50 (moderate effect size)
			No when using the 'prompt-poster' approach to stimulate discussion)	$p \geq 0.05$
Webb 2003 (second study; n=1192)	'Exploratory talk' in science classes	11–12	Yes	$p \leq 0.05$; d=0.50 (moderate effect size)
Villanueva 2010 (first study; n=168)	'Exploratory talk' in science classes	11–12	No	$p \geq 0.05$
Villanueva 2010 (second study; n=675)	'Exploratory talk' in science classes	11–12	Yes	$p \leq 0.05$; d=0.83 (large effect size)
Webb 2010 (n=403)	'Exploratory talk' in mathematics classes	11–12	Yes	$p \leq 0.01$; d=1.41 (large effect size)
Sepeng 2011 (n=176)	'Exploratory talk' in mathematics classes	13–14	Yes	$p \leq 0.05$; d=0.86 (large effect size)
Boschmans 2013 (n=215)	Discussion in supplemental instruction classes for Pharmacy students	Young adults	No	$p \geq 0.05$

Note: p values less than or equal to (≤) 0.05 represent changes about which one can be at least 95% sure that they have not occurred by chance. When the p values are greater than 0.05 researchers generally accept that the changes in score might have been merely due to chance. Where changes are statistically significant, their practical significance, i.e. their effect size, is expressed as a Cohen's d score.

one how to structure an argument, and therefore to organize one's thoughts. Breaking down a larger task into more easily manageable bits is a widely recognized strategy for solving complex problems. Toulmin's argumentation pattern (Toulmin, 2003) specifies in a straightforward way a set of 'rules' for an argument, and how they are related to each other. To make a believable claim one needs data (evidence), warrants and backings. The data must be sound, the warrants (underlying assumptions) that connect the data to the claim must be believable, and the backings (knowledge that already exists) must support the reasons given for what is being claimed. One must also be able to answer all the possible queries, different perspectives and arguments that may be leveled against what is being claimed. Whether this is the best way of explaining far transfer remains moot, and ideas on the mechanics of transfer need further research. Nevertheless, what is clear is that far transfer does appear to happen in terms of developing abstract reasoning via 'exploratory talk.'

7 What Do These Findings Mean?

The Raven's problems represent one form of reasoning, namely abstract reasoning; and abstract reasoning forms a large portion of the intelligence tests that are used worldwide. Psychologists distinguish between two kinds of mental ability, one that can be learned and one that is assumed to be inborn. The learned one is the ability to reproduce a culture's store of verbal concepts through exposure to knowledge and enables one to do well in examinations which require memory. The other is the ability to generate meaningful abstractions from concrete referents (which is essential for thinking scientifically). Intelligence Quotient (IQ) tests measure both types of abilities in equal proportions. In England and the Netherlands IQ scores have been recorded for a long time and it is interesting that they have increased considerably over the last 100 years (Flynn, 2009). Similar trends have been seen in other countries around the world, but currently increases seem to be 'tailing-off' in developed countries while they are still rising in developing countries. Some of the improvements have been attributed to improved literacy and schooling, in other words they have been learned, but what is most interesting is that the greatest gains have been in the non-verbal abstract reasoning aspects of the test.

As such, the transfer effect to abstract reasoning via 'exploratory talk' begs the question 'does 'productive discussion' make children more intelligent?' James Flynn (1987), who collated a great deal of IQ data gathered over decades in 14 different countries, thought about the gains in abstract reasoning over time and came to the conclusion that the answer to this question is probably 'no.' He explains that improved scores on the Raven's test do not mean that people have necessarily become more intelligent. His reason is that, as abstract reasoning tests are so detached from reality, solving them is not the same as solving real-world problems. He believes therefore that the Raven's tests only reflect a correlation with a weak causal link to intelligence, and we should thus say that such tests simply measure abstract problem-solving ability.

Flynn (2009) recognizes that different kinds of reasoning may be appropriate in different circumstances and that using abstractions and logic may not be necessary or of much use for illiterate people who live simple lives. But most people live in a modern world where much of what they experience is mediated by symbols and abstractions. Our thinking relies less and less on memory and more and more on representations, particularly for our young 'digital natives' who seem to be permanently attached to a digital device of one sort or another. Flynn attributes the constant gains in IQ scores of the British population since 1942 to the fact that we use our brains differently as compared to the ways in which our parents, grandparents and great-grandparents used them. As an example of altered brain functioning he points to a study by Maguire et al. (2000) which showed that the brains of London taxi-drivers have an enlarged hippocampus, the area used for navigating three-dimensional space. This and many other studies show that brains *can be changed* by the context in which they are used. (I wonder if young London taxi drivers of today, who depend on Global Positioning Systems [GPS] to get around the city, have enlarged hippocampi and understand the principles of big city navigation in the same way as their older peers who had no access to this kind of technology?).

Flynn, recognizes that the processes which cause brain alteration are most probably complex and that any simple one-dimensional attempt to explain them, such as "nutrition for better brains, better schooling, more stimulated toddlers, etc., will be found wanting" (Flynn, 2009, p. 25). He cites a multiplicity of possible factors, for example different schooling, smaller family size, modern parenting, the rise of a visual culture, more jobs that require problem solving, more leisure and the social interaction between parents, children and peers. In terms of the latter effect (peer interaction) he believes that there has been a decrease in oral socialization between children because of the growing use of modern technology. He attributes the tapering off of IQ scores in adolescents to decreasing oral socialization in Britain over the past decade. If true, the development of abstract reasoning, whether by 'productive discussion' or any other mechanism which promotes the higher order abilities associated with the scientific mind, become worth considering within the rapidly changing contexts of modern life.

8 Making Meaning through Talk

Functional Magnetic Resonance Imaging (fMRI) is able to produce brain scans that show differences between brain activity of literate and illiterate people. Literate people can 'see' the words they hear and convert them into visual representations. These differences can be partially explained by understanding word recognition. When literates listen they not only use their auditory cortex, they use their visual cortex as well (Dehaene, Cohen, Morais, & Kolinsky, 2015), and visual word recognition helps one to analyse meaning in abstraction—an ability that is essential to science (Goody, 1977).

However, while becoming literate in the reading and writing sense has many benefits, it has also been shown to have some negative effects. While literate people get better at analysis, they lose some of their facial recognition skills and their capacity for holistic perception (Dehaene et al., 2015). Wegerif (2016) argues that, while being literate helps us to associate meaning with words, it also creates the illusion that meanings are 'things,' rather than aspects of relationships. He also notes that the idea of meanings as relationships was first put forward by the 'illiterate' Socrates, as recorded in writings by Plato.

Print literacy in particular has led to an ideological dominance of ways of thinking that rely on representation on paper. In oral cultures thinking is situated in contexts and relationships become expressed in real-life dialogue. Text books are written monologues, expressed in words which are objects and which seemingly simply have to be believed. However, the written monologues in print are almost always the products of a larger flow of meaning that is produced as part of relationships, the juxtaposition of ideas, and consideration of counter arguments. Such dialogue implies openness to others in order to allow creative thinking (Wegerif, 2013).

While certainly not arguing against the development of literacy in the traditional sense of reading and writing, or the necessity of developing visual perception for reading, writing and numeracy, the relatively recent shift towards the notion of social construction of knowledge through discussion has been a fruitful and welcomed change (Hodson, 2009). The question is, how can such understandings be developed at home, in school, and in the 'third-spaces' for learning available to children? One such way that has been suggested is via social media.

Rupert Wegerif, one of the pioneers of "exploratory talk" research, has a much more upbeat take on technology than Flynn has, particularly in the social media aspects of the Internet. He sees many affordances in terms of the creative co-presence of many voices via online interaction (Wegerif, 2016). He likes the idea of the 'affinity spaces' (Gee, 2005) that open up between an unknown host of possible respondents when one writes a forum post or blog. He sees a role of modern technology in taking us out of our home space and culture and into dialogue with 'infinite others.' However, an important caveat is that it is the *type of talk* that is important if it is to be productive. It is how people interact with others and the nature of such interactions that are highlighted in studies of 'productive' discussion. As such, the digital age presents many research opportunities for investigating how far transfer effects via 'productive discussion' could be achieved online. Similarly, there are also questions as to what other desirable transfer effects through 'productive discussion' may be achieved. For example, taking the notion of productive discussion into other disciplinary discourse situations seems promising for, among others, issues of sustainability, aesthetics, and alternate worldviews.

9 Pulling Things Together

I have spent the greater part of this chapter building the argument that socio-cultural activities can transfer in terms of the way that we think and understand (make meaning). I dwell on this issue as for too long too many have believed that mathematics and science (in the context of this chapter read 'having a scientific mind') is only for those who are born with mathematical or scientific ability, and that certain abilities, such as abstract reasoning, are an inherited gift. In contrast, there is growing evidence that this is not the case (Kan, Wicherts, Dolan, & Van der Maas, 2013). There is sufficient evidence to suggest that the "holy grail" of education (Resnick et al., p. 11, 2015), namely that people can learn more than just what they are taught, should no longer remain largely aspirational.

I have also emphasised the notion of transfer, and particularly far transfer, as a counterpoint to sociocultural theorists' beliefs that (i) knowledge is not an entity in the head of an individual, (ii) developing knowledge is a process of enculturation in a community of discourse, practice and thinking, and (iii) what is learned is specific to, and grounded in, the situation in which it is learned. In other words they believe that transfer from one context to another simply does not take place. The example of street children who can work very well with money but show no transfer of this ability in school mathematics is often used as an example by those operating in a socio-cultural paradigm (Sfard, 1999).

In contrast, cognitive psychologists believe that individuals learn by moving from naïve understandings to more sophisticated conceptual structures by a process where knowledge may transfer through domain-specific restructuring towards better understandings. In science education personal cognitive and conceptual change as a result of instruction and knowledge restructuring was initially the focus of research to investigate mental representations (Duit, 1999). More recently, with the development of new technologies to better understand how the brain works, investigations have been made as to the plasticity of the brain via increasing the capacity of working memory, understanding the biological bases of learning and memory at a synaptic level, and computational models of brain activation (Rudy, 2013).

Like Flynn (2009), I believe that, as we use our brains differently from how our parents, grandparents and great-grandparents used them, we have probably altered their functioning. The specific example offered in this chapter is that the use of productive discussion improves abstract reasoning (Webb et al., 2016). There are a number of studies which provide measurable evidence of changes in brain activity and structure and an equal number of plausible explanations as to how these changes take place. However, while research based on cognitive psychology principles has produced understandings of the effects of learning on brain functioning, it has not been able to contribute much to the development of successful teaching and learning interventions (Simons et al., 2016). Conversely, sociocultural teaching and learning interventions have provided a plethora of success stories on improved learning (Resnick et al., 2015), but provide few explanations as to how they might work on a cognitive level. As such it seems that the sociocultural approach is most useful for articulating how new ways of thinking are internalized through interactions with others, while a cognitive approach is better for explaining the outcomes.

Unfortunately, there seems to be a 'cultural divide' between researchers who approach education from a sociocultural or a cognitive science perspective (Mason, 2007), with sociocultural approaches to promoting learning currently in the ascendancy. To me, 'cultural divides' between fruitful sources of knowledge and ideas, whether deliberate or simply because one views the world differently, are unfortunate and less than optimal. A probably much more fruitful way of doing things would be for those operating in different research 'cultures' to cooperate much more closely with one another when trying to develop educational strategies and understand the gains made by particular approaches and interventions to teaching and learning.

10 Conclusion

It has been argued in this chapter that thinking is situated in contexts and relationships which become expressed in real-life dialogue to generate meaningful abstractions from concrete referents. This type of thinking is essential for scientific understanding. Evidence that children's ability to think in this way can be developed through meaningful discussion has been provided in this chapter. The evidence provided also shows that one can calculate the level of effect that appropriate training of teachers has on their ability to facilitate such discussion. Over and above the 'statistically significant' findings, there are a plethora of examples of transfer, both quantitative and qualitative, with or without measures of practical significance (effect size), to support these claims. If the 'holy grail' of education is for learners to learn more than just what they are taught and be able to transfer what they have learned in new and different situations, it appears that there can be little argument (pardon the pun) as to whether teacher education in the use of effective discussion under any guise or name (for example, exploratory talk, explanatory talk, collaborative reasoning, accountable talk, argumentation, etc.) is achievable and worth the money and effort required to do so.

It is also highly probable that socio-cultural activities such as productive discussion can also contribute to the "dispositions, values, ethical concerns, aesthetic considerations, and the complex array of skills and mental capabilities in varied domains" (BtSM, n.d.) required for modern day 'scientific thinking'. If so, there should be a multitude of opportunities for socio-cultural and cognitive science researchers to work together to get better understandings of how these attributes form, and to develop and test viable teaching and learning strategies towards this end.

I think they should do just that. I also think that perhaps it is time for more of us who are engaged in any form of education to sit down and productively discuss these issues. We all need to be part of guiding and accelerating the promotion of the 'types of thinking', understandings, insights and dispositions required of 'scientifically minded' citizens if we are to be able to collectively meet the threats to the survival of the human species and the world as we know it.

Acknowledgements

I must thank Jan and Muriel Visser for their patience and guidance as they helped me work out what it was that I was trying to say in this chapter. I also want to acknowledge the influence of the discussions held at the BtSM colloquia I attended on my thinking. In particular, I fondly remember the gentle, measured conversations with Jan Visser while walking up and down the soft hills of Lembang, Indonesia, towards the erstwhile Dutch astronomical observatory, which was the venue for the 2013 BtSM colloquium. I must also thank my friend and colleague, Jesse (Bill) Whitlow of the Psychology Department at Rutgers University, Camden, New Jersey. Without his insights I might still be not much the wiser in terms of cognitive explanations as to why we saw startling results when facilitating exploratory talk with disadvantaged children in urban townships and rural areas in South Africa.

Note

1 For those interested in seeing examples of what 'productive discussion' looks like, there are a number of YouTube sites where you can see variations in process. The first ones that came to hand when I looked are the following sites, but if they are not what you want you will have landed amongst pointers to a great deal more: https://www.youtube.com/watch?v=uHBzRhgQQ4k; https://www.youtube.com/watch?v=uC3AfFkLyi4

References

Ackerman, P. L., Beier, M. E., & Boyle, M. O. (2005). Working memory and intelligence: The same or different constructs? *Psychological Bulletin, 131*, 30–60.

Barnett, S., & Ceci, S. (2002). When and where do we apply what we learn? A taxonomy for far transfer. *Psychological Bulletin, 128*, 612–637. doi:10.1037//0033-2909.128.4.612

Boschmans, S.-A. (2013). *Teaching pharmacology: Issues of language and learning in a multilingual classroom setting* (Unpublished PhD.). Nelson Mandela Metropolitan University, Port Elizabeth, South Africa.

BtSM. (n.d.). *The building the scientific mind colloquia.* Retrieved June 22, 2017, from http://learndev.org/BtSM%202005-2015.html

Chinn, C., & Anderson, R. (1998). The structure of discussions that promote reasoning. *Teachers College Record, 100*, 315–368.

Dehaene, S., Cohen, L., Morais, J., & Kolinsky, R. (2015). Illiterate to literate: Behavioural and cerebral changes induced by reading acquisition. *Nature Reviews Neuroscience, 16*(4), 234–244.

Duit, R. (1999). Conceptual change approaches in science education. In W. Schnotz, S. Vosniadou, & M. Carretero (Eds.), *New perspectives on conceptual change* (pp. 263–282). Amsterdam: Pergamon.

Flynn, J. R. (1987). Massive IQ gains in 14 nations: What IQ tests really measure. *Psychological Bulletin, 101*(2), 171–191. doi:10.1037/0033-2909.101.2.171

Flynn, J. R. (2009). *What is intelligence: Beyond the Flynn effect.* Cambridge: Cambridge University Press.

Gee, J. P. (2005). Semiotic social spaces and affinity spaces. Beyond communities of practice language power and social context. In D. Barton & K. Tusting (Eds.), *Beyond communities of practice: Language, power and social context* (pp. 214–232). New York, NY: Cambridge University Press. http://dx.doi.org/10.1017/CBO9780511610554.012

Goody, J. (1977). *Domestication of the savage mind.* Cambridge: Cambridge University Press.

Greeno, J. G. (2006). Learning in activity. In R. K. Sawyer (Ed.), *The Cambridge handbook of the learning sciences* (pp. 79–96). New York, NY: Cambridge University Press.

Hodson, D. (2009). *Teaching and learning about science: Language, theories, methods, history, traditions and values.* Rotterdam, the Netherlands: Sense Publishers.

Kan, K.-J., Wicherts, J. M., Dolan, C. V., & Van der Maas, H. L. J. (2013). On the nature and nurture of intelligence and specific cognitive abilities: The more heritable, the more culture-dependent. *Psychological Science, 24*(12), 2420–2428.

Kane, M. J., Hambrick, D. Z., & Conway, A. R. A. (2005). Working memory and fluid intelligence are strongly related constructs: Comment on Ackermain, Beier & Boyle. *Psychological Bulletin, 131*, 66–71.

Keefer, M., Zeitz, C., & Resnick, L. (2000). Judging the quality of peer-led student dialogues. *Cognition and Instruction, 18*(1), 53–81.

Lipman, M., Sharp, A., & Oscanyan, F. (2010). *Philosophy in the classroom.* Philadelphia, PA: Temple University Press.

Maguire, E., Gadian, D., Johnsrude, I., Good, C., Ashburner, J., Frackowiak, R., & Frith, C. (2000). Navigation-related structural change in the hippocampi of taxi drivers. *Proceedings of the National Academy of Sciences, 97*(8), 4398–4403.

Mason, L. (2007). Introduction: Bridging the cognitive and sociocultural approaches in research on conceptual change: Is it feasible? *Educational Psychologist, 42*(1), 1–7.

Mercer, N., & Littleton, K. (2007). *Dialogue and the development of children's thinking: A sociocultural approach.* London: Routledge.

Mercer, N., Wegerif, R., & Dawes, L. (1999). Children's talk and the development of reasoning in the classroom. *British Educational Research Journal, 25*(1), 95–111.

Michaels, S., & O'Connor, M. (2002). *Accountable talk: Classroom conversation that works.* Pittsburg, CA: University of Pittsburgh.

Monaghan, F. (2005). Don't think in your head, think aloud: ICT and 'exploratory talk' in the primary school mathematics classroom. *Research in Mathematics Education, 7*(1), 83–100.

Oberauer, K., Schulze, R., Wilhelm, O., & Suss, H.-M. (2005). Working memory and intelligence—Their correlation and relation: Comment on Ackerman, Beier, & Boyle. *Psychological Bulletin, 131*, 61–65.

Osborne, J. (2010). Arguing to learn in science: The role of collaborative, critical discourse. *Science, 328*(5977), 463–466.

Raven, J., Raven, J. C., & Court, J. H. (2003). *Manual for Raven's standard progressive matrices and vocabulary scales: Section 1 general overview*. San Antonio, TX: Pearson.

Resnick, L. B. C., Asterhan, C. S. C., & Clarke, S. N. (2015). Introduction: Talk, learning, and teaching. In L. B. Resnick, C. Asterhan, & S. Clarke, (Eds.), *Socializing intelligence through academic talk and dialogue* (pp. 1–12). Washington, DC: AERA.

Reznitskaya, A., Kuo, L., Clark, A., Miller, B., Jadallah, M., Anderson, R. C., & Nguyen-Jahiel, K. (2009). Collaborative reasoning: Dialogic approach to group discussions. *Cambridge Journal of Education, 39*(1), 29–48.

Rudy, J. (2013). *The neurobiology of learning and memory*. Sunderland, MA: Sinauer Associates, Inc.

Sepeng, P. (2011). *Grade nine second-language learners in township schools: Issues of language and mathematics when solving word problems* (Unpublished PhD). Nelson Mandela Metropolitan University, Port Elizabeth, South Africa.

Sfard, A. (1999). On two metaphors for learning and the dangers of choosing just one. *Educational Researcher, 27*(2), 4–13.

Shayer, M., & Adey, P. (2002). *Learning intelligence: Cognitive acceleration across the curriculum from 5 to 15 years*. Maidenhead: Open University Press.

Simons, D. J., Boot, W. R., Charness, N., Gathercole, S. E., Chabris, C. F., Hambrick, D. Z., & Stine-Morrow, E. A. L. (2016). Do "brain-training" programs work? *Psychological Science in the Public Interest, 17*, 103–186. doi:10.1177/1529100616661983

Toulmin, S. (2003). *The uses of argument*. New York, NY: Cambridge University Press.

Trickey, S., & Topping, K. (2004). Philosophy for children: A systematic review. *Research Papers in Education, 19*(3), 365–380.

United States Common Core Standards for English Language Arts. (2010). Washington, DC: National Education Association.

Villanueva, M. G. (2010). *Integrated teaching strategies model for improved scientific literacy in second language learners* (Unpublished PhD). Nelson Mandela Metropolitan University, Port Elizabeth, South Africa.

Vygotsky, L. S. (1978). Interaction between learning and development. In M. Cole, V. John-Steiner, S. Scribner, & E. Souberma (Eds.), *Mind in society: The development of higher psychological processes*. Cambridge, MA: Harvard University Press.

Webb, L. (2010). *Searching for common ground: Developing mathematical reasoning through dialogue* (Unpublished PhD). Nelson Mandela Metropolitan University, Port Elizabeth, South Africa.

Webb, P. (2003). *Initiating classroom discussion in science classrooms: Practical work, conversational texts and prompt posters* (Unpublished PhD). Curtin University of Technology, Perth, Australia.

Webb, P., & Treagust, D. (2006). Using exploratory talk to enhance problem-solving and reasoning skills in grade-7 science classrooms. *Research in Science Education, 36*(4), 381–401.

Webb, P., Whitlow, J., & Venter, D. (2016). From exploratory talk to abstract reasoning: A case for far transfer? *Educational Psychology Review, 29*(3), 565–581. doi:10.1007/s10648-016-9369-z

Wegerif, R. (2013). *Dialogic: Education for the Internet Age*. New York, NY & London: Routledge.

Wegerif, R. (2016). Applying dialogic theory to illuminate the relationship between literacy education and teaching thinking in the context of the internet age. In S. Brindley, M. Juzwik, & A. Whitehurst (Eds.), *International perspectives on dialogic theory and practice* (pp. 1–21).

Wilkinson, I. A. G., Murphy, P. K., & Binici, S. (2015). Dialogue-intensive pedagogies for promoting reading comprehension: What we know, what we need to know. In L. B. Resnick, C. S. C. Asterhan, & S. N. Clarke (Eds.), *Socializing intelligence through academic talk and dialogue* (pp. 35–48). Washington, DC: AERA.

INTERMEZZO 8

Facing Life's Biggest Questions

Jan Visser

Reflections inspired by Premana Premadi: 'Seeking to know and understand the self'

I can't find another word. Extraordinary! Never before have I been involved in a project that took so long to come to fruition and that counted with so many unpredictable hurdles. Even more extraordinary is the effort made by all who were involved, together with me, to persevere against the odds and to overcome the difficulties encountered. I'm talking about the making of the book on *Seeking Understanding: The Lifelong Pursuit to Build the Scientific Mind*. This brief intermezzo is to honor all those involved.

Some never made it to the end. I already mentioned, in Intermezzo 6, Harry Kroto, who died in 2016 from complications of amyotrophic lateral sclerosis (ALS). At that time we were still in the middle of composing a chapter on his work, based on existing documentation. I had been interested in Kroto's work because of his multifaceted interests. A similar case was the American hematologist Leo Vroman, who combined his scientific interest with graphic art and poetry (in Dutch and English). He was 98 when I contacted him. He felt too old to write, but we would work something out on the basis of interviews, focusing on the intersection of science and poetry. He died at age 99 on February 22, 2014. I heard the news the night I was going to phone him. A third case was the philosopher Hubert Dreyfus. We first met in January 2015 in his office at UC Berkeley. I had been interested in Dreyfus's philosophical insights and particularly also in the dialogic way in which he 'taught' his classes. I still managed to do a video recorded interview with him while he was briefly in Paris in December 2016, but he died in April 2017, aged 87. It never came to a written contribution. Interview material for both Kroto and Dreyfus, though, will be available on the YouTube channel for this book.[1] Unfortunately, no interview material is available for Vroman.[2]

Questions of life and death concern us all. They become more prominent when we grow old or when we face serious challenges to our health. In addition to the three cases mentioned above, nine contributing authors to this book faced such challenges during the five-year period of its gestation. All cases are different. So are the ways in which we all deal with the challenges we meet.

 | DOI: 10.1163/9789004416802_016

The mind, our astuteness at seeking understanding, can then be a powerful attribute in facing life's biggest and most profound questions.

Astrophysicist and cosmologist Premana Premadi is afflicted with the same condition as her scientific colleagues Kroto and Hawking were. She elucidates in her chapter on 'Seeking to know and understand the self" how she deals with it, highlighting the integral nature of the mind as it seeks to understand both the universe and the self, seeing them in a similar light.

Note

1 Building the Scientific Mind YouTube channel, https://www.youtube.com/channel/UCNx_SL357aoLl7ihqVHXqwQ
2 An example of how Vroman's mind could connect scientific writing with poetry can be found in a 2009 article in the journal *Materials* on 'When blood is touched,' available at https://www.mdpi.com/1996-1944/2/4/1547/pdf

CHAPTER 9

Seeking to Know and Understand the Self

Premana W. Premadi

1 Introduction

I am an exceedingly fortunate person and I am ceaselessly grateful for it. I was born to parents who themselves were born to families with—as I came to gather late in life—serious family values. My great grandfather from my father's side was one of the first few Indonesian (or Javanese, they called it then) medical doctors. He practiced in many places in Java, and for a number of decades, he dragged his large family with him to wherever he was stationed. My father's mother never left the island, but she was fluent in high Javanese language as well as Dutch, and also read and wrote well English and French. She taught herself classical piano and then taught my father. She died in her forties, long before I was born, but stories of her live in me like a legacy. I suppose her younger sisters, with whom we, as a family, enjoy close relationships, bear some resemblance; they all speak multiple languages with ease, even now when they are all very advanced in age and are challenged by dementia. My grandmother from my mother's side was a primary school teacher. She graduated from a teacher preparation school where she was the only female student. A large proportion of our relatives from this side of the family are school teachers. My mother's father worked in the pharmacy of a large mental hospital up on a hill in the countryside, and the family was used to helping patients reconnect with society. It is obvious that, for all of them, education was not just a series of training events to master certain professional skills, but more of a personal enrichment, which in turn enriched their way in serving their environment.

My good fortune is faithful to me as the Moon to the Earth. I remember being aware of it after I left my family home to attend university, when I realized that the values I packed from home sustained me far, thanks to good fortune. I am blessed to have met, been closely acquainted, with family friends, relatives, teachers, students, and even strangers, who have provided me with a rich loving and nurturing environment. I feel they all contributed to sculpting me.

Living alone for an extended length of time in a couple of foreign countries, and traveling alone in many more, made me learn about who I am, the place and people I came from, my cultural and spiritual background, perhaps more than what I learned about the hosting country and people. Foreignness

 | DOI: 10.1163/9789004416802_017

is such a relative and reflective notion. The process of self-identification requires multiple references and willingness to reflect with a sincere and open mind. Failing to do so would result in a biased view of oneself and of the rest of the world. This course of learning covered almost my entire early adulthood. Although the wide exposure is undoubtedly enriching, I admit that to have to deal with sharp dissimilarities could be overwhelming as it forced me to reexamine what I believed was true. I just mindfully executed what had already been cultured in the brain and evaluated the incoming responses, and then recultured my brain as I deemed appropriate. Being mindful is like working on a fertile ground that offers you the freedom to choose what best to cultivate considering the demand and the season. It invites you to apply your knowledge on how nature behaves. After a while, to be mindful is almost instinctive.

While mindfully venturing out into the world, I still let my naïveté take the steering wheel through most of the journey. Not because I think that ignorance is bliss, but because I wish to quench what I desire to know, so as to avoid that ideas I already have get in the way of the further development of my knowledge and understanding. It rolls out the multi-dimensional landscape of curiosity-necessity-knowledge, and points to where I stand. Only then the tentative knowledge has a calibrated value for me. The landscape first serves as reference against which I do a fierce self-evaluation, and afterwards it serves as bearings when I decide to embark onto my next journey. I find this meta-method helpful as I am rather conscious about not wanting to fall into the trap of accepting the ubiquitous information as knowledge, and about the dark gap between knowing and understanding. Unsolicited information is overwhelmingly present and only proves to be quite capable of disorienting the mind in too many instances. The responsibility which a scientist has to produce and reproduce scientific ideas and works, explore new territories, examining new ideas, and indeed to assist the public in filtering information, has made me seriously mind that gap, and consequently, undeniably scrupulous.

When I stack together all the pictures of life in the world seen from various points of view and projected against various backgrounds, I notice something remarkable: the clear commonality in what humans all over the world hold as supreme virtues and what makes them superior on the face of the Earth.

2 To Teach Is to Learn

In retrospect, while it is fairly conceivable that my personal desire and my work responsibility converge, I do not think it is entirely coincidental, and thus count it as a blessing. My profession as a cosmologist with a tenured faculty position is definitely one of my greatest blessings. When I look around, I realize how fortunate I am to land a job that not only suits my personality, but also generously caters to my deep interest in the universe and all it comprises, and my genuine longing to be a part of the journey of knowledge and a part of the dynamic of the universe itself. Once trained as a scientist, a person perceives oneself and the world with an attitude that bears witness to freedom of inquiry, and is equipped with some measuring tapes; so yes, it is somewhat methodical.

Being a scientist is not an end in itself. Being a scientist is like having a large door open for you to step out and explore the deep and grand universe, and to decipher the enigma of Existence. But then you need to come back home, sit down and make sense of it all, and share with people what you have discovered. Science does not belong to the scientific community alone, it belongs to the civilization. It is like our ancestor the hunter-gatherer; you go out to look for sustenance, you return to digest, and you must give something back to your environment. You repeat it every day to survive. Except that scientists survive on scientific knowledge.

Astronomy and astrophysics develop by straddling a constructive interference of two waves: serendipitous discovery and increasingly methodological search. The former reflects the vastness of the still uncharted universe to which our eyes and mind open in amazement. The latter, driven by the development of our knowledge, and by more educated questions, teaches us to be meticulous and functions as reality check for our bold theoretical speculations. This spirit is so ingrained in us because our objects of study in these fields are too far beyond our physical reach, let alone to be under our control. Astrophysics presents itself as the ultimate inverse problem that pushes the power of mind to an ever receding limit.

It is curious that my mind seems to automatically translate almost every mental input into geometrical entities and trace a web of traffic contours from one point to another. This mind craft is not a deliberate act to solve any particular problem necessarily, but just pops out as an instinctive desire to see how things fit in the whole picture and how they relate to one another. It is a busy mapping process as the scheme evolves, dynamically taking into consideration numerous self-critiques infused by observations on my surroundings, readings, conversations, and also quiet personal reflection.

Working as a cosmologist is truly a luxury as I am presented with a limitless spacetime for my mind to roam. The universe is truly a home for me. I can indulge myself with any celestial object or phenomenon, or even just an idea, to think about and work on. At the same time it is a private cocoon so vast and quiet and beautiful into which I retreat from this clamorous world. I suppose everyone should reserve a quiet moment on a regular basis, to, quite literally, quiet the mind.

The values that I learned from my family and from the various encounters I have with people and their stories might very well be the driving spirit that brought me to choose my profession. I love to share. Once I ascertain how vast the universe is and how beautiful the physical processes are that create its beauty, all I want to do is share those feelings and insights. Teaching is one obvious way to share, but there are certainly more. The requirement to teach properly, to deliver well, put me into a perpetual process of learning, which not only prepares me to be a teacher, but enriches me with insights and capabilities which proved indispensable in other aspects of my personal development. I am pleased to also share these personal empowerments as I see fit. Teaching cosmology is not just a matter of elucidating the physics of the expanding universe, but to make my students see the universe as their larger home, which comes into view and into our comprehension through the evolution of our thinking. Understanding that we are a part of this revealed universe consequently allows a developing view of ourselves. We have better knowledge of ourselves because we understand our environment better (see also Greenberg, Chapter 2, this volume, regarding the crucial role that knowledge of context plays in creating meaning).

As long as one is driven by love there seems to be abundant energy and infinite inspiration that could be poured into the work. Teaching involves multiparty communication and is commonly embedded in a certain system in such a way that the success of it is not entirely up to the teacher (see also Webb, Chapter 8, this volume). The question is how does one maintain the passion to keep going when the going gets tough? Here the love needs to be supported by a solid conviction of what one thinks is the right way to go. An example for it would be a set of well thought ideas, clear ideas, the mastery of which one truly desires to see as the outcome of one's teaching. One then commits oneself to create the best learning experience possible for the pupils. It is the rationale of the profession. The clear mind supplies reason to the tired love to keep going. This principle must be resilient, not stiff, and must be allowed to evolve, such that at any given time it serves as a reliable guide and support when frustration dims the sparks of love.

Life is full of intersections. We are to make decisions all the time. Some are less forgiving to mistakes than others. But what actually leads to a good decision? Geared by good knowledge, the power of analysis and the ability to predict are definitely crucial in making decisions. Lessons from experience, one's own and others', are valuable. All these tools need continuous sharpening to optimally function. Everyone must be trained to construct a mental map of systems to be considered and have a good idea of how each component in the system logically relates to another. The process of analysis begins with this map. Problem-based learning offers ways to exercise that.

Learning the hard way offers deep lessons by actively engaging and mindfully immersing oneself in the series of experiences one goes through (although some experiences, arguably, can only be understood in remote hindsight) to identify and meet challenges. This is perhaps a desirable route for those who are tough, but the hard drills are not for everyone. Institutionalized educational systems should arrange for a balanced combination of easy and hard learning experiences such that the process is beneficial to all. The very hard ones correlate with specific objectives and perhaps can be used for professional placement purposes. Not leaving anyone behind in education ensures better citizen participation in all sectors. Taking up the challenge to solve a community problem as a personal responsibility encourages people to activate and thereby develop their personal capabilities, particularly in collaborating, as opposed to just relying on others to perform the task, an attitude not uncommon in large communities. Often we do not realize what is missing until we need it. Clearly the network of individual positive attitudes and capabilities builds social capital.

Science is a human endeavor that yearns to discover the nature of the world, from the tiniest to the largest, and its fundamental governing laws. It asks big questions that often drive one along strenuous paths, only to relay the quest to the next generation which proposes some reasonable answer. This answer is subsequently scrutinized by the following generation which will then develop it into a more plausible answer. And so the pursuit continues. Along the way it sows the seeds of numerous more specialized sciences that grow in their own right and beauty. Telling the history of a particular scientific discovery is something I delight in doing while teaching and I use it in any context. Without meaning to be presumptuous, I consider that cosmology sits at the apex of human knowledge, in the sense that every branch of knowledge eventually, via a shorter or a longer route, connects to cosmology. More impressively, people feel personally related to the cosmos, and find cosmology an uplifting science. Of course, this is not entirely surprising, considering that humans have marveled at the cosmos ever since they set eyes on the heavens. But from the grand story of modern cosmology that connects quarks to galaxies—with objects the size of humans hanging precariously, in an infinitesimal slice, in the middle of the known scale of the physical world as a whole—emerges new awe and meaning as humans are now included in the story. People long for a reason- and evidence-based explanation of how they relate to the cosmos. Cosmology is the grand tapestry *par excellence* woven out of all relevant sciences that brings to light magnificent scenery that grows sharper with age. This revelation renders cosmology a grand entrance to science for educational purposes. High school teachers who have become familiar with cosmology report that they feel much more motivated in teaching science knowing how we and the rest of the universe relate to one another, and how the bits and pieces of science that they have been teaching come together. They now teach science as a great story of nature, not merely a long list of cold and detached facts. They experience a sense of liberation, of being rescued: learning cosmology liberated the teachers from the bound and pain of having to teach something they don't fully understand. Such a profound experience makes them realize that being a good learner is a necessary condition for being a good teacher. Hearing teachers make these comments truly made my day.

Because cosmology may, for a significant fraction of the human population, have an impact in some form on feelings of spiritual transcendence, there needs to be a constant reminder that cosmology is a physical science, a product of the human mind. It is therefore subject to endless examination. It is neither magical nor prophesied. In my cosmology classes students are trained to argue the causal relation of each physical process, involving whatever phenomena under scrutiny, and to identify the conditions and limits of its validity. They are compelled to comprehend the hard physics, only to realize that, in the end, all scientific knowledge is tentative. This should be interpreted as an open invitation to contribute to the development of science, as nothing in science is final. Stories about how many great scientists of time past persevered before they triumphantly arrived at their *magnum opus* encourage the participation of young people who initially thought science is for geniuses only.

The exposure to the richness of the universe, to its enormously wide spectrum of structure and scale, and the various degrees of complexity in its subsystems, plus the training in causal logic, construct a scientific mind which can reliably explore other multi-parameter realms. Someone working on the nature versus nurture debate in galaxy evolution might find a similar debate in human cognitive development exciting and inspiring, not necessarily because the two systems are thought to share some common working principle, but because the logical machinery evolved in the mind and the open attitude formed by being widely exposed to varied phenomena make paying mindful attention spontaneous. Well exercised, the power of thinking can enhance the beholder's existence far beyond the domain of the physique.

3 Stamina: Physical, Mind, Mental, Spiritual

With the mind continuously occupied throughout my waking hours, and while the heart perpetually yearns to give, I must act so that something gets materialized. Time and again fortune lends its good hand to propel me forward a significant distance in this pursuit.

I have lived with ALS (Amyotrophic lateral sclerosis) for the past seven years. ALS, also known as MND (Motor Neuron Disease), impacts one's life progressively and seriously. Even now that this ALS is surely gaining the upper hand, the wheel is still turning. However, since it would be naïve to say that ALS cannot divert you from reaching your dream, you must go back to the drawing board and revise the strategy. Here is my story.

After being diagnosed, the first step I took was to reevaluate everything on my life's roadmap in terms of importance, relevance, and value to me, to my loved ones, my work, my institute, my country, my et cetera, et cetera. At first and second and third attempts all things remained important, relevant, and invaluable to all parties. Nothing, no project, wanted to budge. That is until my physical condition plunged further, which forced a project or two on the roadmap to be erased or merged or delayed indefinitely. On the other hand, there are long-term projects that got squeezed into a smaller package and rushed into the present. Before getting this ALS I had never realized that I could be so defiant. Certain objectives in life remain though; in fact they are only made more prominent by the approaching time limit and decreasing energy budget. I began to sympathize with the frequent political headaches at top government level all over the world, where decisions have to be made within limited time-frames, while juggling conflicting priorities, for I am now weighing both the pros and cons in my own head under time pressure, having no one to consult. But then, every challenge pushes us to open new doors to tinker with a plethora of possible solutions. We learn from a great variety of contexts how taking up challenges makes way for development. Unlike the word 'problem,' which has a darker tinge, the word 'challenge' feels more like a call to engage our entire system to meet it. Without being asked, family and friends, all who know me well, identify with some of the projects on my roadmap. They would like to be part of them. In short, I am receiving and accepting help. Learning to balance help received and the desire to still do things on my own as much as possible turned out to be an enlightening experience. Not just for me, but for all around me. It is a wonderful opportunity that we all embrace.

Learning to delegate is a challenge entirely new to me. To delegate is not the same as breaking the task into pieces and distributing them to a team working towards an agreed-upon target. To delegate is not only to make someone else do something on your behalf, it is more of trusting that this someone would do the task as well as you would, if not better. For someone who had been pretty much self-reliant, this is far from easy. Trust encompasses a wide spectrum and is built upon both merit and generosity. The weight of the objective must rest on the quality of the work, not on my feeling.

In practice, one tends to put gratefulness far above anything else. We refrain from subjecting kindness to criticism or lament. Yet, 'kind' and 'correct' do not operate in the same playing field. When we need to get the job done right (and perhaps with a certain style), is it fair to let kindness take the backseat? It is a serious juggle that teaches me to be always careful, mindful, and perceptive. Empathy is reciprocal. I could succeed with one person, but have a trying time with someone else. There are strands in the fiber of human relations which are never meant to be generalized. Nonetheless, I truly feel fortunate to have this opportunity to experience a unique, but ambiguously hard to cherish, manifestation of human interdependency.

ALS brings the word 'limit' to my face. Limit no longer presents itself as a challenge one wishes to overcome as part of personal development. It is now a solid barrier within which I am obliged to remain. The challenge is now to make the best out of this limitation, and in fact to make friends with it and regard it as my personal bodyguard rather than a frightening prison officer. Its manifestation is in admitting one's regime of strength. Several years ago I could still play the piano for a couple of hours straight. Two years later I could only play for less than half an hour, forcing me to stop just before my fingers would become stiff and my wrists painful. Within that thirty-minute interval I could then still produce a bit of music albeit at a rate of just half the required tempo. The objective of playing the piano thus became exclusively personal enjoyment. The hard lesson was that extending beyond my limit only caused a cruel cut in that enjoyment, with a repercussion permeating beyond physical discomfort towards agony and self-pity. I discovered that my physical stamina dictated my mental stamina. Through this I also came to realize that the spiritual fulfillment from playing the piano depended only loosely on my technical prowess but more on how successfully I addressed the spiritual longing that brought me to play the piano that day in the first place. My demand from myself has become simpler, and thus clearer, as my current top priority is to maintain my mental well-being. I doubt that it will ever recoil to a lower rank, for now that I realize having mental well-being as the center of command, my life enjoys a new balance of the trio: physique, mental, and spiritual. The ability to think well in all situations proved crucial in holding up the whole system.

With ALS one's physical world relentlessly shrinks. This is the world one can physically access, the very world in which most of one's freedom is exercised. When you move around in a wheelchair the world looks remarkably different from when you walk. The field of view is obviously altered, and this promptly stimulates the mind to get busy recalibrating all measurements. The mind strives to make up for the loss in the physical freedom. The challenge is to make sure that this non-physical world, the world occupying one's mind and soul, does not shrink and yet remains examinable. I try to keep it in check by creating some kind of mental mapping from an idea in my non-physical world to some set of observables in my physical world. Those observables are the common terminologies available for others and me to communicate with. I think of the way I think. It is a rational conversation within myself which is not easy to share with others. I do not mean to keep it to myself, but I should not expect others to be able to fully relate to it as they do not share my condition. I just need to safeguard my mind for it is my only constituent that I can still have full control of. Hopefully for a much longer time.

4 Cosmological Innervation

One of the things we do in studying the universe is collecting a large set of data of galaxies, where each of them is carefully identified with all its unique properties and further studied to be fully understood. Presently just a small percentage of the vast body of data finds immediate use. But in time it will, and with the acquired understanding, we will be able to refine our reconstruction of the history of our universe.

My outward interest in the cosmos is now complemented by an inward one. Unavoidably, I look at my getting ALS from the point of view of a scientist. I find striking similarity in cosmology and neurology in its unknowns and complexities. I wish I could fulfill all the protocols involved in getting my blood and skin sampled for the laboratory to generate my induced pluripotent stem cell so that the misfolded proteins and mutated genes in me could be identified. This would enable us to model plausible mappings from the mutated gene(s) to the affected neurons, and therefore, perhaps, yield a description of my particular variant of Motor Neuron Disease. The collection of such identifications derived from a large sample of patients would generate a library on which we can rely for future reference. The availability of such a data base would stimulate like-minded scientists to engage in further research and develop deeper understanding of the disease. As many variants as possible must be identified in full to build a model that might resemble something fundamental, the common underlying condition and process, that constitute the cause of MND. Until that is the case, finding a cure for it remains largely dependent on a fair amount of guesswork.

Although I do not believe, and certainly do not hope, that the cause is as diverse as its manifested variants, one needs to prepare for a radically different approach in case it was.

This line of thought is too raw in the common scientific logic and it therefore appears mundane at best. My wish is to have a good understanding of neuroscience, neurology, microbiology, biochemistry, or whatever the subject is categorized into, for I truly desire to be the competent scientist who takes her own self as a scientific object. To just consider myself a helpless and thoughtless patient is taking me nowhere but down, and worse, it is a waste of potential patient participation in the progress of MND science.

From a rounded perspective, I wonder whether having this ALS is a blessing in disguise to me personally. People who love me dearly find it difficult to agree on this. Here is my take. I was 46 when the diagnosis knocked me off. Before this vicissitude my life was brimming. I worked fervently and played seriously. I have been so very fortunate to have met many and various kinds of people, and seen many places around the world. I have been exposed to the vastness of life, great and small, high and low, complex and simple, dark and illuminated, endowed and impoverished. I wish more people are at least as lucky as I have been.

Now at 53, I am feeling more and more of the physical challenges, which occasionally extrapolate to the psyche. Although with increasing effort, I am still breathing on my own, eating through my mouth, indulging my palates with all the richness of taste, speaking my mind, and am still up and about, even traveling from time to time. Living with ALS long enough provides me with a unique and invaluable opportunity to do a thorough rational observation of how the progression of the disease causes alterations in me in the course of time, directly and reflectively using responses from my environment. I have been made exquisitely aware of what changes, and therefore, in contrast, but equally invaluable, of what remains invariant. All of this, I reckon, must serve a purpose. Or at least that is how I like to think of it. Here I am, a scientist with a take on the Aristotelian final cause. I just give it a go if it could lend a boost to a positive and feasible proposal.

My inner world of music, geometrical and colorful images, of words, has apparently produced a patterned dynamics in my mind that perpetually aspires to create the best harmony for each point in spacetime, i.e., for each circumstance. The more worldly manifestation of this seemingly abstract effort is in the exhaustive elaboration of the scientific mind in analyzing the multilateral communication among various perspectives and in weighing the various cause-and-effect trajectories in order to come up with a sound decision. The above exercise yields my individual evaluation sheet. This evaluation of my having ALS, combined with my serious take on language and communication, the crucial importance of a good network of people and resources, sparks a new dream, my dearest: to establish an ALS support system in Indonesia—a dream that has meanwhile materialized. Somehow my whole person constitutes a complete cause for this endeavor, and therefore the final cause argument.

I never ever ask why I am afflicted by this incurable, disabling, and invariably fatal disease. Instead I do ask through what process I became afflicted, what are the effects on my body and my life, and what are the risk factors, all of which are valid scientific questions. Furthermore, I ask what I can do to make a positive difference, however miniscule, along the way to discover the cure and to hold up the quality of life of ALS patients. Of course being a patient does not make me an authority in the subject, but surely it makes me an authentic case, provided I present the case via a rational approach. I thereby hope that my story would be a contributing piece of the puzzle in the ALS mosaic. Reflectively, even though I might know my story well, I would only have a better comprehension of it when I know where my story stands in reference to the big picture.

It is quite compelling that, like the story of the universe, the highly complex nature of MND renders the construction of its big detailed picture a never ending task. Yet, just the same, we strongly believe that it is only through rigorous analysis that we might unfold the whole story and thus discover a solution. In practice, as we keep our noses to the grindstone, new parameters are being revealed. Science always starts with simple bookkeeping. The emergence of those new parameters calls for additional dimensions in the bookkeeping, a revision of our simple-minded starting model, which could reveal a huge miss in the fundamentals. One must balance the need for having an open mind towards serendipitous discovery with a scrutinous attitude to be on the lookout for pitfalls. ALS is an interesting scientific object, and is indeed presented in an exuberant line of research works, yet it is heavily burdened by the human psyche. Here, unlike in cosmology, the agony of not knowing cannot be contained in the laboratory and observatory, but it travels far into the patient's home, shaking the fabric of family life.

Fortunately the lack of local medical support is compensated, albeit partially, by information resources available on the Internet in the form of formal and informal medical references and communication forums among ALS patients and their caregivers in a number of developed countries. Equipped with this set of references I venture to do some analysis to identify my current position on the ALS landscape and predict my future trajectory, within an acceptable margin of uncertainties and errors. This corridor of prediction allows me and those near me to prepare ourselves reasonably well, mentally and otherwise, for things to come. The ability to avoid cataclysmic shocks hugely helps in keeping life at the best possible quality by managing stress levels and avoiding unnecessary astronomical financial expenditures. To be resilient requires knowing how to prepare. For me personally, it goes deeper than that. This personal learning experience of pushing rational power to work in areas where it is not traditionally championed, presents me with a new and refreshing outlook toward human mental capacity. Not only can we empower our rational capability, we can also engage it to other elements of human attributes, such as the spiritual element, and empower them in a controlled chain reaction sort of manner. Ratio and spirit enlighten each other. They then conspire to set and achieve a higher goal, unfathomable through the lens of just a single element.

5 Cultivating Hope

ALS is a physical illness that has the capacity to brutally corrode the mental. It is so easy to slip into negativity and to plunge into bitterness. Having ALS has made me slow down significantly which results in a substantial decrease in productivity in various areas of my work. On the other hand it gives me more time to be in tranquil solitude and to contemplate. It gives me time to remember the past, to awaken feelings that have been left dormant for too long, and just to wander and pick up what I have been missing and ignoring. We all know of an advice that tells us to stay positive by training our mind to only pay attention to positive signals. However, we know negativity exists. We should not neglect it lest it becomes uncontrollable undergrowth. We need to discern those negative signals, learn of their behavior, and then train our mind not to let them dominate. This way we do not only acknowledge the presence of the counterpoints, but more importantly, we are driven to reason why the positive signals are considered positive, and vice versa, through the impact analysis of each on the well-being of our mind. That is, we train ourselves to pick up and accentuate positive signals on real fertile ground, not on artificial flatness. The difference between signal values at an instant of time may not always be significant, but if evaluated over a longer period, one sees a definite change towards a more positive value. The presence of this positive gradient is also something that I consider as a crucial element in making me a happy person, for it is a mathematical representation of hope. Even if things are not good now, knowing that there is an indication that things will be better tomorrow would make me content now. This is not just an empty consolation, but a mindful strategy to cope with real day-to-day situations which are never ideal, and to make the best of today and rejoice, and to always be eager to work for a better tomorrow.

When the physique is failing, the mind—in its mental and spiritual dimensions—joins forces to act as the last defense to prevent a climactic breakdown of the whole person. It is as dramatic as it sounds. When a certain physical disability results in the loss of freedom, the impending frustration and depression are not exaggerations, particularly if one's dignity is compromised.

They are the seeds of those negativities that we need to acknowledge and manage. I suspect this willingness to acknowledge, and then manage, whatever we consider as drawbacks is what distinguishes responsible living from being careless or apathetic. It starts in the person before getting communal.

This mental exercise needs knowledge and creates knowledge. It sounds like a vicious circle, but I see it as an ascending spiral staircase. Through this process we understand the value of knowledge which, by the nature of the exercise, starts as something very personal. In recent years I realize that the purpose of acquiring knowledge is neither to ensure that we have the upper hand of knowing the answers to certain questions, nor to secure ourselves a place above the dividing line on the social ladder, not either to give us the benefit of being the frontrunner in the human race, et cetera. For me, to have knowledge, in the end, amounts to having a thorough understanding of oneself: a holistic measure of oneself with respect to an encompassing set of references, essential in any circumstance that requires one to decide and act wisely. Wisdom, consequently, is now weighted against the backdrop of those references. True knowledge can successfully guide us towards wisdom and mutual respect for the existence of others. This acquired knowledge will propel me forward, but not without the whole community moving forward as well. A knowledgeable community is then also a compassionate community, for all its members are equally respected. How wonderful that the Cosmological Principle, which states that no point in the universe is more special than others, has something to say to humanity.

∴

I began this chapter by taking a retrospective look at my family values. In the process of writing I came to realize that what I have been doing and involved in is not much more than safeguarding certain values I believe to be worth keeping. I am not preserving the values in a tightly sealed jar, not to be touched or changed, but I am watching in amazement how values follow and control the dynamics of humanity. Combing the threads and zooming in on knots allows me to see the dynamics of the control parameters and how symptoms appear and disappear. The intricate relation between humanity and values then seems to resemble Einstein's field equations: The content shapes the geometry of spacetime, and the geometry governs the dynamics of the content. But then the naughty quantum-y question: Would the world still be a reliable reference for me if the way I look at it influences its behavior? It only confirms my conjecture: I am a part of the whole world, and thereby I matter, and therefore an affirmative and responsible part I should be.

INTERMEZZO 9

A Sense of Beauty

Jan Visser

> Beauty will save the world.
> DOSTOEVSKY (The Idiot)

∵

Reflections inspired by Matthew Colless: 'Beauty in science, science in beauty: The scientific aesthetic as an evolving heuristic'

We live in interesting, but also exceedingly difficult, times, an epoch that is usually referred to as the Anthropocene, i.e., an era in our planetary history in which the impact of human activity on the living conditions on Earth has become a threat to our own survival and that of other species. Never during human history have humans had to deal with challenges of a magnitude we are currently facing. While it may be possible—perhaps, and hopefully—to make rapid adjustments to avoid the worst, the problem situation that is will not go away. We will have to learn to live with it.

The question comes up, should we care? What if we disappear, together with other species, in the next mass extinction? Though we like to believe otherwise, we are simply an accident of evolutionary history. Is there a purpose for our being here? Can we think of a reason why we should care for the survival of our species? I think there *is* a reason, and it's this: It is the essence of life to enhance life. Creating beauty falls in that category. Beauty enhances life. Being careless about our sustained presence on planet Earth would thus be an insult to our ancestors who created the beauty we inherited, making it inaccessible to future human life. Such carelessness goes against the grain of being human. It hurts. The pain we feel at the disappearance of beauty is the signal we receive to save the world (Visser, 2016).

In a world without beauty we don't feel pain. In a world whose beauty we have not learned to recognize and appreciate, we don't feel pain either. Becoming sensitized to specific ways in which beauty reveals itself is like passing a threshold. Getting across the threshold takes time, time to get exposed multiple

 | DOI: 10.1163/9789004416802_018

times to what we may later recognize as beauty. I vividly remember having had that experience as a small child when I first discovered the beauty of wildflowers and much later, as a reader of theoretical physics, when, all of a sudden, I saw beauty in the simplicity with which compact mathematical description can create a unified vision of seemingly distinct natural phenomena. We probably all have such experiences being exposed to new and unfamiliar styles of artistic expression in the visual and plastic arts and in music. Matthew Colless exposes us in his chapter on 'Beauty in science; science in beauty: The scientific aesthetic as an evolving heuristic' to a lesser known domain of beauty.

Reference

Visser, J. (2016). Human learning and the development of mind in the Anthropocene: Reflections against the backdrop of Big History. *Origins, 6*(9), 3–12. Retrieved from https://www.researchgate.net/publication/308612239

CHAPTER 10

Beauty in Science, Science in Beauty: The Scientific Aesthetic as an Evolving Heuristic

Matthew Colless

1 Introduction

This chapter is based on a talk given at the *Building the Scientific Mind* colloquium held at Bosscha Observatory in Lembang, Indonesia, in May 2013. It is not a research paper by an expert in the history and philosophy of science, but rather a speculative essay by a working scientist that flows from personal reflection on aspects of the scientific process as he experiences it.

2 Beauty in Science

What has beauty to do with science? Isn't beauty a subjective aesthetic emotion, far distanced from the objective rigor and logic of science? Well, no. Beauty, harmony, elegance, and related aesthetic concepts in fact lie close to the beating heart of science. Indeed, they are intimately related to the way that the scientific program is carried out in practice by scientists, who are not necessarily more naturally given to objective rigor and logic than the rest of humanity.

The close relation between aesthetics and the application of the scientific method is revealed in the language scientists use about their work. Scientific theories, experiments, observations and models are often commended as being rigorous, logical, insightful, seminal, radical or profound; but equally often they are lauded as beautiful, harmonious, elegant, appealing, splendid or pure. Examples of this are to be found throughout the scientific literature, as illustrated by the handful of quotations from well-known scientists below:

> Science shows us truth and beauty and fills each day with a fresh wonder of the exquisite order which governs our world. (Polykarp Kusch, physicist, cited in Liljestrand, 1956)

> We especially need imagination in science. It is not all mathematics, nor all logic, but it is somewhat beauty and poetry. (Maria Mitchell, astronomer, cited in Mitchell, 1896)

 | DOI: 10.1163/9789004416802_019

> Mathematics, rightly viewed, possesses not only truth, but supreme beauty—a beauty cold and austere, like that of sculpture, without appeal to any part of our weaker nature, without the gorgeous trappings of painting or music, yet sublimely pure, and capable of a stern perfection such as only the greatest art can show. (Bertrand Russell, mathematician & philosopher, cited in Russell, 1945)

> I am among those who think that science has great beauty. A scientist in his laboratory is not only a technician: he is also a child placed before natural phenomena which impress him like a fairy tale. (Marie Curie, chemist & physicist, cited in Curie, 1937)

> The scientist does not study nature because it is useful to do so. He studies it because he takes pleasure in it, and he takes pleasure in it because it is beautiful. If nature were not beautiful it would not be worth knowing, and life would not be worth living. I am not speaking, of course, of the beauty which strikes the senses, of the beauty of qualities and appearances. I am far from despising this, but it has nothing to do with science. What I mean is that more intimate beauty which comes from the harmonious order of its parts, and which a pure intelligence can grasp. (Henri Poincaré, mathematician & scientist, cited in Poincaré, 1908)

> Some people would claim that things like love, joy and beauty belong to a different category from science and can't be described in scientific terms, but I think they can now be explained by the theory of evolution. (Stephen Hawking, physicist, in a National Public Radio interview, 2013)

That final quote is particularly telling, and an inversion of it forms the basis of my thesis: I argue that science is not disjoint from aesthetics, and can be described (if only heuristically) in aesthetic terms; moreover, I think this observation can be explained by evolutionary theory.

3 Aesthetics and the Working Scientist

One way to approach this point is to consider the prevalence of aesthetic language used in describing science and ask the question, what do these terms mean to the scientists using them? Why do they have emotional content? Why do researchers value these qualities? And, why does this matter?

It seems that the experience and pursuit of scientific research has, for most researchers, an acute aesthetic component, either explicit or implicit. We (for I certainly include myself) find beauty in many aspects of the research enterprise, including the special attraction of our particular field of our research, the specific problems we choose to address, and—perhaps especially—the effective solutions that we find to those problems.

In my own case, I was drawn to astronomy by two distinct aesthetic pleasures: first, by the sensational beauty of the night sky and the universe beyond, and, second, by the intellectual beauty of the theories that allow us to understand the universe. The first of these aesthetics is closely related to a deep sense of awe and wonder, to mystery and surprise; the second is clearly based on an equally deep desire for sense and meaning, and a delight in order, pattern, and understanding. The see-saw tension between these two aesthetics—seeing mystery resolve into meaning, and that new understanding generating yet more profound questions, requiring yet deeper probing—that is the exquisite, tantalizing pleasure of the scientific aesthetic. For scientists, a beautiful solution is one that not only answers the greatest number of old questions, but the one that also generates the greatest quantity and quality of new questions.

In the course of my research career in astrophysics and cosmology I have often been guided by my sense of the most beautiful solution to a problem or the most harmonious interpretation of the data, and I believe this aesthetic reaction is a common experience for many scientists, if not all. My speculation, developed here, is that this aesthetic sense is an adaptive heuristic that good scientists develop (and great scientists have to a marked degree) because it is effective in guiding them in complex situations to identify the right questions and the right answers in their efforts to solve difficult problems.

4 Thesis

The thesis of this chapter is based on the lived experience for many scientists of the power of an aesthetic response for identifying and solving important problems. I hypothesize that this sense of scientific beauty is in fact a vital and effective adaptive heuristic that the human intellectual apparatus develops in an evolutionary response to the challenging fitness problem of selecting amongst competing hypotheses and paradigms in the course of normal scientific research. Furthermore, I argue that scientific revolutions are generally accompanied by aesthetic revolutions in the definition of scientific beauty, and that which of these revolutions (the scientific or the aesthetic) happens first varies from case to case. Finally, I speculate that in recognizing the existence and nature of this heuristic mindset there is an opportunity to train and develop the scientific sense of beauty and make scientists more effective at understanding the world.

Note that at no point am I arguing that science is a wholly subjective process or understanding. That it is partly subjective (subject to personal comprehension and attitudes, to the current state of knowledge and cultural biases, and in other ways) is undoubtedly true; but I am nonetheless a pragmatic realist. I find the assumption of an objective truth that lies beyond all the filters we impose, and have imposed on us, to be a viable working hypothesis that, if perhaps it cannot be proved, has neither yet been disproved. As a practical matter, I find such a belief to be a valuable defense against the lazy thinking and special pleading that subjectivism too often falls prey to. However, these philosophical issues are a digression, and not germane to my main thesis, which is entirely open to being addressed by the scientific method.

5 Historical Examples

I begin by setting out some simple (and from a historical perspective, greatly over-simplified) examples from my own field of astronomy that illustrate the way in which a sense of beauty informs scientific theorizing, and how major changes in the scientific paradigm are sometimes preceded by changes in the scientific aesthetic and sometimes force changes in that aesthetic. More extensive coverage of this history and its scientific implications can be found in, e.g., North (1994).

5.1 *Ptolemy, Copernicus & Kepler*

For classical astronomers of the Ptolemaic school, the sun and planets had to move in circular, uniform orbits about the Earth because it was only natural that Man should be at the center of the universe, and that all objects must proceed in unvarying courses following that most perfect of forms, the circle. This, clearly, is a theory that is strongly influenced by a deep sense of what is 'beautiful', right, and proper.

Observations, however, disagreed with this theory, and those natural philosophers who were scientists (i.e. those who submitted their theories to the judgement of observation and experiment) were obliged to 'save the phenomenon' (and the underlying aesthetic principle) by introducing additional circular motions, in the form of epicycles and deferents. In this case the aesthetic principle guided the revision of theory in the face of disagreeable data—but a choice was forced between the simplicity and the circularity of the orbits in the original theory. The complexified Ptolemaic theory chose circularity over simplicity (and of course retained the Earth as the center of the universe).

Nicolaus Copernicus championed a scientific revolution that, initially at least, had a stronger aesthetic than evidential base. He argued the case for a heliocentric rather than a geocentric solar system not because the heliocentric model produced better predictions of planetary motions but because it simplified the nature of those motions, reducing the number of complexities in the model.

With Tycho Brahe providing more precise new observations of planetary motions, Johannes Kepler continued the Copernican revolution by noting that the simplest description of the planetary orbits was obtained if they were modeled as ellipses with the Sun at one focus. Moreover, this aesthetic push for simplicity also yielded a better, more precise match to Brahe's observations.

A simplified view of the Copernican revolution is thus that an aesthetic evolution, based on emphasizing simplicity of description over the supposed perfection of circularity, drove a shift in the scientific paradigm: a heliocentric system with simple elliptical orbits is more beautiful than a geocentric system with complex, nested, circular motions. Of course, the ultimate justification of the new aesthetic rested on the improved match to observations and the greater predictive power of the theory.

The degree to which this simplified view of the historical events is justified is certainly arguable. And even to the extent that it is correct, it is debatable to what degree the aesthetic arguments provided a conscious, rather than subconscious, motivation of the main actors. However, for the purposes of my argument, that is essentially irrelevant, since their degree of self-awareness does not make their aesthetic motivations any whit more or less powerful.

In the language of evolutionary processes, what happened was that the concept of scientific beauty mutated, found a selective advantage, and outcompeted the old aesthetic by leading to a theory that was better adapted to observational reality. I will come back to the case for using such evolutionary language later.

5.2 *Kepler & Newton*

It is important to note that the new model of Copernicus and Kepler, while undoubtedly scientifically superior to the Ptolemaic model, is still wholly descriptive—there is no deeper explanatory level underlying Copernicus' heliocentrism or Kepler's elliptical orbits than Ptolemy's geocentric nested circular orbits.

Thus, while Kepler's three laws of planetary motion provided a conceptually simpler and ultimately more accurate phenomenological description of the observations, they have strictly limited predictive power (or 'depth'): they apply to all planetary orbits, but do not anticipate or explain other phenomena. Consequently, although they represent a major revolution in understanding the world, they are nonetheless a relatively weak scientific paradigm.

Isaac Newton's development of his theory of gravity was fundamentally different, because instead of just supplying a better *description* it gave a unifying *explanation* having predictive power not only for Kepler's laws but also for a wide variety of other phenomena (such as tides).

Newton's laws provided a profound and radical new scientific framework that led to perhaps the most important generalization that the concept of scientific beauty has ever undergone. Henceforth scientific theories did not just require the apparent, superficial, beauty of providing a better descriptive framework; they also had to have a 'deep beauty' that proceeds from 'explaining' at some more fundamental level the connections between disparate phenomena.

In this case the concept of scientific beauty did not lead the theoretical development, but rather was vastly extended and amplified by it, and consequently obliged to meet a much more profound measure. This new conception of scientific beauty has (in a broad sense and with much additional adaptation) continued to apply within the scientific enterprise ever since Newton.

6 Restating the Thesis

The evolution in aesthetic understanding outlined in the previous section provides a close analogy for the gradual adaptation of researchers to the shock of a new discovery or paradigm shift.

Initially, new facts and ideas are viewed in terms of previous experience, old facts, and old paradigms. At this stage the new fact or paradigm seems 'ugly' or 'discordant', and there is emotional and intellectual misunderstanding and confusion. Over time, however, researchers (either individually or as a population) adapt their thinking and perspectives in order to accommodate the novelty; the underlying aesthetic changes, and the fresh information is placed in a new ecology of ideas in which it seems natural, right, and proper—and, indeed, beautiful.

Thus, to restate my thesis: 'beauty', in the context of science and technology, is a heuristic mental construct that is continually adapted by an evolutionary process to better select true models of the world. The scientific aesthetic—an acquired sense of fitness—evolves in order to function effectively in the new order of things.

This is very explicitly a restatement of the thesis within an evolutionary view of the scientific process as a whole, and of the scientific aesthetic in particular. In a very general sense, evolution is the adaptive propagation of populations of entities through feedback involving selection of the fittest (see, e.g., Mayr, 2001). Here the 'entities' are scientific models and hypotheses, the 'feedback' loop is embodied in the scientific method (observe, hypothesize, test, repeat), and the 'selection of the fittest' is provided by testing the models and hypotheses against reality through experiment and observation.

7 An Example of an Evolving Aesthetic

Let me give an example of how the scientific aesthetic is currently evolving at a fundamental level in my own field of cosmology. Again, the history is greatly (perhaps overly) simplified—see North (1994) and Singh (2004) for more historical detail and Peacock (1999) for a technical introduction.

In 1917, Einstein introduced the cosmological constant into the equations of General Relativity due to his aesthetic preference for a static, eternal universe. Perversely, it subsequently emerged that Einstein's 'static' universe is unstable to any perturbation and must either expand or contract. By 1929, Edwin Hubble's observations of galaxies had shown that the universe is in fact expanding, in the sense that galaxies are generally moving away from each other and that more distant galaxies are moving away from each other faster. According to physicist George Gamow, Einstein realized that he could have predicted this expansion of the universe and considered the introduction of the cosmological constant his greatest blunder.

After many scientific twists and turns, the evidence today for the Big Bang paradigm is overwhelmingly strong. This paradigm provides a quantum physics version of an *ex nihilo* beginning for the universe. It also allows a choice of scenarios for the ultimate fate of the universe: a 'heat death' (perhaps better called the 'big chill', since the universe ultimately cools to absolute zero) if the density of matter in the universe is less than a critical value, or a 'big crunch' (in which the expansion eventually stops and the universe contracts, ending in a 'gnaB giB'—a Big Bang in reverse) if the density is greater than this critical value.

However, if one allows a cosmological constant, the universe can have another fate—not just expanding forever but expanding at an accelerating rate. Variants on this theme posit not Einstein's cosmological constant as the driving force behind this accelerating expansion but a more general (and even less understood) 'dark energy' that might have a different equation of state governing its behavior and might vary with time. In the most extreme versions of the 'dark energy' scenario, the accelerating expansion ultimately rips apart the spacetime fabric of the universe.

Within the many variants of this paradigm, the question remains: what is a 'beautiful' universe? On the grand scale of cosmology (i.e. ignoring all of chemistry and biology!), the geometry, history, and ultimate fate of the universe all turn out to be specified by three energy densities, expressed in units of the critical density: Ω_m, the energy density of matter (from $E=mc^2$, including both ordinary atomic matter and the as-yet-unidentified 'dark matter'); Ω_Λ, the energy density of dark energy (or the cosmological constant); and Ω_k, the energy density in the curvature of spacetime (from Einstein's field equations). Moreover, under General Relativity, these three quantities are linked by the extraordinarily beautiful—and beautifully simple—Friedman equation:

$$\Omega_m + \Omega_\Lambda + \Omega_k = 1. \tag{1}$$

So in fact specifying any two elements of this triad of cosmological parameters determines the third. What sort of universes emerge from different specifications of these quantities, and how are they compared aesthetically?

The most obviously 'beautiful' universes are those in which these quantities take the 'special' values of either 0 or 1. Initially, therefore, cosmologists preferred a universe in which $(\Omega_m, \Omega_\Lambda, \Omega_k) = (1, 0, 0)$—i.e. a universe where matter is all there is, and there is no dark energy (for which there is no good physical theory) nor any curvature of spacetime (a 'flat', Euclidean universe). This universe is very simple and beautiful indeed: only matter exists (which seems unavoidable, since here we are); there is no need for Einstein's *ad hoc* cosmological constant or any other form of the mysterious dark energy; and the universe has an intuitive and straightforward geometry. On the other hand, a truly 'ugly' universe would have no special value for any of these quantities—i.e. it would have $0 < \Omega_m < 1$, $0 < \Omega_\Lambda < 1$, and $\Omega_k \neq 0$.

Over the last 20 years, a host of observations have led to precise measurements for all three of these key cosmological parameters. These measurements have exploited 'standard candles' (like Type IA supernovae) and 'standard rulers' (like the baryon acoustic oscillation scale) to map the changing geometry of the universe over vast reaches of cosmic time. They have also tapped into the rich vein of information encoded in the structure of the matter distribution revealed by the cosmic microwave background radiation shortly after the Big Bang and by the galaxy distribution at more recent epochs. The results of this remarkable scientific program produce a concordance across many methods and measurements, yielding current best estimates (Alam et al., 2017) for these quantities and their uncertainties:

$$\Omega_m = 0.311 \pm 0.006, \Omega_\Lambda = 0.689 \pm 0.007, \Omega_k = 0.0003 \pm 0.0026. \quad (2)$$

Thus the universe is far from 'simply beautiful', but neither is it entirely ugly. The 'beautiful' element emerging from these observations is that the universe appears to be flat (i.e. it has zero—or very nearly zero—curvature). On the other hand, the matter density is not the critical value (i.e. $\Omega_m \neq 1$) and there is a non-zero amount of dark energy ($\Omega_\Lambda > 0$). Worse yet, the dark energy dominates the energy density in matter ($\Omega_\Lambda > \Omega_m$), meaning that the expansion of the universe is in fact accelerating, as observed in 1998 by two teams who won the 2011 Nobel Prize in Physics for their discovery (Frieman, Turner, & Huterer, 2008). As another Nobel Laureate, I.I. Rabi, quipped regarding a different unexpected discovery, "Who ordered that?"

How do we find beauty in the universe when apparently it is neither beautifully simple nor randomly ugly? Is there a deep reason that fundamental physics and the parameters of the universe are the way they are (and must be that way)? Or do we have to undergo another 'Copernican displacement' from the centre of things, recognizing our universe as just one instance within a vastly larger multiverse in which all possible cosmological parameters are represented and we observe the particular values we do because they happen to permit the existence of sentient creatures like ourselves? These are still very much open questions that are currently the subject of vigorous research and debate.

But the answers to these fascinating and important questions are not the focus of this essay. Rather, the questions we ask here are: Will our concept of a 'beautiful universe' change after we gain a deeper understanding of the underlying physics? Or will a more profound aesthetic lead to a deeper physical understanding? I do not know, but my personal scientific aesthetic whispers that the Copernican Principle is more beautiful than its rival, a cosmological version of Manifest Destiny that states our universe is not just special, but unique. That is, my aesthetic leads me to favour models in which there are many more ways than one to build a universe, and where the prodigious diversity of creation is not limited to a single set of physical laws and one specific instance of their outworking. Rather, my aesthetic heuristics lead me (along with increasing numbers of my colleagues) to think that our universe is just one of an effectively infinite number of cosmological universes that span the whole allowed parameter space of fundamental physics (whatever that might be). This principle also tends to favour the Everett-Wheeler many-worlds interpretation of quantum physics (Everett, 1956), which says that each individual universe within this cosmological multiverse proliferates into an essentially infinite number of quantum alternatives. Thus multiplicity upon multiplicity at multiple levels!

In essence, the Copernican Principle states that we are in no way special (let alone unique), and I believe that this ultimate modesty may well be the under-pinning theme of a successful new cosmological aesthetic. But perhaps thinking of the key principle of this aesthetic as 'modesty' is still too anthropocentric a view. Perhaps, rather than our own utter insignificance, the over-arching principle is the overwhelming, prodigious bounteousness of the multi-multiverse view—multiplicity, plenitude, vastness, and infinite possibility may be better descriptors of this further generalisation of the Copernican worldview and its associated aesthetic. If so, then awe and wonder will have a continuing role at the heart of the scientific story.

8 Science in Beauty

The thesis that I have argued here is that the scientific aesthetic is an adaptive response to the evolutionary nature of the scientific process. While scientists remain unconscious of this, the evolutionary response is a 'natural', undirected one, since natural evolution is a random process—the work of a 'blind watchmaker'. But if scientists become self-consciously aware of the process, then in principle, they can direct (or modify) the adaptive process (i.e. the scientific method) to optimize (or at least improve) the outcomes.

Note that this assumes two abstractions in viewing the situation: first, that scientists can sufficiently abstract themselves from the *process* of the scientific method to allow them to *recognize* their own responses to that process; and, second, that scientists can have a sufficiently abstracted and meaningful understanding of the *goals* of that process that *direction* of the process yields better scientific outcomes. While the former assumption seems plausible, the latter is far from clear. However, evolution at a higher level of abstraction can come to our aid: rather than letting science evolve blindly, unconscious and unheeding of the process by which it evolves, we can consciously tinker with the process, modifying the conditions underlying the *process* of scientific evolution as well as science itself. This, after all, is what human selection in plant and animal breeding has been doing for thousands of years—providing unnatural selection in order to guide an evolutionary process in particular directions.

How might this work? The elements of the scientific method that we can change are: (i) *selection*—the process of testing scientific models and hypotheses; (ii) *variation*—the means of generating potential models and hypotheses; and (iii) the *fitness* measure—the metric used to determine which models and hypotheses are selected based on their scientific merit. My perception of science is that we have focussed heavily on the first of these elements and have a formidable arsenal of methods for testing models and hypotheses, but that we have (relatively) neglected the second element, which may well also be susceptible to great improvement by application of the scientific method itself. The most challenging of the three elements to 'improve' is the last, since this might be argued to be non-negotiable: a test of a model's 'fitness' is whether it can make predictions that can be tested, and whether it survives the testing of those predictions. However recent arguments (Ellis, 2014) about whether the multiverse paradigm is a valid scientific model suggest that this notion may need sharpening up, even if it cannot be generalised.

Regardless of these speculations, surely it is time that we applied the scientific method more rigorously to the method itself, and in particular to the messy processes by which it is implemented by actual scientists? Can we intervene to improve the efficacy and value of our naturally-evolved heuristics for scientific beauty and harmony? Might we even be able to replace such heuristics with a scientific model for (scientific) beauty?

9 Summary

In sum:

- Our sense of beauty in science (the scientific aesthetic) is a powerful adaptive heuristic that the human intellectual apparatus develops as an effective evolutionary response to the challenging fitness problem of selecting amongst competing hypotheses and paradigms.
- Scientific aesthetics evolve: they adaptively propagate due to the variation-selection-feedback cycle of the scientific method.
- The scientific aesthetic is a crucial guide for the scientific mind, and we should consider ways in which we might sharpen or otherwise develop this heuristic tool through directed evolution.

10 Epilogue

In any discussion of evolution, it is wise to give Charles Darwin the last word. The famed final paragraph of the *Origin of Species* (Darwin, 1859) is ablaze with a sense of awe, wonder, and scientific beauty, which culminates in this concluding sentence:

> There is a grandeur in this view of life, with its several powers, having been originally breathed by the Creator into a few forms or into one; and that, whilst this planet has gone cycling on according to the fixed law of gravity, from so simple a beginning endless forms most beautiful and most wonderful have been, and are being evolved.

However, for the final thought of this essay, I turn to the penultimate paragraph of *Origin of Species*, where Darwin reminds us that:

> Judging from the past, we may safely infer that not one living species will transfer its unaltered likeness to a distant futurity.

This rule applies to all evolutionary processes, not only to the evolution of biological species and populations by natural selection. Applied to the evolutionary process we call the scientific method, and the theories, models, and paradigms that constitute its species and populations, we can predict that the science of 'a distant futurity' will be 'most beautiful and most wonderful', though perhaps as unrecognizable to present-day scientists as quantum physics would be to Aristotle.

References

Alam, S. et al. (2017). The clustering of galaxies in the completed SDSS-II Baryon Oscillation Spectroscopic Survey: Cosmological analysis of the DR12 galaxy sample. *Monthly Notices of the Royal Astronomical Society 470*, 2617 (arXiv:1607.03155).

Curie, E. (1937). *Madame curie* (V. Sheean, Trans.). London: William Heinemann.

Darwin, C. (1859). *On the origin of species by means of natural selection*. London: Everyman's Library.

Ellis, G. F. R. (2014). Does the multiverse really exist? *Scientific American, 23*, 76–81.

Everett, H. (1956). *Theory of the Universal Wavefunction* (PhD thesis). Princeton University, Princeton.

Frieman, J. A., Turner, M. S., & Huterer, D. (2008), Dark energy and the accelerating Universe. *Annual Reviews of Astronomy & Astrophysics, 46*, 385–432.

Kusch, P. (1956). Nobel banquet speech. In G. Liljestrand (Ed.), *Les Prix Nobel en 1955*. Stockholm: Nobel Foundation.

Mayr, E. (2001). *What evolution is.* New York, NY: Basic Books.

Mitchell, M. (1896). *Life, letters and journals.* Boston, MA: Lee & Shepard.

North, J. (1994). *The Fontana history of astronomy and cosmology.* London: Fontana Press.

Peacock, J. A. (1999). *Cosmological physics.* Cambridge: Cambridge University Press.

Poincaré, H. (1908/1914). *Science and method* (F. Maitland, Trans.). New York, NY: Cosimo.

Russell, B. (1945). *A history of western philosophy.* New York, NY: Simon & Schuster.

Singh, S. (2004). *Big bang: The origin of the universe.* New York, NY: Harper Collins.

INTERMEZZO 10

Making the Unfamiliar Familiar

Jan Visser

Reflections inspired by Lê Nguyên Hoang: Science popularizer is the most important job that does not yet exist: Why modern societies need more science popularizers

There is a particular beauty in mathematics. Mathematics is the language in which the poetry of nature is written. However, many people see a serious threshold that keeps them from acquiring the ability to interact with this particular language environment. Why? We don't know. We can only guess. Lê Nguên Hoang thinks he has an answer.

I first made contact with Lê Nguyên (Lê for short) in October 2012. Less than 140 characters on my twitter account had alerted me to the 'Science 4 All' website[1] Lê had created to make science and mathematics appealing and to bring enthusiasm and excitement to his readers' learning experience. As soon as I had seen the site it had aroused enthusiasm and excitement in *me*. This is someone I must get in touch with, and urgently so, was my immediate thought. A quick search led me to Lê's LinkedIn page. I sent him an invite which was responded to the next day. A rapid exchange of email messages followed, which led to Lê's participation in the Building the Scientific Mind colloquium held the next year in May in Lembang, Indonesia.

'Do you know the name of a famous football player?' was the question he asked to open his presentation at the colloquium. I'm afraid I will have been the only exception among the members of the audience who could not quickly have come up with a name had Lê pressed on, but that's just me. It was likely clear to everyone else that this was mere rhetoric. The next question he asked placed the rhetorical question in context. 'Do you know the name of a famous mathematician?'

The point had been made. There is a huge lack of familiarity when it comes to using mathematics as a tool to help one understand the world and interact with it wisely. What makes people enter territory that is unknown to them? It's the prospect of a treasure that may be found within. That such a prospect is not just wishful thinking, a dream, must be made evident to those who are unafraid to open their mind. And that's exactly what Lê does.

 | DOI: 10.1163/9789004416802_020

It is this same spirit that motivated, many years ago, biologist and astronomer Marcel Minnaert to publish (in Dutch) in 1937, 1939 and 1940 three volumes of richly illustrated text about '*Physics in the open air,*' aimed at encouraging his readers to go out, enjoy nature, explore and understand it. The first volume of this work is available in English.

To explore Lê's work, read his chapter and follow the multiple pointers it provides to his work on the Web. But beware! It may excite you.

Note

1 See http://www.science4all.org/

Reference

Minnaert, M. (1993). *Light and color in the outdoors.* New York, NY: Springer-Verlag. (Originally published in Dutch with the title *De natuurkunde van 't vrije veld. Licht en kleur in het landschap.*)

CHAPTER 11

Science Popularizer Is the Most Important Job That Does Not Yet Exist: Why Modern Societies Need More Science Popularizers

Lê Nguyên Hoang

1 My Taste for Math May Not Be Lost

Teaching, learning and research are hard. Too often do I feel frustrated, when I seem unable to explain a concept simply, when I am lost in some too technical seminar or when my research papers get rejected. Too often do I feel stuck by my blatant limited intellectual capacities. Too often do I end up doubting myself and not appreciating my own work. And worse of all, too often do I forget why I have chosen math and science in the first place. Too often do I forget to simply *have fun*.

Fortunately, there is one particular webpage section that makes me feel so good that simply opening it will immediately make me want to get back in the game and to make science enjoyable. Enjoyable for me and for everyone. I am referring to the YouTube comment section of a documentary[1] on university-level mathematics that I wrote and produced. The documentary is in French, and is entitled *Magie des Mathématiques des Classes Préparatoires*. I cannot help but share a few of the comments I received (translated by me from French—see Figure 11.1), as they are indicative of the emotions that math and science can arouse in those who watched the documentary.

> Your two videos are really super, nice job, I loved it a lot! My taste for math may not be lost. ☺ (Comment by Cédric Moreau)

> A very good job! Congratulations! I really loved your series and I think it's a shame you have so few views, as you would deserve many more ... You have made me discover a new extremely interesting aspect of maths and, thanks to you, I am super motivated to work. So, thanks! ☺ (Comment by Sharkou)

> This documentary is really splendid, bravo! One thing though, it's sometimes too complicated for me XD I study literature, there are certain

 | DOI: 10.1163/9789004416802_021

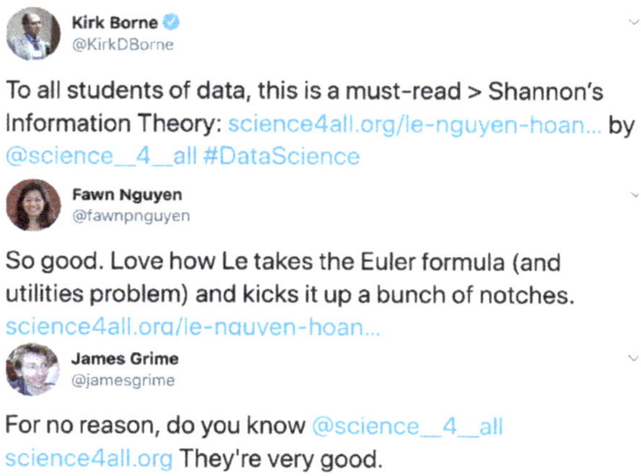

Kirk Borne
@KirkDBorne

To all students of data, this is a must-read > Shannon's Information Theory: science4all.org/le-nguyen-hoan... by @science_4_all #DataScience

Fawn Nguyen
@fawnpnguyen

So good. Love how Le takes the Euler formula (and utilities problem) and kicks it up a bunch of notches. science4all.org/le-nguyen-hoan...

James Grime
@jamesgrime

For no reason, do you know @science_4_all science4all.org They're very good.

FIGURE 11.1 Original tweets of feedback received on the documentary on university-level mathematics: *Magie des Mathématiques des Classes Préparatoires*

> things that may be obvious for a math student, but are not explained enough for me. Still, I discovered plenty of things: I am craving to better understand them now. I have always believed that I wasn't smart enough for math and physics, that there were the math geeks and others, and that I wasn't sufficiently mature to understand. I recently started to question that. Videos like yours help me a lot. Thanks for the inspiration :-) (Comment by Lania Kea)

The last comment struck me in particular. This was exactly the ambitious and seemingly unrealistic goal of my video. I did not mean to teach people any fact. I just meant to convince them that math, *'real math,'* was an inspiring, stimulating and gratifying hobby. I wanted viewers to end up so curious that they would want to go further. And to do so, I purposely made the video a bit complicated and lacking of many technical details, so that people would feel frustrated by not having a full explanation—all the while knowing that I could and wanted to give them a more complete explanation but that limited time was standing in my way to do so. But most importantly, I wanted them to feel that they had in them a natural taste for mathematical elegance, even when their background and math scores might suggest otherwise. That they too could, and should, enjoy mathematics.

In what follows, I will discuss different ways through which other popularizers and I have tried to get people excited about mathematics. I will discuss Science4All and related projects I have completed on the Internet. Then, I shall describe a lecture I once gave at the 2015 MIT *Splash* event, and how it exemplifies what I regard as quality popular science. Next, I will stress the crucial role that I think science and math popularization must play in societies, and I will discuss some features of quality math popularization. Then, I will review today's state-of-the-art in online science popularization. I will also discuss recent developments I have observed or been involved in and I will try to extrapolate to see where these developments may take us. I will conclude by attempting a depiction of an optimistic future for mathematics and science learning, and by saying a few words about why I think that massive funding of science popularization seems instrumental to get there.

2 *Science4All* and My Online Outreach

I launched the website Science4All.org back in April 2012. I had recently started my PhD at that time. My goal was to create a platform of science popularizations. I counted on my friends to populate the website with entries on topics they had classes in or done research about. With hindsight, that was way too optimistic. One thing I quickly learned was that it is extremely hard to get PhD students to do additional work. But, while I was harassing my friends to help me make Science4All active, I published myself numerous articles on the website.

I vividly remembered many of my friends in quite technical areas like pure mathematics repeatedly telling me that their research topics could not be popularized. It thus became a challenge for me to popularize high-level topics; and early on already, I tackled technical topics such as *duality in linear programming*,[2] *P versus NP*[3] or *Fourier analysis*.[4] I am not sure I would recommend my readers to start with these articles though. I do not think they are that bad *per se*, but, without even realizing it, I was on a steep learning curve. More importantly, I started to enjoy it. From that point on, Science4All became almost solely my website. I became happier and happier with what was going on there. And the more I posted new articles, the more I enjoyed writing popular science.

I have selfish reasons for this. First, there is the obvious fact that writing such articles adds personal visibility on the Internet, and, today, I proudly refer curious people who want to know more about science to my Science4All articles. After all, if you are reading this chapter right now, it is solely thanks to my wide web presence. Second, the only way to really learn something in depth that I know of is to write (or make a video) about it. Since I started Science4All, I have learned dramatically more, both in terms of quantity and quality, than I have in the twenty-five years before that (even though I am supposed to have been a "good" student).

Third and finally, I just love it. I just love putting pieces of science and mathematics together in larger, more comprehensive and more meaningful wholes in ways that have hardly ever been done before. Some people think that popularization is a degraded version of research-level science. It is not. At least, quality popularization is not. Quality popularization, as I have found out, is, and should be, a highly creative endeavor, just like research is. It is about questioning previous popularization work and improving on it. It is about drawing the limits of previous analogies and coming up with more relevant ones. It is about shuffling disconnected pieces of knowledge and structuring them in such a way that all pieces fit together perfectly, as part of a wider and more grandiose puzzle. This also means that I think that it takes years of experience to become a great science popularizer, and that only too few have had more than half a bite of this experience.

But evidently, these selfish reasons to do popularization are nowhere near the main and most exciting perk of being a science popularizer. By far, what excites me the most about popularization is, as you have probably guessed, affecting the way people regard science and mathematics. Nothing tops this feeling that I am submerged by when, after someone has read or watched me, he or she feels happier and more excited about science and mathematics.

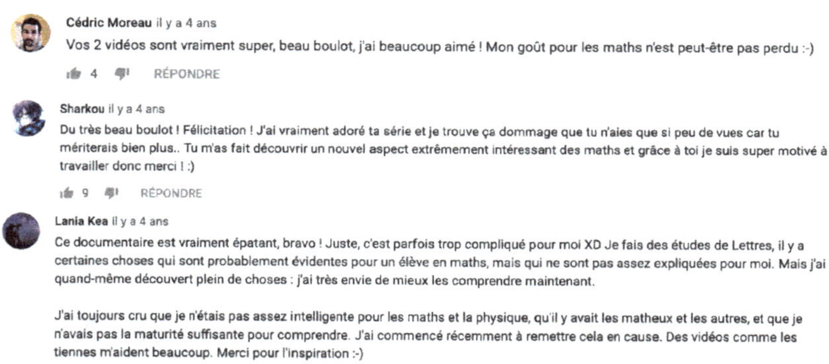

FIGURE 11.2 Original tweets of feedback received on some Science4All articles

By December 2015, Science4All.org had over 400,000 views, with an average of 10,000 per month. The most popular entries are on *Euler's identity*,[5] *entropy*,[6] *conic sections*,[7] *water in trees*,[8] *information theory*[9] and *Galois theory*.[10] I am proud of the diversity and technical sophistication of these top entries, as well as of the many positive instances of feedback they have received. A few tweets about Science4All articles that speak for themselves are presented in Figure 11.2.

3 Alternative Media

I have been leaving Science4All.org a little bit aside since March 2015. There is a good reason for this. I started to produce a new series of videos on YouTube. As I expected, there was once again a steep learning curve. Growth started pretty slow. But by September 2018, the (French) Science4All YouTube channel had gathered over 140k subscribers and 7 million views. These numbers are overwhelmingly huge. They correspond to centuries of teaching in classrooms!

There are arguably different reasons for this success. First, I think that videos (and especially music!) are a great way to communicate the excitement (or frustration) I feel when I am doing mathematics. Second, there is already a large community of successful science YouTubers. Third and finally, direct vocal communication allows to develop a more intimate relationship with my viewers than was the case of Science4All.org. Indeed, it has become clear to me that the lengthiness of my Science4All articles prevents people from reading them regularly. They seem to be more of a once-in-a-while thing to follow.

In particular, I have developed a slightly different approach to science popularization on my YouTube channel. Instead of jumping between different topics, as I did on Science4All.org, I try to build things up, through series of around 20 videos, to a deeper understanding of some broad and major theory of mathematics or physics. As of December 2015, the first series, which is a thorough exploration of Einstein's theory of general relativity, has been finished, while the second series on logic and math foundations is being launched. Even though I do not know whether this approach will eventually be as successful as my Science4All articles or as the more successful YouTube channels, I am extremely excited about this. I have myself found it a bit frustrating that, in Science4All articles, I always had to limit myself to 4,000 words—Science4All articles have been growing in size through time and some topics I later tackled like homotopy type theory were in fact presented through series of articles. Hopefully, through 20 episodes, I can be much more complete on a given topic, as I think was the case with general relativity … I do not know to what extent my hopes are justified, but I do know that I love making these videos and that some viewers like watching them. At the end of the day, this is what matters to me.

One last thing I want to mention about my online work is the two-hour documentary on university-level mathematics that I mentioned in the introduction. This project spanned through 2014. I first wrote a script in late 2013, and got feedback from many friends. In the summer of 2014, I shot the documentary and interviewed a dozen people working in math-related areas, from researchers in pure mathematics to engineers in hospital management. In the fall, I edited the videos and eventually released the documentary on YouTube, with a success I did not expect. Within a year, the documentary got over 140,000 views and 162 viewers even took the time to respond to a very interesting survey[11] which unveils what makes such documentaries interesting to them. And, as I said in the introduction, the comment sections have become my little paradises that help me feel that what I do matters, i.e., that, time and time again, I successfully changed people's approach to mathematics; that I got to make them feel like they could appreciate mathematics. Or, at least, quality popularization of mathematics.

Through the viewers' comments, I often feel that their mind repeatedly had had moments where things clicked. That they had had these indescribable 'aha' moments. Or that they had had one of those intrigued moments where they said to themselves "*that's funny*," as they spotted one of those things that, as Isaac Asimov would say, "*heralds new discoveries*." Sadly though, through articles or videos, I do not get to witness or share these exciting and privileged moments …

4 Teaching at *Splash*

However, later, I got the chance to teach to wonderfully curious high school students at the 2015 MIT *Splash* event.[12] I got to live multiple 'aha' moments in real time. It was amazing. Amazing for me. And amazing for them.

In particular, in the middle of a four-hour lecture on the theories of gravity, I got them to feel for themselves that gravity was not a force. I kept repeating that this was Einstein's happiest thought, and I bet that they felt Einstein's excitement through my own excitement. I do not know if you have ever experienced it, but, trust me, Einstein's happiest thought is a very, very happy thought to have. It basically says that a falling man, or an astronaut in a space station, does *not* feel gravity. In fact, I argued that no one ever feels gravity. I asked them to experience this by themselves, to live it and to be convinced by it through their own senses. As they were sitting on their chairs, they could feel that all they really felt was an upwards push from their chairs. It took time to realize it and to reject everything about Newtonian mechanics that they had been taught or had heard of. But in the end, they could feel it. They could feel that all they felt was the upwards push of their chairs. But gravity? Nope.

So, I asked, why the hell do apples fall? That was, after all, the question of my lecture, and it's such a basic question that it does not feel like a question a student—let alone a professor!—seems to be allowed to ask. But once we know that Einstein claimed that gravity is not a force, it suddenly becomes a wonderful question—perhaps the most wonderful question I have ever encountered! Why do apples fall? I could feel that they wanted to answer: "*gravity*" or even "*science*"; and in fact, they did. But I could feel that they could feel that these were unsatisfactory answers. Words are not explanations. They started to realize this. And I could feel that they were on their way to Einstein's happiest thought. But that they were struggling to follow Einstein's footsteps. Who wouldn't? It was a long moment, with alternations of long silence and unclear suggestions. It was a hard and frustrating moment, where they felt lost and unsure about everything. How can such a basic question as the falling of the apples have no explanation to them?

I really enjoyed this moment. I think that, with hindsight, they now enjoy having lived this moment as well. It was, I would argue, a moment when they were doing real, top-level science. A moment of deep investigation, mistrust of common sense and high creativity. I have no doubt that all the students present in the room now have at least some comprehension of Einstein's otherwise mysterious quote:

> The years of anxious searching in the dark with their intense longing, their intense alternations of confidence and exhaustion, and the final emergence into the light—only those who have experienced it can understand it.[13]

At last, after ten minutes or so (a lifetime when you're lost!), as I gave small hints away, a student said half-hesitantly: "*The ground is accelerating upwards …*" Yes! YES!

They were still lost though. It made no sense. "*Doesn't this mean that the Earth is expanding?*" Yes! It does! I was so excited at that point that they got excited. Not much made sense yet. But slowly, through questions and remarks, it started to kind of make sense. Slowly, but surely, I made them feel why it made sense to Einstein, when in 1907, he came up with the infamous equivalence principle,[14] and that, finally, he could explain Galileo's law of falling objects that we had discussed earlier,[15] without involving any mystic gravitational '*force*' or any cosmological coincidence such as the equality between inertial and gravitational masses. What is more, Einstein's happiest thought wonderfully explained why helium balloons bend the way they do in an accelerating car![16] In fact, as they started to realize it, our everyday world can be explained without any gravitational force. All that is needed is the upwards acceleration of the ground. Fantastic!

In the rest of the lecture, I then explained why, because of spacetime curvature, the Earth does not actually expand, and how spacetime curvature allowed to make sense of planetary and cosmological scales. I also insisted on the Historical quest of Einstein for the equation of general relativity. In particular, I stressed the uncountable mistakes that Einstein made all along his journey. In the end, I know that the students did not leave the room with a full understanding of gravity—I doubt general relativity can be learned in 4 hours! But I could see that their eyes had brightened all along the lecture, and that they ended up very excited about gravity in particular and physics in general. And to see this, to feel this, was a feeling I do not know how to top.

5 Science and Math Popularization Matters

If there is one thing that I have learned by doing science popularization, it is that science popularization matters. It matters a lot. I would even claim that it matters much more than science knowledge or science investigation. In this section, I will try to argue so by focusing on the science I know best: mathematics.

Here is my claim. I believe that anyone will be good at mathematics if and only if he or she (really) loves mathematics. As the quote of Steve Jobs—which I gave at the beginning of my two-hour documentary on university-level math—goes:

> The only way to do great work is to love what you do. If you haven't found it, keep looking. Don't settle.[17]

My equation is then very simple. If, in order to do great mathematics, the only thing that people need is a love for mathematics, this is what I need to deliver; I will make my viewers fall in love with mathematics. But I guess that this raises two questions. First, why would it be that people are good at mathematics if and only if they like it? Second, how can we make people fall in love with mathematics? Perhaps, a third question that could be raised is why people need to be good at mathematics, but I will not address this question in length here. I think that it suffices to ponder the increasing importance of rigor in today's economy to be able to give an appropriate answer to this last question.

Let me address the first question now, and stress the intimate relationship between being passionate and gaining skills. Anyone familiar with the activity of doing mathematics knows that ideas often seem to come out of nowhere. The great French mathematician Henri Poincaré describes this phenomenon particularly well, as he told the story of how he repeatedly got struck by ideas at the least expected moments. He mentions one of these moments in the following excerpt from his book *Sciences et Méthodes*.

> When we arrived at Coutances, we got into a break to go for a drive, and, just as I put my foot on the step, the idea came to me, though nothing in my former thoughts seemed to have prepared me for it .... (Poincaré, 1914, p. 53)

But where did this idea come from? Where do great ideas come from? Can anyone thereby come up with great ideas? Louis Pasteur once said that, "in the fields of observation, chance favors only the prepared mind,"[18] while Einstein added that "curiosity, obsession and dogged endurance, combined with self-criticism, have brought me to my ideas."[19] For Poincaré himself, the biggest ideas came from a long subconscious work, that his conscious busy mind had initiated. This concept seems widely familiar to most mathematicians, as illustrated by the following quote of the British mathematician Andrew Wiles, well-known for having proved Fermat's last theorem.

> I carried this problem around in my head basically the whole time. I would wake up with it first thing in the morning, I would be thinking about it all day, and I would be thinking about it when I went to sleep. Without distraction, I would have the same thing going round and round in my mind.[20]

In my documentary *Magie des Maths*,[21] I asked interviewees about their greatest aha moments that seemed to come out of nowhere. While I was hoping for a couple of nice stories, I was nevertheless surprised by how often their minds had clicked in unexpected moments. Under the shower, while walking the dog, or in the metro, it was often at unlikely locations that new ideas had come to them.

Let me take a bold step and generalize this. I would argue that it is this very ability to reflect in unlikely locations, consciously or not, that makes the difference between a good and a bad (math) student. Moreover, it takes a great desire to understand and a real appreciation for a topic to constantly keep in mind a problem we want to understand or solve.

From my personal experience, the best students, especially in mathematics, do not seem to be hard-working students. They do not necessarily spend hours at their desk working out again and again the same problems. Instead, their minds often seem to be elsewhere, as they constantly try to make sense of what is taught to them, and of what isn't. I know that I myself was this kind of student. Back when I was an undergrad, I used to work on math homework with a couple of friends in late afternoon. Every now and then we got all stuck, and had to call it a day. Well, they called it a day, and I thought I did as well. But, more often than not, it was then, on my way back home, in the metro with no book nor sheet of paper with me, that I would crack the problem. The next morning, they would be surprised that I had cracked the problem, especially given that I hardly ever actually work at home, while they were revising the lessons of the previous day all evening. But, at the risk of sounding pretentious, I really hardly worked at home—I vividly remember playing soccer video games a lot. Like the prototypical mathematician, I was fundamentally lazy.

But keeping in mind an intriguing math puzzle was something I could easily do, despite being lazy. This is because pondering a nice math puzzle has never felt like work to me. I love having an intriguing question I can keep my mind busy with, especially when this question seems to be about an elegant underlying mathematical structure. In fact, if the puzzle is really interesting, it will be hard to prevent me from keeping it in mind!

An example that stands out in my mind was a first-year college set-theory one-line theorem. The professor had given the proof of it, but, even though I perfectly understood every line of the proof, I had no global understanding of it all. Somehow, this particular one-line theorem kept bothering me until, literally a year later, I finally made sense of it. And I would argue that such long-lasting thinking was the main—perhaps only—difference between me and less successful students. I can stay on a math problem for a very long time and without a real effort. And this is *not* because I like thinking—nothing bothers me more than having to split a bill! This is because I love *mathematical* thinking. I love asking new unexpected questions I had never even thought could be asked. I love cracking a well-posed problem that seems like it should have a nice answer. I love drawing new connections between seemingly unrelated concepts. I love mathematics.

I would argue that most mathematicians have the same sort of thought process — we probably became mathematicians because we have this thought process. And I would argue that this thought process, this excitement for mathematical elegance, is instrumental to be comfortable with abstract mathematical objects and concepts. Mathematicians are not sharper because they were born to get things right away; they are sharper because, while you were showering and had an absent mind, they were cooking and pondering a mathematical object they have just discovered, hence preparing themselves for any question that could be related to this mathematical object. But they never pondered the mathematical object in that purpose; they pondered the mathematical object because they could not help but ponder it. They pondered it, because they loved pondering the mathematical beauty of this object. And I would argue that this deep appreciation for mathematical beauty is an unavoidable requirement for the understanding of mathematics. More than that, I would claim that it is almost sufficient to go far in the learning of mathematics, at least much further than where most undergrad students go.

But here is the trouble. Here is the big trouble. Most students and most grown-ups have absolutely no idea of what I mean by 'mathematical elegance.' When asked to define this concept, the English physicist Paul Dirac answered that it is

> [a quality that] cannot be defined, any more than beauty in art can be defined, but which people who study mathematics usually have no difficulty in appreciating. (Farmelo, 2009, p. 301)

Of course, many others, including *Wikipedia*, have tried to define mathematical elegance by reducing it to a combination of a few key principles. But it is one thing to learn this non-rigorous definition. It is quite another thing to feel it.

Also, I would go one step further than Dirac and claim that anyone 'really' studies mathematics if and only if he or she is able to appreciate its elegance. I am convinced that a love of mathematical elegance is the key to quality mathematical education. As Dirac said, it is necessary. And it seems almost sufficient to me. Once someone truly loves mathematical elegance, he or she will see it everywhere and ponder it at length. Plus, given today's widely available resources throughout the web, such people will have no trouble finding out more about the beautiful mathematical objects that they cannot take their eyes off.

6 Different Approaches to Math Popularization

Having said that, I now turn to the second question. How can the love of mathematics be taught? One thing that will be obvious to anyone who has tried to teach mathematics to non-mathematicians is that this is no easy task. These days, so many people somehow proudly claim that they hate mathematics, or that mathematics is not *their* thing, or that they do not have a 'mathematics brain.' But I cannot stress enough the fact that the people who do say that have absolutely no understanding of the mathematical beauty that Dirac did not even bother to try to define for them. How can they even think that they do not like mathematics if they do not even know what mathematics *really* is? It would be like listening to the sound of a screwdriver and claiming that music is not enjoyable!

Evidently, most of the trouble is most people's mathematical experience, which, in so many ways, is absolutely not an expression of, or a contact with, mathematical elegance. In fact, in many respects, school mathematics seems to forbid any real appreciation of mathematical elegance. On the one hand, much of the curriculum is based on the learning of repetitive procedures, such as the steps that must be followed to perform a multiplication. On the other hand, mathematics is the selective discipline *par excellence*. There might be good reasons for this—math scoring may be regarded as a more objective judgment. However, I cannot help but protest this mistreatment of mathematics education.

Regarding the first point, any mathematician would agree with me if I say that mathematics is *not* the ability to perform a repetitive procedure. In fact, like computer scientists, mathematicians are often said to hate to repeat the same task twice. This is, after all, what I really meant when I said that mathematicians are lazy. Instead, mathematics is about consistently addressing new problems that our current skills cannot yet solve at once. Or, about addressing old problems in different, more insightful ways.

For instance, one thing that I have only recently noticed is that I, and many of my college-graduated friends, perform additions in a way that is different from the way in which we have been taught (and that is still widely being taught) in school. Namely, instead of going from right to left and starting with the units, we begin our computations from the left, which has the nice and practical advantage of quickly yielding a good estimate of the result. In fact, more often than not, I do not complete the computation and I am happy enough with the rough estimate obtained by combining the left-most digits. Anyways, the point is that my college-graduated friends and I have never been taught this addition procedure; we have found it out on our own. We have understood why the procedure we were taught worked, and how we could improve on it. We have been doing actual mathematics, while the less successful students were stubbornly and repeatedly taught to act like computers by their teachers and parents, and might even have been scolded at for having tried alternative methods.

I believe that much work is needed in the design of math curricula to erase a lot of (sometimes useless) mathematical knowledge, and to replace it by mathematical *activity*, *exploration* and *understanding*. As I am not a teacher myself and have scarce contact with kids, I will not venture further in a proposal of a math curriculum replacement. But I know through the Internet many (probably excellent) math teachers who would agree with most of what I am saying here. These are the people who are probably the most able to come up with curricula that allow students to really experience mathematics and to get a taste of mathematical elegance. One of them is Alec Resnick, of Sprouts & Co.[22] Resnick and his colleagues are working on the development of dramatically different approaches to high school education, proposing a gradually more project-oriented education, with a diminishing set of common-facts-to-be-learned. Feedback from their experience will undoubtedly be very interesting.

Now to the second point. Can mathematics *not* be the main discipline of selection? Unfortunately, au contraire, I even suspect mathematics to become an increasingly dominant vector of selection, as our world is being more and more automatized, and is being more and more submerged in data. Besides, I cannot help but think that the extreme rigor that mathematics teaches is an essential weapon to face the swarm of poor, unreliable and misleading data that the Internet is throwing at us. It seems that those who master this extreme rigor and are able to structure the complexity of the real world will necessarily be in high demand in the future.

So, if mathematics must remain the discipline of selection, how can it still be enjoyable? I only know one answer to this question: math popularization. Or, more generally, mathematics outside of school. For kids maybe, but for parents mostly. Here is why. *The greatest sin of mathematics education is that it is consistently associated with kids.* Too often do people think that kids should do their math homework while parents go to work. Too often do parents hire a (young) tutor rather than learn with their kids. Too often do grown-ups think that they are *too old* for this stuff. Too often do they (happily) consider that mathematics is over for them. But then, if parents have a poor excuse to bypass mathematics, why on earth do kids have to do it? If kids are told to grow up, and if grown-ups do not do mathematics, how can we expect kids to want to do mathematics?

This is why my science and math popularization is aimed at grown-ups rather than kids. Mothers and fathers, big brothers and big sisters, uncles and aunts, grandfathers and grandmothers, these are the people that I want to be, or to become, math enthusiasts. I want all of them to be able to talk about mathematical elegance, about this unique feeling of figuring things out and about this excitement of asking intriguing deep questions. I am convinced that if kids were surrounded by an environment that really embraced mathematics, they would undoubtedly be themselves excited with manipulating new mathematical objects and asking new mathematical questions. They would love mathematics; and they would learn mathematics.

I am not alone here. In fact, much of the successful online math popularization is aimed at grown-ups. So, how do they get their audience to fall in love with mathematics? From what I have observed, there are several different ways to do so. For instance, *Martin Gardner, Matt Parker* or *Mickaël Launay* tend to play around with bits of factoids, such as *the 8 Queen Problem.*[23] This is sometimes called recreational mathematics. The problems one interacts with are of little importance to research mathematics, but they do develop a sort of nerdy intimate connection with basic objects of mathematics like numbers. The great advantage of this approach is that it is much easier to get the audience to engage and think about the problems in question. They can more easily crack them or ask relevant questions. A drawback I see is that recreational mathematics often turns into a collection of unrelated puzzles.

On the other hand, other math popularizers, such as *Marcus du Sautoy, Brian Greene* or *Edward Frenkel*, prefer to discuss the big ideas of mathematics. They do not hesitate to discuss the really big questions such as the Riemann hypothesis, string theory or the Langlands program, and often go through the major developments of their disciplines throughout history. Of course, these two ways of popularizing mathematics are absolutely not disjoint, and someone like *James Grime* can both play with sexy numbers and discuss Nash's embedding theorem. One huge advantage of the big-idea approach is that you can talk about mathematics with a bird's-eye view. It allows to connect popular mathematics with some concepts that will be or have been encountered at school, and provide explanations for their appearance in history (which, too often, seems neglected in classrooms). Finally, it allows to introduce the biggest and most exciting open questions that today's greatest mathematicians are trying to tackle. This means that we often get to see popularizers discussing the bits of mathematics that they are so excited about that they literally spent their lives trying to understand them.

It is definitely this latter approach that I prefer to follow, as I have been reviewing the history of algebra and analysis in my two-hour documentary, and as I have been retracing Einstein's footsteps through the theory of general relativity in my YouTube series. Prior to that, through my Science4All articles, I was already happy to have managed to go much more in-depth than most popularization work does, which allowed me to underline and connect the big ideas of mathematics.

Still another approach to mathematics popularization is in a more dramatized or romanticized manner. For instance, the BBC documentary on Andrew Wiles' proof of Fermat's last theorem was staged in such a way that we could follow the rollercoaster chain of emotions that Wiles had to go through. Similarly, Fields medalist Cédric Villani recently published *Birth of a Theorem*, which tells the story of how he came up with a complete proof of the exponential Landau damping in analytic regularity beyond the linearized model. Quite surprisingly, this book features several copy-pastes of Villani's email exchanges with his collaborator Clément Mouhot, as well as excerpts from their papers, but with (almost) no explanation whatsoever of the mathematics that appears in these copy-pastes. Evidently, the focus is absolutely not on the mathematics itself, but on the mathematical quest. This book also resonates with some of Poincaré's outreach work, where Poincaré wanted to insist on the thought process that led to the proof of a theorem rather than on the theorem or the proof themselves. In a significantly different style, I should like to mention the book *The Simpsons and their Mathematical Secrets* by Simon Singh, which gives bits of mathematics through the more catchy theme of the famous animated sitcom.

At this point, dear reader, you might be surprised that I have not related math popularization with the focus on math applications, even though I myself am an applied mathematician! Somehow, it seems that applied mathematicians are less active in mathematics outreach than some of those active in pure mathematics. But there may actually be a good reason for this. Namely, while we may care about applications and while the audience may think it cares about applications, the really exciting bit of mathematics, the mathematical elegance, does not often have to do with applications. Therefore, while it may be useful to introduce mathematical concepts by mentioning actual or potential applications, at least when we want the audience to taste mathematical elegance, it seems that, more often than not, the eventual focus needs to be on some more abstract mathematics.

In fact, I believe it to be wrong to be focusing on applications rather than the mathematics underlying the applications, especially if the purpose is to make mathematics exciting. As Feynman put it while speaking about physics: "Sure, it may give some practical results; but that's not why we do it."[24] Applicability often adds to mathematical elegance, but it is hardly ever the core of mathematical elegance.

So, which of the above approaches is the best to popularize mathematics? First, I should point out that I do not suggest that the above list is exhaustive.[25] But I would say that none of these approaches should be seen as *the* best way. Rather, it is the combination of the different approaches that makes mathematics popularization enriching and efficient. In fact, the variety of approaches to mathematics popularization provides an insightful illustration of the variety of ways in which mathematical elegance becomes apparent. I was even surprised to rediscover this as I was interviewing guests for my *Magie des Maths* documentary.[26] Some loved Fourier analysis; others hated it. Some found Euler's identity lovely; others found it unremarkable. But all claimed to love mathematics! Evidently, the contrast was particularly enjoyable between the geometer and the number theorist on the topic of algebraic geometry. For the geometer, algebraic geometry was created by algebraists who "claimed it to be geometry (but it isn't)." For the number theorists, algebraic geometry had taken geometry to greater horizons where the focus shifted from basic triangles to more fundamental geometrical objects. In the end, I myself, like all mathematicians, have clear preferences for some areas of mathematics rather than others. For instance, I do not really care for numbers—which always surprises people!—while I find probability theory very intriguing …

But in the end, does any of this work? Did I, or anyone else, get to actually change anyone's perspective on mathematics? I think we did. I think we are changing people's approach to mathematics. And do not forget, it is not only about getting students excited about mathematics. A mother once messaged me that her son shared with her an article I wrote about soccer balls and that *she* loved it. I have been imagining a dinner conversation between mother and son about the shape of the 2018 World Cup ball ever since ... I also often get feedback from teachers who find the resources I put online very useful and enriching. I will not claim that they will become better teachers by reading me, but if they can find a little bit of extra excitement, that surely cannot hurt! And, through the previous sections, I hope to have convinced you that this popularization work has had dramatic impacts on at least some of my viewers (and on myself).

7 Science Popularization Online

Fortunately, I am not the only one to engage a wide audience with mathematics and science. Now, the Internet is huge and it seems hopeless to list what is happening out there. So, while I will try to review what I know and find worthy to mention here, this section is to be regarded as highly incomplete.

Evidently, I need to start with YouTube, which, in the last five years, has completely upset the landscape of science popularization. The leading figure is probably *VSauce*, a channel run by Michael Stevens, with 1.4 billion views and 13 million subscribers. This is overwhelming. If *VSauce*'s viewers formed a new country, it would be bigger than Switzerland! Perhaps equally surprising, despite its background in psychology and literature, *VSauce* does not hesitate to tackle highly technical topics such as the EPR paradox or the Banach-Tarsi paradox, in amazing detail and with stunning success.

Not so far away, we find many wonderful channels such as *Asap Science, Crash Course, Smarter Every Day, Minute Physics, Veritasium, Scishow, Vlog Brothers, Khan Academy, CGPGrey, Ted-Ed, Numberphile, In a Nutshell, ViHart* and so many more. These channels have over 1 million subscribers and are big enough so that their hosts can live from their YouTube incomes. But they are merely the tip of the iceberg. Behind these hugely successful channels, there is a myriad of growing quality channels, a few of which have been regrouped by PBS Digital Studios, like *Space Time, Physics Girl* and *BrainCrafts*. In fact, many of my personal favourite channels are (unfortunately) much less successful. They include *Math Tipping Point, Looking Glass Universe* and *NakedScientists*. In addition to this, the YouTube ecosystem features institutions like *NASA, the Royal Institution* and the *Museum of Mathematics*.

It is probably relevant to say at this point that all these channels are growing, and many of today's less successful channels will likely become huge in a near future. The Internet is moving, and it is moving fast. Quite impressive as well is the growth of websites like *IFLScience* and *AlertScience* that gather bits of science from all over the web in a more news-following way. Meanwhile, the web is fortunate to feature blogs from top-level scientists like *Terence Tao, Sean Carroll* or *Dick Lipton*. Finally, I should mention amazing discussion platforms like *Reddit, Quora* or *StackExchange*, where many professors of prestigious universities take the time to answer specific questions asked by the website users.

So far, I have only mentioned here English-speaking science popularization, partly because they are the most successful ones as they can reach a wider audience. But many YouTube channels in other languages are rapidly growing as well. Among Spanish-speaking channels, I will mention *Ted-Ed Español, MinutoDeFisica* and *Derivando*.

Since I am French myself, I actually know better the French-speaking science popularization community, which has undergone a dramatic growth in 2015. Part of this community was already organized through a website entitled *Café des Sciences*, which regrouped most of the greatest French-speaking science blogs throughout the web. However, in 2015, several of these bloggers shifted to YouTube and their audience grew at a much faster rate, like *E-Penser, Dirty Biology* and *MicMaths*, and groups like *Vidéothèque d'Alexandrie* and *Vidéoscience* started to structure the numerous emerging new channels.

But, still within the French-speaking community, perhaps what is most innovative was the creation of *Facebook* and *Google* groups which facilitated communications and collaborations between online science popularizers, including the introduction of a science popularization peer-review system. I have personally been very excited to be involved in this peer-review system. On the one hand we, as a science popularizer community, can better avoid blatant mistakes in our scripts. On the other hand, it allows science popularization reviewers to set new standards of rigor. I am convinced that this will lead to popularization of greater quality, and, as well, build and strengthen a wonderfully excited emerging community of science popularizers and science enthusiasts.

Finally, there are as well more institutionalized approaches to science popularization that allow researchers to make impromptu contributions, like *World Science U, Plus Maths* and *Image des Maths*. While these are quite successful, I find the quality of the articles published there somehow irregular, as many entries are written by researchers who have had little outreach experience prior to these occasional contributions. Nevertheless, I am a great admirer and a regular reader of these resources.

8 A Sneak Peek into the Future

As the saying goes, *it is hard to make predictions, especially about the future.* But I will nevertheless venture an optimistic guess. It seems to me that more and more brilliant and motivated people are coming to the Internet and are structuring the immensity of human knowledge. Evidently, *Wikipedia* is the prototypical example of this. Who would have guessed back in 2000 that the most reliable source of information a mere 15 years later would be a free and open-source website, with so many entries on so many different topics, and whose authors are basically unknown and not remunerated? I wouldn't have even dared to imagine that this would be possible. And yet, *Wikipedia* does exist, and I must be checking it a dozen times every day.

But I expect *Wikipedia* (and things like the *Khan Academy* or the *MOOCS*) to be merely the first step towards a more exciting learning experience through the Internet. I hope that more and more web surfers will open the Internet in the hope of having a new exciting science and mathematical experience. They may have a few favorite guides through these journeys—like today's leading YouTubers—and these guides would help them discover streams of interesting questions to ponder further, and would redirect them towards more specialized guides or resources. Along the way, the web surfers may meet other web surfers, and exchange thoughts about their understanding. Maybe they could even refer one another to other resources. In other words, web surfers would then be like couch surfers traveling the world, with some guidance here and there, some great encounters along the way and plenty of souvenirs to come back home with. To some extent, this is already a little bit the case for some of us on the Internet.

But, I also expect this online culture to spread into schools, universities and local communities. I think that, at this level as well—in fact, perhaps even more than on the Internet—the popularization of science must span and grow, both in terms of quantity and quality. From a very practical standpoint, this means that I am hoping that significantly more funds will be allocated to science and mathematics popularization. These days, almost all science and math popularizers need to have a different main activity, like research, teaching or something else. Besides, few of these popularizers would say that popularization is their main activity. But I hope to have convinced my readers that this makes little sense to me personally, as, for me, science and mathematics popularization seems way more important than research and teaching.

In 2016, I was hired by the École Polytechnique Fédérale de Lausanne (EPFL) to take over an online video teaching project called *Wandida*. The goal of this YouTube channel is to provide brief videos on key university-level concepts in mathematics, physics and computer science. Since then, I have launched other projects at EPFL that are more popularization-oriented, including another YouTube channel called *ZettaBytes* that features interviews of EPFL researchers, and the *Swiss Alkindi* cryptography competition for secondary school students. Since then, I also have had the chance to collaborate with many different institutes on producing more popularization content.

More recently, in 2018, I published a book entitled *La Formule du Savoir* on Bayesianism. This modern philosophy of science argues that probability theory should be viewed as the foundation of all knowledge. Interestingly, this has strong implications on how to think, teach and learn.

While the valorization of science popularization is slowly improving, I would argue that we are still extremely far from what it should be. After years of struggle, I am still in a precarious situation—albeit I am doing financially fine. But I am arguably an exception. I have been lucky. The overwhelming majority of my colleagues still cannot afford to make science popularization their full-time job.

Yet I hope to have convinced you that math and science popularizer should become a proper job in a near future. I hope to have convinced you that it may well be the most important job that does not yet exist.

Notes

1 La Magie des Mathématiques des Classes Préparatoires, https://www.youtube.complaylist?list=PL8ovs-QtxcNzwAlSfYOLvc9QOtsPWZUht

2 Duality in Linear Programming | Science4All Article, http://www.science4all.org/le-nguyen-hoang/duality-in-linear-programming/

3 P versus NP: A Crucial Open Problem | Science4All Article, http://www.science4all.org/le-nguyen-hoang/pnp

4 Fourier Analysis: Signals and Frequencies | Science4All Article, http://www.science4all.org/le-nguyen-hoang/fourier-analysis/

5 The Most Beautiful Equation of Math: Euler's Identity | Science4All Article, http://www.science4all.org/le-nguyen-hoang/eulers-identity/

6 Entropy and the Second Law of Thermodynamics | Science4All Article, http://www.science4all.org/le-nguyen-hoang/entropy/

7 The Beauty of Ellipses, Parabolas and Hyperbolas | Science4All Article, http://www.science4all.org/le-nguyen-hoang/ellipses-parabolas-hyperbolas/

8 The Amazing Physics of Water in Trees | Science4All Article, http://www.science4all.org/le-nguyen-hoang/the-amazing-physics-of-water-in-trees/

9 Shannon's Information Theory | Science4All Article, http://www.science4all.org/le-nguyen-hoang/shannons-information-theory/
10 The Revolutionary Galois Theory | Science4All Article, http://www.science4all.org/le-nguyen-hoang/galois-theory/
11 Google View Analytics Form of *Magie des Maths* documentary, https://docs.google.com/forms/d/1asI9wJ1jrnMWjpWaO29XvORSIK-ZQDoG_Ivt5m-It5M/viewanalytics
12 MIT *Splash* event, https://esp.mit.edu/learn/Splash/index.html
13 See https://history.aip.org/exhibits/einstein/emc1.htm
14 Are astronauts weightless? Einstein's Happiest Thought | Science4All 14, https://www.youtube.com/watch?v=9s2tObpO_rU
15 Don't heavier objects fall faster? Galileo's Thought Experiment | Science4All 12, https://www.youtube.com/watch?v=Vz9q-TuBrgo
16 Is Gravity an Illusion? | Space Time | PBS Digital Studios, https://www.youtube.com/watch?v=NblRo1hHK6U
17 Steve Jobs's Commencement address at Stanford University, delivered on 12 June 2005. See https://news.stanford.edu/2005/06/14/jobs-061505/
18 Lecture, University of Lille (7 December 1854).
19 http://www.phnet.fi/public/mamaa1/einstein.htm
20 http://www.pbs.org/wgbh/nova/physics/andrew-wiles-fermat.html
21 La Magie des Mathématiques des Classes Préparatoires (2/2)—L'analyse | Time 40:40, https://www.youtube.com/watch?v=DRciHKXWneA&start=2440
22 Sprouts & Co., http://thesprouts.org/
23 The 8 Queen Problem—Numberphile, https://www.youtube.com/watch?v=jPcBUoZ2Hj8
24 See e.g., http://www.legacy.com/news/culture-and-trends/article/richard-feynman-and-the-wonders-of-science
25 Klara Kelecsenyi's thesis (2009) gives a more thorough analysis of mathematics popularization, although, much has happened since 2009, especially on the Internet, http://spectrum.library.concordia.ca/976628/1/NR63411.pdf
26 Op. cit. (see note 1).

References

Farmelo, G. (2009). *The strangest man: The hidden life of Paul Dirac, mystic of the atom.* New York, NY: Basic Books.

Poincaré, H. (1914). *Science and method* (F. Maitland, Trans.). London/Edinburgh/Dublin/New York, NY: Thomas Nelson and Sons.

INTERMEZZO 11

It's Not Just a Right; It's an Obligation to the Future

Jan Visser

Reflections inspired by Michael Spector: 'Fostering inquiry, reasoning and critical thinking'

People like to believe what they like. It's called wishful thinking. When scientists feel inclined to steer their research in the direction of what they believe should be its outcome, it is called researcher bias. When politicians want you to believe what they like and try to sway your opinion without providing the assumptions and evidence based research outcomes on which their own belief is based, it's called deceit, which is not rarely accomplished these days by spreading fake news. When commercial interests drive consumer behavior in the sole interest of profit making for the few, thereby limiting the public's options to choose, it should be called infringement of the right to exercise one's free will. Should you feel comfortable to be part of a religious community in which you grew up and should you then shield yourself from being exposed to and seeking to understand the religious feelings of others, you are shortchanging yourself. When you, living in a multi-party democracy, decide to consider only the arguments of your preferred party or ideology, you are prone to self-deception. If you live in a dictatorially governed state, it requires courage to challenge authority. Yet courageous acts occur and they have often resulted in desirable change. We all admire such manifestations of courage.

There are multiple ways and countless circumstances in which we are at risk of fooling ourselves or deceiving others involuntarily. Moreover, it would seem that deliberate deceit has increasingly become an accepted way of influencing people for purposes such as profit making or the building of power bases. The potential consequences of deceit are serious, particularly in the perspective of how, as a planetary community of currently 7.7 billion people, we should be able to evolve in ways that allow us to coinhabit the planet harmoniously. Deceit undermines the development of mutual trust within and among communities and societies. It's exactly trust that we need most to invent a different future.

This has obvious implications for how we configure the learning landscape. It should be an environment that offers multiple and diverse opportunities for people to learn along the lifespan, starting at infancy, involving, among other

© KONINKLIJKE BRILL NV, LEIDEN, 2020 | DOI: 10.1163/9789004416802_022 things, the family, the school, broadcast and social media, living and work environments, civil society participation and constructive involvement in democratic development. In all these spheres there should be a strong focus on improvement of effective use of our mental faculties, allowing us all to think critically. Mike Spector addresses this question in his chapter on 'Fostering inquiry, reasoning and critical thinking.' It is not just the right you have to be inquisitive, to think critically, and to take active control of yourself and the world of which you are part; it is also an obligation to future generations who will build new forms of cohabitation of planet Earth on what we leave behind for them to inherit.

CHAPTER 12

Fostering Inquiry, Reasoning and Critical Thinking

J. Michael Spector

1 Introduction

The argument presented in this chapter is that higher-order reasoning should be considered a critical 21st century skill and, as a consequence, inquiry-based learning and argumentation should be integral parts of a 21st century education addressed at every level, beginning in primary school and continuing through secondary school, college and subsequent learning experiences. Currently, emphasis on engaging in higher-order reasoning rarely occurs in primary and secondary schools and somewhat sporadically in higher education contexts. The emphasis in primary and secondary contexts primarily involves readings and discussions *about* higher-order reasoning rather than having learners become engaged in higher-order reasoning activities. As M. David Merrill has said in many conference presentations and implies in his *First Principles of Instruction* (2002, 2013), people learn what they do. This principle implies that learning activities should be aimed at applying and practicing the targeted knowledge and skills. As a corollary to Merrill's activation principle is the notion that people are unlikely to learn things that they do not practice and apply. As a consequence, failure to develop higher-order reasoning skills can result in a population that becomes easily swayed by extreme views and is intolerant of those who appear to be different or who have different beliefs and values. In short, there are ethical and social implications of neglecting higher-order reasoning skills. Integrating inquiry-based learning activities early and often and developing argumentation skills are therefore critical aspects of education in the 21st century.

2 Habits of Scientific Reasoning for a Tolerant World

In many places around the world one can find people with divergent views. In many of those places, tensions have grown considerably and taken the form of hateful language and even violence. Tolerance and an openness to considering alternative views and perspectives seems to be at all-time lows in many

 | DOI: 10.1163/9789004416802_023

places. Advocacy seems to be displacing evidence. Anger seems to be displacing thoughtfulness. Are these trends real? Are they acceptable? If not, what can be done?

Earlier in my career, I thought that reason and logical argumentation could and would enlighten those suffering from the throes of dogmatism, prejudice and intolerance. I thought that helping others to develop a scientific attitude could help cure the ills of a society suffering from intolerance. Those efforts, mainly in higher education settings, have met with only modest success. My conclusion is that the appropriate habits of mind—namely, those associated with scientific reasoning—were not developed in primary and secondary education. As a consequence, this piece is an extended argument for targeting the development of higher-order reasoning skills, especially in primary and secondary settings.

The trend in higher education seems to be towards developing highly skilled workers for a competitive global marketplace in which information and communication technologies play a critical role. That trend bends towards an economic perspective that favors such things as productivity and revenue, from an enterprise and corporate perspective rather than from an individual citizen perspective. Such a trend is likely to result in further widening of the digital divide that separates those with appropriate skills and jobs from those without those skills and jobs. As gaps between the favored and those left behind widens, the tendency to only see those with whom one associates will increase widening divisions within society. An unintended outcome, then, is likely to be a society of selves and self-interests rather than a socially cohesive society of families and communities. The habits of mind that need to begin developing early might include those of (a) thinking from alternative perspectives, (b) understanding the views of those with different backgrounds and experiences, and (c) respecting—even embracing—the rich diversity that exists naturally in and among humans around the world.

On a personal note, I had the benefit of a father who was a Rabbi who taught me by example and by asking difficult questions and then analyzing my naïve responses. I once asked him about what it meant to be a Rabbi as I was then considering that possibility. His response was that a Rabbi was essentially a teacher—someone whose voice was one of encouragement and comfort, who listened to others before offering opinions or views, whose eyes examined all sides of an issue and reflected something new to others, whose behavior served as an example and whose hands were always available to help others succeed—someone who never turned away from another in need.

3 Definitions

The key terms involved in the following discussion include: (a) argumentation, (b) critical thinking, (c) educational ethics, (d) higher-order reasoning, (e) inquiry-based learning, and (f) scientific attitude. A short notional definition for each is offered next, with the understanding that these terms are interrelated and interdependent, in this author's mind.

3.1 *Argumentation*

An argument, in this context, is a collection of statements, some of which are offered in support of another. The supporting statements are often called premises, and the statement being supported is often called the conclusion. Rather than teaching concepts and facts, there should be some emphasis on using those facts and concepts to develop evidence in support of a variety of additional claims (conclusions). Suppose, for example, that the year is 1905 and you have been studying astronomy. You have been taught some facts, such as there are 8 planets in the solar system. You also know some concepts and principles, such as Newton's laws of motion and Kepler's law of planetary motion. In other words, the concept of gravitationally determined orbits is within your set of knowledge and beliefs. As an amateur astronomer, you have been observing the paths of the eight known planets and keeping a log of their orbital location. As you review the logs, you notice that Neptune and Uranus are not exactly where they would be expected to be according to Kepler's law of motion. These are additional facts. Now there is a problem. Your logs could be mistaken. Newton and Kepler could be mistaken. Or, there is an additional fact not yet known that accounts for the discrepancy. Considering that Kepler's laws of planetary motion are so accurate most of the time with regard to most of the planets, it seems unlikely that Newton and Kepler are mistaken. You check your logs and record additional data for several more years being somewhat curious and having no social life. These data confirm the discrepancies in the orbits of Neptune and Uranus. You then ask yourself what additional fact could account for those discrepancies. Suppose there was a very distant orbiting object as yet unobserved in the solar system. How massive would it need to be and where should it be located to account for those discrepancies? Being skilled in algebra and calculus, you make a few estimates and then go looking in the night sky for the object you believe is causing the discrepancies in the orbits of Neptune and Uranus. Amazingly, you find an object that is later named Pluto. This example can be formulated as an argument:

Premises:
- Kepler's laws of motion accurately predict the orbits of the planets.
- Kepler's laws of motion are based on the effects of gravitational attraction between massive objects in accordance with Newton's laws of motion.
- Neptune and Uranus are not exactly where they should be according to Kepler's laws of motion.
- If there were another orbiting object, somewhat smaller and quite distant from the sun, it would account for the discrepancies in the motion of Neptune and Uranus.

Conclusion: There is another massive object orbiting in the solar system.

This argument can be simplified and examined more closely to make a point about the nature of arguments. Suppose the first two premises are simply taken as givens; established facts that are in no need of explicit statement in the argument. We then have a simplified argument form that looks like this: (a) Premise 1: Discrepancies apparently exist in the orbits of Neptune and Uranus; (b) Premise 2: If there is an as-yet unobserved additional orbiting object, then those apparent discrepancies can be explained; (c) Conclusion: There is an additional massive object orbiting in the solar system. Further simplification yields an argument form like this: (a) Y; (b) If X, then Y; (c) Therefore Y. (My apologies for the simplified rendition of the discovery of Pluto; I am not an astronomer. I am just someone who occasionally thinks about thinking.)

While this particular form of argument led to an important scientific discovery, it is worth noting that it is not a conclusive form of argument (or, expressed differently, it is not a deductive argument). As suggested previously, there are other possibilities, however unlikely, that could account for Y: the apparent discrepancies in the orbits. So, even arguments that do not establish a conclusion with certainty (i.e., non-deductive or inductive arguments) can be useful and can guide inquiry and investigation. The point for now is that an argument consists of some statements (the premises or evidence) that are offered in support of another statement (the conclusion). Some arguments establish a conclusion with certainty and some only suggest a probable conclusion. In both cases, the evidence needs to be linked as strongly as possible to the conclusion. The question now is: When and how might argumentation activities be introduced into primary and secondary curricula and learning activities? This is left as a take-home exercise for primary and secondary teachers and the odd instructional designer who might be reading this chapter. Presumably, one might begin with much simpler arguments that might involve everyday occurrences and phenomena rather than science. One might also consider blending in activities that take place outside a school or classroom and that go beyond a curriculum oriented towards mandated education standards.

3.2 *Critical Thinking*

Critical thinking can be traced back at least to the time of Socrates and the pre-Socratic philosophers in western civilization (circa 450 BCE). According to the evidence presented in Plato's early dialogues, Socrates engaged in what was recorded as dialogues with one or more other interlocutors, most often with regard to ethical issues, including whether or not it would be right for Socrates to escape from prison and avoid the death penalty that had been imposed on him (Cooper, 1997; see Plato's dialogue entitled *Crito*[1]). Crito tries to convince Socrates that escaping would be the right things to do, and Crito has already imagined an apparently risk-free escape plan. Socrates listens and reflects and questions, and then arrives at the conclusion that escaping would violate things that Socrates holds higher than himself, namely, the laws of Athens and a guiding voice that Socrates reports hearing.

Glaser (1941) defined critical thinking as involving three aspects: (a) an open attitude to considering a range of possibilities to which one can relate, (b) familiarity with logic and logical argumentation (similar to the discussion in the previous section), and (c) skill in applying logic and logical argumentation. Taken together, these three aspects of critical thinking involve a persistent willingness to take into account evidence, to seek clarity, to recognize and question assumptions (recall the two premises in the former section that were taken as givens but briefly questioned), and to consider or evaluate the strength of the evidence and its relationship to the conclusion. Scriven and Paul (1987) add to the process the notion of disciplined conceptualization, analysis, synthesis and evaluation, emphasizing clarity, accuracy, precision, consistency, relevance, sound evidence, and fairness.

Given the Socratic model and the definitions sketched above and other definitions, one can conclude that critical thinking is a much more complex concept than argumentation. While an argument might be at the core of a critical thinking activity, that activity extends the argument to reflection about unstated assumptions, implications of accepting the conclusion, the consistency among the various pieces of evidence, and more. The concept called 'critical thinking' is broad and somewhat fuzzy. Given that learning a concept involves (a) learning critical attributes or a definition of the concept, (b) being shown examples and non-examples, and (c) identifying examples and non-examples (Merrill & Tennyson, 1977), it might be appropriate to briefly sketch an example and a non-example.

Here is a brief sketch of an example of critical thinking. The conclusion being questioned is whether and to what extent human activity is causing climate change. This is a matter involving causal claims about the physical world. Causal claims are generally established by non-deductive arguments—that is, with more or less probability. What might count as relevant evidence? First, is the global climate actually changing? There are many factors to consider; to simplify, only temperature trends will be considered (simplifying is not necessarily a good practice when conducting a rigorous inquiry, by the way). According to NASA,[2] 2016 was the warmest year on record for planet earth. In fact, 2016 was the third year in a row to set a new record for global average surface temperatures. According to the same source, globally-averaged temperatures in 2016 were 0.99 degrees Celsius above those in the mid-20th century and 1.1 degrees above those in the late 19th century. Is that a significant change? What has happened with regard to the net global temperature since that time? A warming trend appears to be accelerating, which might lead one to look for possible causal factors such as CO_2 emissions, electric power consumption, population growth, rainfall patterns and changes, etc.[3] The process just described might constitute an example of someone thinking critically about the possible causes of climate change.

Here is a brief sketch of a non-example of critical thinking. The conclusion being discussed is whether and to what extent lowering taxes on the wealthy will cause an improvement in economic productivity and job growth. Again this is a matter involving a causal claim and can only be established with more or less probability. However, suppose the person who advocates that conclusion is unwilling to consider any evidence that might support or fail to support the conclusion. Discourse is then cut short; the claim that lowering taxes on the wealthy is, in effect, not refutable from the advocate's perspective. Unwillingness to question assumptions, examine evidence, and consider alternatives are strong indicators of non-critical thinking.

Two additional comments are in order with regard to critical thinking. First, critical thinking is relatively rare as it takes effort and persistence and an open attitude. People are often busy and have become habituated to simplifying shortcuts and quick decisions about many things. Simplifying is part of human nature. According to a constructivist epistemology, people create internal representations (e.g., mental models) to make sense of things experienced—especially things that are new or puzzling. A model is necessarily a simplification of the thing it represents. A critical aspect of critical thinking is deciding when to begin extending, elaborating or exploring a natural simplification that one has made. Indeed, creating a habit of routinely making explicit one's mental models, and then reflecting on and questioning them is a habit of mind worth developing. The second comment has to do with the nature of non-critical thinking, to which everyone is prone. Non-critical thinking all too often becomes the basis for prejudice, intolerance and hateful behavior. Another take-home exercise is to identify instances in which you or someone you know has been intolerant on the basis of unwarranted prejudice (this exercise may be much too easy to complete in the current political climate).

3.3 *Educational Ethics*

Given the tendency to simplify and maybe become intolerant, there might be a connection between critical thinking and ethical behavior. Perhaps, there are ethical principles to which educators, instructional designers and others who support formal and informal learning should adhere. Earlier (Spector, 2005) I proposed an Educratic Oath that is intended to parallel the Hypocratic Oath, which represents basic principles for medical practice. The Educratic Oath specifies the following ethical principles: (1) do nothing to harm learning or the learner; (2) improve learning and instruction to the best of one's ability; (3) base instruction on evidence; (4) share the principles of instruction that work with others; and (5) respect individual rights at all times. What seems to be happening more and more frequently around the world is that lack of respect for those who appear to be different or who hold different beliefs and values is increasing, and, with an increase of intolerance there seems to be an increase in hateful behavior and violence.

A core argument of this chapter is that teaching young persons to think critically and to engage in higher-order reasoning is a long-term, systemic way to address such intolerance. As indicated in the beginning, this may well be a naïve notion. Regardless, educators do have basic ethical obligations and social responsibilities which should not be overlooked in efforts to build scientific minds. One example of inadvertently creating harm is introducing a particular educational technology in a situation in which some but not all students have access and the requisite knowledge to make effective use of that technology. Less well-prepared students may fall farther and farther behind, which is one kind of harm. Further disadvantaging those who are already disadvantaged, in spite of good intentions to improve learning for others, should not be considered an acceptable consequence of technology innovation and integration in classrooms (in the less than humble opinion of this older than average author). Moreover, there is a general trend among educational technologists to become infatuated with a new technology and believe that it will work for all learners in nearly every circumstance. This is a form of non-critical thinking previously described. A further example of misplaced confidence in a technology to solve serious learning challenges pertains to the use of social networking in a problem-based learning activity. In some cases, the symptom of the problem is poor performance on projects associated with a representative problem. What can happen is that some groups may only interact with each other and the thoughts and beliefs of the most articulate or forceful group member may be reflected. In such a case, project performance may or may not improve; however, the underlying problem of individual understanding may not be addressed at all. It is not necessarily the technology (in this case, social networking) that might improve learning, understanding and performance. It is the use of a technology by skilled educators that might improve learning. It is not about the technology—it is about the learning.

3.4 *Higher-Order Reasoning*

Higher-order reasoning is often considered co-extensive with critical thinking skills, which is a reasonable way to proceed. The two are separated in this chapter as a way to distinguish the things that one does (the activities associated with critical thinking), and the cognitive and non-cognitive processes involved when engaging in those activities. One way to think about the relevant internal processes involved in critical thinking is to begin with Gagné's (1985) nine events of instruction. For example, for the first event—*gain attention*—what happens internally in the learner is the activation of neural receptors; the third event—*stimulate recall of previously learned knowledge*—involves retrieving items from long-term memory and bringing them into working memory; *enhancing retention and transfer*—the ninth event—involves strengthening retrieval and activation with additional cues and strategies. Moving into the explicitly cognitive domain, one might consider Bloom's revised taxonomy (Anderson & Krathwohl, 2001) as a way to represent higher-order reasoning abilities. In that taxonomy, cognitive processes are made explicit in the form of such verbs as remember, understand, analyze, and evaluate. Indeed, many have argued that analyze, evaluate and create (the top half of the taxonomy) represent higher-order reasoning abilities, whereas remember, comprehend and apply (the lower half of the cognitive hierarchy) are lower-order and more common reasoning abilities. In addition, one might argue that metacognition (awareness of one's own cognitive processes) is a higher-order reasoning ability. One can also argue that understanding (a term used in the lower half of the revised taxonomy) is often used to represent an integration of all levels of that taxonomy. In a sense, the term 'deep learning' now used in the artificial intelligence community as well as by educators is such a high level, integrative term similar to what is being suggested here for the term 'understanding.'

Another way to consider the logical flow being represented in this argumentation model is this: the goal is to develop higher-order reasoning abilities or what could be called understanding or deep learning. That goal is consistent with a need to have a citizenry that understands science along with a highly productive STEM (science, technology, engineering and mathematics) workforce, and it is desirable from the perspective of social responsibility and ethical behavior that could be found in a STEAM (includes arts and the humanities) curriculum. In order to attain that goal, there is a need to develop critical reasoning skills. An instructional approach that seems well-suited for developing critical reasoning skills is inquiry-based learning.

3.5 *Inquiry-Based Learning*

In the USA, one can trace inquiry-based learning at least as far back as John Dewey (1910a, 1910b). More recently, a number of pedagogical approaches have emphasized things related to inquiry, including *problem-based learning* that puts small groups of learning in problem solving situations (Barrows & Tamblyn, 1980), *task-centered instruction* that orients learning activities around realistic problems (Merrill, 2002, 2013), and model-facilitated learning in which learners are gradually confronted with problems of increasing complexity and first asked to explore alternatives and later asked to develop policies to guide decision making (Milrad, Spector, & Davidsen, 2003). Inquiry-based approaches design learning activities that involve solving a challenging problem or investigating a complex question. Jonassen (2011) developed an excellent instructional guide to support problem solving and inquiry learning in which different approaches are linked to different kinds of problems and investigations. Jonassen also presents a model for integrating argumentation into inquiry and problem solving, which is quite rare within the instructional design community.

A general way to think about inquiry learning goes as follows. Inquiry involves having questions and seeking answers. Having a question is different from simply asking a question. Asking a question is a simple grammatical exercise; in English, switch the location of the verb and subject in the sentence and put a question mark at the end. Having a question is not so simple. Having a question involves (a) admitting that one does not know the answer (an admission of ignorance, which might take some time and effort to uncover), (b) committing time and effort to exploring the question and finding potential answers (wanting to know and understand and following through on that desire), and (c) being open to alternative explanations and willing to question one's assumptions (recalling the discussion of critical thinking skills). In an inquiry-based learning approach, the primary task of the educator is to help students *have* questions (which naturally implies openness on the part of the educator to questions that students may already have to start with). Giving answers is not the focus in inquiry-based learning, although giving answers and presenting facts and concepts constitutes most of primary and secondary education. Having a question involves doing something, and people generally learn what they do (the mantra for today—repeat it at home twenty times). Gagné often said that the job of an educator and instructional designer is to help people learn. The inquiry-based approach transforms that task into helping people have questions.

3.6 *Scientific Attitude*

What is it to have a scientific attitude? Some might say that it involves considering beliefs and statements that could potentially be shown to be false or mistaken as those beliefs and statements are worth investigating (for those keeping notes, I am thinking about Karl Popper's *Conjectures and Refutations*). If the person holding a particular belief or making a certain statement is not willing to consider that it could be wrong, then there is little to be gained by pursuing discourse if one believes that continued discourse is unlikely to reach a mutual understanding with regard to basic facts and beliefs. This could be called the *walk-away strategy*. The drawback is that no progress in terms of understanding or openness occurs as a result. I have walked away many times, but cannot recommend this strategy.

What happens all too often in and out of school is that having questions is not encouraged or supported or facilitated. It takes an exceptionally patient parent to keep encouraging a child to have a question and explore possible answers. What might happen on account of new technologies with conversational interfaces is that the role of helping a child explore a question could shift to an intelligent conversational agent with access to large and reliable datasets. However, that technology is not yet able to help a child pose or rephrase a query in a more focused or meaningful way. Some parents and teachers can and sometimes do so. For example, consider a child witnessing thousands of monarch butterflies in North Texas. An inquisitive child might ask where did those butterflies come from. A computer agent might reply that they were once caterpillars, focusing on a particular interpretation of 'come from.' A more knowledgeable parent or teacher might say, in reply, Canada. The curious child may then say "Really?" A parent or teacher who wishes to encourage inquiry might say, thinking ahead to further questions, "Yes—Canada, which is about a thousand miles from here. Subsequent questions might then include" (a) How far can a butterfly fly in one day? (b) How long do butterflies live? (c) What do butterflies eat during their migration? (d) Do they follow the same route every year? (e) Why do they migrate? (f) Do they return to the same place on the return flight? Such questions can span multiple disciplines including entomology, ecology, and possibly economics.

A somewhat different approach involves thinking about what underlies thinking and reasoning. For this task, it is worthwhile considering Ludwig Wittgenstein's (1922) works. In the *Tractatus Logico-Philosophicus*, Wittgenstein observes at remark 2.1 that we picture facts to ourselves. He fails to note that we also picture things that are not factual to ourselves. Nevertheless, the remark that we picture facts to ourselves is the kernel notion in a naturalistic or constructivist epistemology—we create internal representations to make sense of things that we experience that are new, unusual or otherwise puzzling, as noted previously. The last remark in the *Tractatus* is this: what we cannot speak about (clearly) we must pass over in silence. As it happened, there was much that Wittgenstein wanted to speak about, as is evident in his *Philosophical Investigations*, published after his death (Wittgenstein, 1953). In that posthumous work, he introduced the notion of language games. Not only do we create internal representations to make sense of new and puzzling things, we talk about those representations with others. We engage in language games. Such discourse represents a re-representation of the internal representations that are basically hypothetical entities that no one ever directly observes. These two ideas—creating internal representations and engaging in discourse about those representations—are the basis of a socio-constructivist epistemology that is prevalent in some educational and philosophical circles. As it happens, such a perspective also informs an inquiry-based learning approach and can be considered the foundation for developing critical thinking skills.

Next to consider in properly situating critical thinking, inquiry learning and higher-order reasoning are the works of Sextus Empiricus—namely, *Outlines of Scepticism* (Annas & Barnes, 2000).[4] After describing what it is to be a skeptic, Sextus argues that being a skeptic will result in peace of mind and tranquility. Is that not odd? How to achieve such intellectual peace and tranquility? The answer is that having questions and seeking answers will ease a troubled mind; holding on to dogmatic beliefs will eventually create mental disturbance and unhappiness. The evidence may not support that particular conclusion. Indeed, one could argue that a state of uncertainty and cognitive dissonance may be desirable in terms of promoting serious and deep inquiry (Festinger, 1957; Quine & Ullian, 1978).

Becoming a skeptic involves recognizing that many (possibly most) beliefs and statements are open to investigation. A dogmatic position results when one believes something with certainty and without considering evidence or alternatives when there exists a basis for doubt or further investigation; a variant of dogmatism occurs when one believes that there is no possible answer to a question or issue. In both of those cases, no further inquiry can occur and the discourse is cut off (recall the walk-away strategy mentioned earlier). A skeptic, however, is someone who recognizes that further investigation is possible even when the available evidence apparently favors one or another of the many possible beliefs or positions. A skeptic, in the world of Sextus Empiricus, is not held captive by his or her beliefs. Socrates came close to living the life of a sceptic in the sense that he seemed willing to consider and re-consider many of the conclusions which he previously held.

4 A Framework to Develop Critical Thinking Skills

Based on the prior discussion, what is next is a presentation of an initial framework that might guide teachers, instructional designers and parents in the development of activities to support the development of critical thinking skills. An initial step is to help instill a habit of asking and then having questions early in a child's development.

The first part of the framework therefore involves the general notion that helping students to have questions should be a teacher's primary task. As previously stated, the job of a teacher is to help students have questions—not to help them ask questions or give them answers. To have a question involves (a) admitting that one does not know something, which is a process rather than a singular event, (b) committing time and effort to exploring and finding potential answers, (c) being willing to consider alternative perspectives and approaches, and (d) being willing to revisit the entire process again and again. All too often students are taught to have answers rather than to have questions. Learning how to have questions should be considered a basic skill, akin to reading Plato's *Symposium*, writing love letters, and calculating π (I happen to love transcendental numbers). If this argument is accepted, then inquiry should be introduced early and often in primary and secondary education.

The second part of the framework I am suggesting should support a wide variety of inquiries—that is to say, a general essential structure that can support many cases of having questions. One such structure that might fit into this framework (supporting inquiry, reasoning and critical thinking) involves considering argument forms, as previously discussed. Recalling the previous discussion, an argument can be considered a collection of statements (premises) offered in support of another statement (the conclusion). A form of scientific reasoning can be mapped onto this notion of an argument form. Logicians typically distinguish deductive arguments from non-deductive arguments since

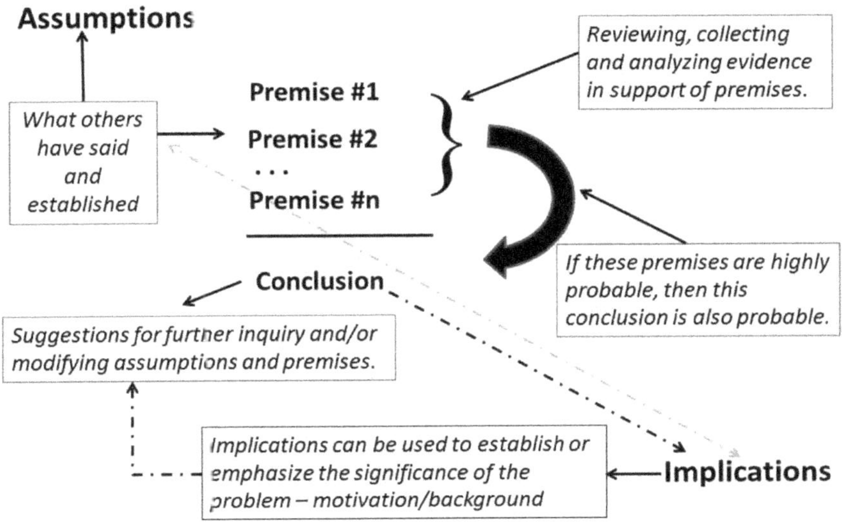

FIGURE 12.1 A general framework for inquiry and critical reasoning

the criteria for evaluating them are different. That distinction is not necessarily relevant to this brief and tenuous excursion into inquiry learning. Figure 12.1 (which I have developed for use in some of my courses with graduate students where it arrives much too late in their intellectual development) depicts the general form of an argument. The reason for introducing such a framework early in a child's education is that it can establish a habit of mind—namely the habit of thinking of the adequacy of the evidence, the habit of identifying unstated assumptions that might also merit inquiry, and the habit of looking at the implications of what one believes. If those habits are established early in a child's education, then there is a remote possibility that more children will begin to think scientifically. Perhaps there is a remote possibility that when those children become adults that they will be free from dogmas that create tension, confusion and strife within society. There is a remote possibility.

5 Concluding Statement

I cannot pretend to have found answers or even useful paths to follow. There are limits to human knowledge and understanding. One cannot, for example, love all and only those persons in this room who do not love themselves. Nor can two. Contradictions (statements that are necessarily false) and tautologies (statements that are necessarily true) form the extreme limits of understanding; neither contradictions nor tautologies provide any knowledge or understanding. Then there are possibilities—the domain of what can be meaningful and understood. With regard to possibilities, there are several varieties. There are bare possibilities, such as the possibility that the number of grains of sand on Waikiki Beach, Hawai'i at a particular point in time is an odd number (I quit counting at 13). There are also practical possibilities, such as the possibility of significantly reducing hunger and violence in a particular part of the world. Another practical possibility is that the children of today may indeed become skeptical inquirers (i.e., non-dogmatic investigators) and critical thinkers. That last practical possibility, however, is not likely to occur without the efforts of many of those involved in such efforts as the Building the Scientific Mind enterprise.

Notes

1 Located at http://classics.mit.edu/Plato/crito.html
2 See https://www.nasa.gov/press-release/nasa-noaa-data-show-2016-warmest-year-on-record-globally
3 See https://climate.nasa.gov/ for a more detailed analysis.
4 See also http://www.skeptic.com/ for more about skepticism.

References

Anderson, L. W., & Krathwohl, D. R. (Eds.). (2001). *A taxonomy for learning, teaching and assessing: A revision of Bloom's taxonomy of educational objectives*. New York, NY: Longman.

Annas, J., & Barnes, J. (Eds.). (2000). *Sextus empiricus: Outlines of scepticism*. Cambridge: Cambridge University Press.

Barrows, H. S., & Tamblyn, R. (1980). *Problem-based learning: An approach to medical education*. New York, NY: Springer.

Cooper, J. M. (Ed.). (1997). *Plato: Complete works*. Indianapolis, IN: Hackett.

Dewey, J. (1910a). Science as subject-matter and as method. *Science, 31*, 121–127.

Dewey. J. (1910b). *How we think*. Boston, MA: D. C. Heath.

Festinger, L. (1957). *A theory of cognitive dissonance*. Stanford, CA: Stanford University Press.

Gagné, R. M. (1985). *The conditions of learning* (4th ed.). New York, NY: Holt, Rinehart & Winston.

Glaser, E. M. (1941). *An experiment in the development of critical thinking*. New York, NY: Columbia University Teachers College.

Jonassen, D. H. (2011). *Learning to solve problems: A handbook for designing problem-solving learning environments*. New York, NY: Routledge.

Merrill, M. D. (2002). First principles of instruction. *Educational Technology Research & Development, 50*(3), 43–59.

Merrill, M. D. (2013). *First principles of instruction: Identifying and designing effective, efficient, and engaging instruction*. San Francisco, CA: Wiley.

Merrill, M. D., & Tennyson, R. D. (1977). *Concept teaching: An instructional design guide*. Englewood Cliffs, NJ: Educational Technology.

Milrad, M., Spector, J. M., & Davidsen, P. I. (2003). Model facilitated learning. In S. Naidu (Ed.), *Learning and teaching with technology: Principles and practices* (pp. 13–27). London: Kogan Page.

Popper, K. (1963). *Conjectures and refutations: The growth of scientific knowledge*. New York, NY: Routledge & Kegan Paul.

Quine, W. V., & Ullian, J. S. (1978). *The web of belief* (2nd ed.). New York, NY: Random House.

Scriven, M., & Paul, R. (1987). *Critical thinking as defined by the National Council for excellence in critical thinking*. Retrieved from http://www.criticalthinking.org/pages/defining-critical-thinking/766

Spector, J. M. (2005). *Innovations in instructional technology: An introduction to this volume*. In J. M. Spector, C. Ohrazda, A. Van Schaack, & D. A. Wiley (Eds.), *Innovations in instructional technology: Essays in honor of M. David Merrill* (pp. xxxi–xxxvi). Mahwah, NJ: Erlbaum.

Wittgenstein, L. (1922), *Tractatus logico-philosophicus* (F. P. Ramsey & C. K. Ogden, Trans.). London: Kegan Paul.

Wittgenstein, L. (1953). *Philosophical investigations* (G. E. M. Ascombe & R. Rhees, Eds.). Oxford: Blackwell.

INTERMEZZO 12

How Long Can We Still Wait and Who Takes Responsibility?

Jan Visser

Reflections inspired by Martin De Wit: 'The shifting mind of economics'

It is January 16, 2019. A new study has just come out yesterday, published in the journal *Nature Communications* (Smith et al., 2019). Quoting from the abstract, the results of the study show that

> if carbon-intensive infrastructure is phased out at the end of its design lifetime from the end of 2018, there is a 64% chance that peak global mean temperature rise remains below 1.5°C. Delaying mitigation until 2030 considerably reduces the likelihood that 1.5°C would be attainable even if the rate of fossil fuel retirement was accelerated. Although the challenges laid out by the Paris Agreement are daunting, we indicate 1.5°C remains possible and is attainable with ambitious and immediate emission reduction across all sectors.

In slightly simpler words: If from now on we decide to retire each and every fossil fuel powered contraption at the end of its useful life (different for power plants, aircraft, vehicles, ships, etc.) and then replace them with zero-carbon alternatives, we will still only have a 64% chance that we can limit global warming to 1.5 degrees, the most ambitious target of the Paris Agreement. Looking around, I see a world in which it seems highly unlikely that such action, as suggested in the study, will be implemented. Headway is being made in some sectors, but not enough or nothing at all is accomplished in others. More is thus necessary and the authors of the study call for radical action across all relevant sectors. Such radical action implies fundamental change in attitudes and mindset in humans across the globe.

The study is new, but the underlying issues are not. In fact, already in the 1950s and particularly 1960s, warning signs were given off, notably in the writings of Arthur Koestler (*The Ghost in the Machine*, 1967), and Paul Ehrlich (*The Population Bomb*, 1968), of impending disastrous developments. Several

 | DOI: 10.1163/9789004416802_024

similarly worrying studies during the past three decades are cited in my introductory chapter to this book.

Mere tinkering with the current reality, trying to avoid the pain involved in making necessary changes, will not be enough. A fundamental shift of mind is required in multiple domains of exploration, research and thinking. Martin de Wit addresses this issue in the context of economic thinking in his chapter on 'The shifting mind of economics.' He notes that "a posture that is open to learning and a commitment to an objective reality that is subjectively approached" is required to engage fruitfully with the shifting economic mind. The same posture will be indispensable in any other area of necessary change.

Waiting to see what happens is not an option. We must all act now and take responsibility.

Reference

Smith, C. J., Forster, P. M., Allen, M., Fuglestvedt, J., Millar, R. J., Rogelj, J., & Zickfeld, K. (2019). Current fossil fuel infrastructure does not yet commit us to 1.5°C warming. *Nature Communications, 10*, Article 101. https://doi.org/10.1038/s41467-018-07999-w

CHAPTER 13

The Shifting Mind of Economics

Martinus Petrus De Wit

1 Introduction

The financial and ecological crises are still creating havoc on the planet. The emerging risks, costs and benefits of relatively rapid climatic changes as well as other planetary boundaries on economies and society are challenging economists on how to think about natural resources and the environment. Yet, it was the financial crisis that has led to a shock and disbelief by the main proponents of financial deregulation, not only because of the scale of its impact, but mostly because it challenged economists' belief in rational market actors and the self-regulating forces of the market (Andrews, 2008). Mainstream economists have focused on including the price of environmental damage and resource use and the need for adequately defined property rights in their theories and models to achieve objectives of sustainability. Such a form of 'enlightened rationality' is perceived to be achieved, in some cases, with the help of governmental regulation, but essentially within the framework of self-regulating markets. Despite long traditions in dissenting schools of economic thought, it was through the financial crisis that rationality and self-regulating markets are now challenged on a more fundamental level (INET, 2009).

This chapter examines how the financial and ecological crises are contributing to a discernible shift in economic thinking. The pursuit to better understand and respond to complex financial and ecological realities is tied up in a continuous process of learning. Effective learning is facilitated through the willingness to ask penetrating questions and expose hidden assumptions in any mode of thought. Such a philosophy of learning, as elaborated on in this book, is employed here to expose the non-neutrality of economic theorizing on real-world crises and to argue for an openness to change in the discipline.

Alternative schools of economic thought have been around for decades, but is the mind of economics really shifting? What are the effects of a shifting economic mind, not only on economics as a discipline, but also on policy and economic life? If the economic mind is changing, what kind of thinking is shifting and what kind of thinking is untouched? How does the shifting economic mind affect real-world economic policy and life? The purpose of this chapter is to start addressing these questions by reflecting on the effects and depth of

 | DOI: 10.1163/9789004416802_025

the changes that are taking place in economic thinking in the aftermath of the financial crisis and amidst the ongoing ecological crisis. To achieve this purpose the chapter makes use of resources borrowed from theories of change in corporate management and from the philosophy of economics, aiming to contribute to a critical, yet realistic, discussion on the state of, and change in, the economics discipline. The chapter is presented in three sections: first, a description of the shift in economic thinking that is taking place, second, a reflection on the effects and depth of this change and, third, a discussion on how to engage fruitfully with such shifts.

2 The Shifting 'Economic Mind'

The term 'economic mind' is used sparingly in the economic literature, mostly limited to economic psychology (Furnham & Lewis, 1986) and the history of economic thought (Rutherford, 1998; Dorfman, 1946, 1949, 1950). Singer's (1950) question in response to Dorfman whether the economic mind needs to be understood only as thinking about economics or whether it also includes the decisions of people in the economy, remains relevant. In this chapter the focus is primarily on economic thought and how such thinking may affect policy and practice. The economic mind is broadly defined as 'how economists think,' or as the mainstream, dominant and generally acceptable way of thinking for an economist. The mainstream 'economic mind' informs how curricula are developed and how students are trained to 'think like economists,' how manuscripts are evaluated for publication in mainstream economic journals, what is expected from economists in business and the marketplace, and how economists usually inform strategy and policy. John Kenneth Galbraith's inclusive maxim that economics is a useful form of employment for economists is a valuable heuristic, but too broad for our purposes; instead mainstream economists are here defined as those who hold to the dominant 'economic mind.'

Although many have proclaimed that the financial crisis is over and the ecological crisis can be met through existing economic structures and actors, several have argued otherwise. By way of example, one critical review has pointed to the inability of mainstream economics to anticipate a crisis of such a magnitude, the lack of intellectual diversity in the field of economics, as well as the focus on mathematical elegance and equilibrium at the expense of the real world (Heise, 2014). The economic discipline has basically responded by broadening what is tolerated within the discipline, all of which is now loosely defined as 'mainstream economics.' Ecological questions were included in standard economic theory initially through developments in the field of environmental and natural resource economics and since the late 1980s through the heterodox-orientated ecological economics (De Wit, 2016; Carpintero, 2013). Complex systems science has also increasingly influenced economic theory and modeling (Foster, 2006; Arthur, Durlauf, & Lane, 1997). Foster (2006) speculated that a "… great economic problem [needed to arise] … to further advance the complex systems perspective in economic science" (p. 1087), and it seems that this is happening in the aftermath of the global financial crisis. Whether this will affirm Colander, Holt and Rosser's (2008) view that these developments will "create a new orthodoxy centered on a broader complexity vision" (p. 497) remains to be seen.

The Institute for New Economic Thinking (INET)—initiated in 2009—has been at the forefront of reshaping economics in the wake of the financial crisis. It has played a key role in questioning the assumptions of rational expectations theory, efficient markets and predictable human behavior, and in accepting the inherent complexity and uncertainty of economic and financial systems (INET, 2009). One of its main founders, George Soros, reflected on the question whether economic phenomena can be predicted by universally valid laws:

> I contend that they cannot be, because the phenomena studied by economics have a fundamentally different structure from natural phenomena. The difference lies in the role of thinking. Economic phenomena have thinking participants, natural phenomena do not. The thinking of the participants introduces an element of uncertainty into the course of events that is absent in natural phenomena. The uncertainty arises because the participants' thinking does not accurately represent reality. (Soros, 2010, p. 3)

For Soros, financial markets cannot be treated as a form of physics, but they need to be seen and studied as a form of history. He questions the lack of realism in economic models regarding human behavior and instead works from the assumption that the human condition is fallible and that perfect information is a myth. These ideas are not new and have a long tradition in social science and economic thought. John Kenneth Galbraith (1987) already started his book on the history of economics with the statement that "[t]here can be no understanding of economics without an awareness of its history" (p. 1). Herbert Simon observed the limits of knowledge and cognition with his theory on bounded rationality in the 1970s, taken further with the rise of behavioral and complexity economics in the 1990s and beyond. The realization that economics cannot be studied in an ahistorical fashion and that humans have cognitive limits and are not able to maximize their well-being, accentuated a shift from deductive to inductive reasoning in economics. Brian Arthur (1994) commented: "In other words, when we cannot fully reason or lack full definition of the problem, we use simple models to fill the gaps in our understanding. Such behavior is inductive" (p. 407).

According to Thomas Kuhn in his book on *The Structure of Scientific Revolutions* a theory holds until too much contrary evidence accumulates after which the theory is discarded and a new one emerges. For Kuhn, whole 'paradigm shifts' happen when anomalies start to occur and accounts of reality become irreconcilable. The development of science is not only objective, but also relies on subjective interpretations of reality. Although Kuhn interpreted his idea of paradigm shifts to the physical sciences only, economists have applied this idea to mark shifts to Keynesianism and Monetarism. In development theory the swing from government-led to the market-orientated policies of the 'Washington-consensus' and beyond have been portrayed as Kuhnian 'paradigm shifts' (Gore, 2000). In the early 1970s the rise of radical neo-Marxian economics against the backdrop of a complex of unresolved social problems was seen by some as a much-needed switch to a new paradigm, while for others such as Robert Solow, conventional economics just needed better articulation in response to solving such problems (Worland, 1972). A similar development can be seen today. It is still a matter of intense debate whether a combination of complexity and behavioral economics is producing a shift in mainstream economics or be usurped in it. Although the financial crisis did provide a jolt to the complacency of the mainstream, true paradigm shifts in the Kuhnian sense are not readily occurring. There are signs that the dominant center of economic knowledge has become more self-critical, but a true change in economic thought—that is "a shift in the axiomatic, methodological and epistemological openness of the discipline"—is unlikely (Heise, 2014, p. 83). Colander, Holt, and Rosser (2004) on the other hand, argue that economics is undergoing a fundamental shift in its methods away from rationality, selfishness and equilibrium towards "... dynamics, recursive methods and complexity" (p. 458). These changes cannot be interpreted as a Kuhnian paradigm shift though. The authors prefer to speak of a Lakatosian shift (after the philosopher Imre Lakatos):

> ... changes do not lead to sudden paradigm shifts, but instead lead to cumulative evolutionary changes that ultimately will be recognized as a revolutionary change. The changes leading to this ex-post revolution were initially accepted within the profession only gradually .... (Colander, Holt, & Rosser, 2004, p. 488)

One key difference in the Kuhnian and Lakatosian approaches to change is whether it comes from outside or within. Coming from within would suggest that there is some general acceptance of method and epistemology. In mainstream economics, modeling remains a central core, whether in orthodox or non-orthodox approaches. Simulation, integrated ecological-economic, evolutionary, complexity and agent-based modeling approaches are now used by mainstream economists, signaling that important shifts from more static, neoclassical modeling approaches may be taking place. Social science relationships are now also modeled on biological and psychological relationships and not exclusively on physical relationships (Davis, 2008).[1] It must be noted that although the philosophers Kuhn and Lakatos are valuable in conceptualizing types of change, the application of their work to shifts in economic thinking is not straightforward and remains controversial.

On the basis of the discussion so far it can be observed that a shift is occurring in economic thinking, not only in academia, but also in institutes and high-powered individuals in finance and economics. The changes occurring in economic thinking have deep historical roots and their increased acceptance has been perpetuated by the financial crisis. The next section examines the depth of this shift and reflects on its possible effects.

3 Reflections on the Shifting Mind of Economics

3.1 *Effects of Change*

Questions remain whether the shifts are peripheral or whether a critical mass of economic minds has shifted; how fundamental these shifts are; and whether these shifts will have any practical impact on minimizing the risks of another financial crisis and the ongoing ecological crisis.

Such questions may benefit from employing alternative theories of change emphasizing structures, processes and directions (Haferkamp & Smelser, 1992), but also intangibles such as beliefs and social relationships (Huy, 2001). Huy's ideal types of change intervention could help guide an analysis of change as follows: the time perspectives and the anticipated pace of change, analytical frameworks or diagnostic models to measure change, the goals of change, intervention theories, roles and typical identities of change agents and the change tactics employed. According to this typology, the approach of changing economic thinking through networking and teaching as outlined in the previous section has a long-term perspective, the pace is gradual, it is focused on shared norms and psychology with a goal of enhanced learning through actions such as probing, revealing and facilitation. The limitation of such interventions is that cognitive change does not always lead to sustained behavioral change and seldom leads to a wider realization in the corporate world (Huy,

2001). These limitations also apply to the lack of impact on economies and in societies. What this short analysis reveals is that without normative and more powerful changes aimed at the wider structures and processes, systemic change is an unlikely outcome.

The evidence regarding economic policymaking does seem to support such a cautious statement on the likely effects of change. The shifts in economic thought as documented in the previous section are still mostly limited to (some) academic institutions and think tanks and the implications for economic policymaking are in their infancy. In the aftermath of the financial crisis, Dolphin and Nash (2012) observe that:

> ... orthodox economics remains the only game in town for economic policymaking. The vast majority of mainstream economic policy debate continues to oscillate between the arguments of free market economists and neo-Keynesians about how best to restore global growth and prevent future crises. (p. 9)

Despite a sea change in what is constituted as 'mainstream economics,' most of economic policy by governments and global financial institutions is derived from a particular way of doing macroeconomic modeling, namely dynamic stochastic general equilibrium (DSGE) modeling. DSGE is based on a preconceived set of axioms, simplified mechanical relationships and a deductive logic (Kay, 2012). Such models are usually validated with a focus on consistency in mathematical modeling and rigor. The assumption that macroeconomic behavior is aggregated from theoretically robust microeconomic foundations of rational actors, has "... not really succeeded in deriving all macroeconomic propositions from fundamental hypotheses on the behavior of individual agents," (Janssen, 1998, p. 6) and is increasingly challenged by the contributions of evolutionary models and behavioral economics. The important point for our purposes is that competing macroeconomic models used for economic policymaking must be empirically validated.

It can be concluded that the effects of the shifts in economic thinking taking place are mostly theoretical. Their full effect on economic policymaking still has to become manifest.

3.2 *Depth of Change*

It is in times of crisis that some economists are becoming dissatisfied with their discipline, compelling them to ask themselves questions and reflect upon the philosophical underpinnings of their field—including regarding epistemology, metaphysics, ethics, ontology and aesthetics—leading them to challenge what is perceived as 'meaningful discourse' in economics. Hands (2008, p. 14), in a review on philosophy and economics has called the rejection of philosophy in positivist economics "the aberration of the 20th century." Such rejection of philosophy by economists was the fruit of a self-conscious development of 'scientific economics,' meaning a narrow, positivist idea of what real scientific inquiry is. This tendency, which was particularly dominant from the 1930s to the 1950s, has started to change. Since the mid-1980s the relation between economics and philosophy has, according to Hands, seen a renaissance. Not only the one-sided philosophy of natural science, but also ethics, ontology, metaphysics and aesthetics, are providing ideas to economists.

The first philosophical question to be discussed here in relation to economics is the method or grounds of knowledge. The implicit assumption that truth claims need to submit to the scientific method, whether inductive or deductive, is a pre-supposition that has governed mainstream economics for a long time. The shift in economic thought from deduction to induction does not examine the shared "… tendency towards 'naturalism' in epistemology and philosophy of science" (Hands, 2008, p. 6). Naturalism starts with the idea that science informs how theories of knowledge should be developed. By definition, in such a view, the possibility of a normative philosophy that explains what can and cannot be done in science or any economic imitation of science (whether physical, biological or psychological), is ruled out. In such a view there is also no room for an epistemology that operates on the basis of a different ontology than what is presupposed by a theory of cognition. Descartes placed the foundation of all knowing in the knower's mind. Kant restricted knowledge to sense experiences and left room for faith, but strictly separated these spheres. The Kantian synthesis of a phenomenological world of rationalist metaphysics and empiricism and the noematical world, which is beyond our capabilities of rational understanding, has pervaded Western philosophy and thought, including economic thinking. Reason and faith, facts and values, immanence and transcendence have all been placed in their own sphere and this division *a priori* excludes an empirical explanation of reality in its totality. Such a position has led to the untenable and incoherent outcome that the different spheres of life are each operating on their own logic, for example, one morality for economic (work, business) life and one for private life. In contrast, knowing can also be thought of as an act of making sense of life as a whole. A focus on the whole person and the person in relation to others, is another philosophical position overcoming the exclusive focus on thinking and theorizing as a means of understanding and responding in reality. Kant, for instance, famously believed that atomistic Newtonian physics was the only physics there possibly is—a clear *a priori* belief in opposition to the more relational

Einsteinian physics. In this vein of thinking a typical response to crises would be to revisit and possibly revise the methods or grounds of knowledge.

The second philosophical idea economists are paying increasing attention to is a focus on morality and ethics. Although there is a long tradition in the line of Kant and Hume's thought in positive economics arguing to keep facts and values separate in the theory of rational choice, most philosophers do consider it a normative theory. Values are forthcoming out of factual measurements on how individuals are thought to behave. The new economic thinking as described in this chapter improves on what is seen as facts, by incorporating experimental results on human behavior showing that "moral beliefs matter to decision-making in experimental environments" (Hands, 2008, p. 9). New advances in the field of economics and psychology help to measure subjective experiences of people and has spawned a field in economics dedicated to human happiness, in contrast to the utilitarian morality of welfare economics. In all these new strands of thinking, satisfying the preferences and well-being of the individual remains the underlying moral philosophy, but now only in closer relation to others and more empirically measured than what standard rational choice theory was able to offer. Values are not separated from facts as in Kantian morality and in the classical physical foundations of rational choice theory, but are now placed on biological and cultural foundations (Hodgson, 2012).

A moral response to the financial and ecological crises would therefore involve a promotion of certain moral values and moral practices, although proponents would differ considerably depending on their foundations of value. One example of a moral response to the financial and ecological crises is to place an emphasis on "… the centrality of persons acting in cooperation" (Donohue-White, Gabrill, Westley, & Zúñiga, 2002, p. 71) and acting on this insight by including this "personalist insight into economic modeling" (p. 84).

Ethical reflection on any economic theory or system is needed, including those systems that include moral and personalist defenses of human liberty. Whether more depth in the shift of economic thought—as portrayed in a shift of human morality—is ultimately desirable in economic thinking is not a new debate in moral philosophy. The question needs to be considered if the satisfaction of individual well-being—whether seen from a *homo economicus* perspective or as pertaining to persons acting in cooperation—is a fruitful response to the financial and ecological crises. Such a question has traditionally been addressed by a Rawlsian emphasis on social, communitarian or public rather than individual preferences, an approach that has also been used to accommodate ecological norms (Sagoff, 1986; Cobb & Daly, 1989), and/or by pointing to other desirable goals for humans (Sen, 1985; Nussbaum, 2011).

These approaches are unified in their rejection of a sole scientific basis for ethics.[2] In response to the crises, the market mechanism is generally accepted, but not always and everywhere. Sen (2009), with an appreciative nod to Adam Smith and the welfare economist Pigou, states that "[t]he present economic crisis is partly generated by a huge overestimation of the wisdom of market processes, and the crisis is now being exacerbated by anxiety and lack of trust in the financial market and in businesses in general ..." (p. 3). Sen furthermore argues that "[t]here is a critical need for paying special attention to the underdogs of society in planning a response to the current crisis ..." (p. 5). Such a more critical acceptance of markets is copied by ecological economists who are attempting a more "... deliberative approach to human development which embraces a strong sustainability perspective" (Pelenc & Ballet, 2015, p. 36). In such views, the satisfaction of individual well-being—no matter how well it might be morally articulated—is not a solution to crises. Normative and deliberative intervention is required through a variety of organizational mechanisms in response to crises.

The third philosophical question of relevance to a meaningful discourse in economics is ontology. The question has been asked by contemporary philosophers of science whether mainstream economics is a science grounded in reality or a pseudo-science leading to delusion (Lewens, 2015). Kant argued that we cannot know reality; we can know only phenomena, organized in our conceptions of things which provide *a priori* categories of understanding. For Kant, metaphysics is outside the grasp of human intellect and beyond the reach of science. For the empiricist David Hume ontology was a misnomer. Hume argued that objective natural laws are some form of belief, as only appearances of things as they are perceived by humans can become part of knowledge. Hume reasoned that reality that is independent of our perceptions can have no bearing on ontology. In such a Humean framework, has economics transgressed into the noematical? In Kant's typology one would ask: Is economics more like religion than science? Is economics, despite all its talk of scientific objectivity and value-neutrality, in fact a value system that has been imposed on us?

Mäki (2008) summarized the views in economics regarding scientific realism and ontology and argued that in the social sciences the existence of social reality cannot be conceptualized as independent of the human mind. Scientific realism relevant to economics suggests that "... much of the social world is characterized by non-causal science-independence" (p. 1). Economic thinking, according to Mäki, shapes people's beliefs and worldviews which in turn shape economic policies and ultimately the economy. This does not mean that beliefs about the world, or the world itself, are socially constructed. Neither does it mean that the mere act of finding agreement will lead to action that leads to social change, nor that the "social world itself is ... socially constructed by the modeling practices of economists" (p. 9). For realists, the acceptance of economic models needs to be grounded in the properties of the real world. Acceptance of these models must not be just a "... matter of power and persuasion" (p. 9). Tony Lawson (2003), for example, observed that the social world is an open system, limiting the use of the closed system approach of deductive, mathematical models. The choice of a model is dependent on its suitability to investigate certain social phenomena. The ontological response to crises would thus be one where methodological deficiencies are addressed (Lawson, 2003).

The discussion on ontology in economics as summarized here has proceeded on the basis of the assumption that the human mind is the gateway to reality with no category for the transcendent. The ontology Hume rejected was one where reality was perceived independent of empirical facts, while the ontology described by Mäki and Lawson is an ontology in opposition to social reductionism. For realists, truth matters and they have a healthy respect for facts, but such a terminology rich in varied meanings already suggests that it cannot be applied without critical reflection on both the transcendental and the social world. Hence, the rise of *critical realism* which rejects Hume's empiricism, has developed into many branches of its own. One notable development is in theological critical realism, providing a ground for a rational debate across Kant's wall of separation between faith and reason (Walker, 2017). An open, fallible and reflective two-way relationship is proposed that should not be closed prematurely through a focus on either empirical reductionism or moral idealism (De Wit, 2016).

The fourth area of philosophy that is of importance for economic discourse is aesthetics. The word is derived from the Greek αἰσθητικός (aisthetikos) which means 'pertaining to sense perception'. Despite an obsession with beauty in modernity, economics as discipline has come to think of beauty as "being scarce and tradable" (Hamermesh, 2011, p. 9). Beauty is perceived as something instrumental, an attribute or quality that helps to achieve socio-economic objective(s) of happiness. In mainstream economics, beauty remains firmly in the eye of the beholder, measured by the preference for beauty by individual agents acting in their own self-interest. In contrast to such an instrumental view of beauty, its pursuit is often associated with metaphysical or transcendental concepts of the good and the truth (Chandrasekhar, 1987). In this vein, the Newtonian-inspired economics of Adam Smith was founded on the idea of order, which in turn was shaped by an ancient cosmology of harmony (Orrell, 2012). Adam Smith, especially in his earlier writings, favored beauty or taste as a standard of judgement in response to the rather harsh empiricism of thinkers such as Bacon, Hobbes, and Locke (Thomson, 1965).

Yet, nature has been observed as not perfect and not always beautiful. Duncan (2011) wrote in his book *The Calendar*, a book on the failed 5000-year long quest to perfectly align the clock to the heavens, that nature is not perfect. Likewise, the abstract beauty of mathematical elegance representing economies is far removed from a complex reality full of novelty and surprise. The aesthetically pleasing, the beautiful, and the elegant cannot be assumed to be equal to the truth. Paul Krugman commented that the economics profession went astray "... because economists, as a group, mistook beauty, clad in impressive-looking mathematics, for truth." (Krugman, 2009, p. 1). One can even speak of a 'Paradox of Beauty' in mainstream economic science: aided by aesthetically elegant theory and models, the outcomes are often ugly and disharmonious. Nor can beauty be perceived as equal to the good. Einstein in his later years, for example, stated that aesthetically elegant abstraction can lead to a loss of ethical life and human dignity. A response to the financial and ecological crises, as informed by this discussion on aesthetics, would argue for a more realistic sense of aesthetics accounting for imperfection and brokenness, in economics (De Wit, 2013).

In summary, the philosophy of economics produces a set of ideas that can assist in analyzing the depth of change taking place in the discipline. The rejection of economics modeled on classical physics has opened up fruitful avenues of engagement with philosophical themes.

4 Engaging Fruitfully with Shifting Economic Minds

How are economists prepared to engage fruitfully with the ongoing changes in their discipline? The choice of the word *fruitfully* is deliberate; standard dictionary definitions are that it produces good results, which is beneficial or abundantly productive. Such a definition signals openness and growth and would require first, and foremost, to move beyond a position of denial.

Second, it would be helpful to understand what types of financial and ecological crises responses are already ongoing. In an earlier contribution, I proposed stylized categories of responses to assist in the debate, namely an instrumental response, a structural response, a moral response, and an ontological response (De Wit, 2011). The instrumentalist believes that the financial and ecological crises can be dealt with through an increase of greener consumption. A structural response focuses on normative interventions in the dominant models of capitalism to achieve objectives such as socio-economic justice and/or ecological sustainability. A moral response sees the root of the financial and ecological crises as a lack of moral qualities, such as greed. An ontological response sees the root problem as a misunderstanding of reality and our misguided efforts of modeling reality. Such typologies may be helpful in organizing the material within various economic schools of thought in responding to crises, but needs to be expanded for contributions from the philosophy of science and economics, notably in epistemology and aesthetics.

Third, the way economists organize themselves in engaging with change matters in the development of economic thought. A logical way of responding to change is to emphasize process, but it cannot be done without a meaningful conversation on what counts as scientific evidence and what counts as valid morals and ethics.

Fourth, this raises the question what information will count as being valid in economic reasoning. The question of economic methodology, defined as "the principles underlying economic reasoning" (Backhouse, 2008, p. 1), remains open in such a process, but to what extent and on what basis? In my opinion, neither focusing *solely* on perfect and objective facts of reality, nor *solely* on the subjectivity of values will provide a fruitful way forward. There is an objective reality independent of human perception and preference and there will be subjectivity in every inquiry of that reality. As human persons we have senses, cognition and emotions available to us to make sense of the reality of the world. A posture of being critical yet realistic on the evidence we use, of being deeply engaged and open to learning, would help in comprehending and fruitfully contributing to the shifting mind of economics.

5 Conclusion

A discernible shift is occurring in economic thinking, not only in academia, but also in institutions and high-powered individuals in finance and economics. The changes occurring in economic thinking have deep historical roots and their increased acceptance has clearly been perpetuated by the financial and ecological crises. The approach to change economic thinking through networking and teaching can be characterized by having a long-term perspective, a pace that is gradual, being focused on shared norms and psychology with a goal of enhanced learning through actions such as probing, revealing and facilitation. The effects of such shifts in economic thinking have not measurably permeated into economic policy and practice.

Deeper change is happening in economic thinking when the grounds and methods of knowledge are revisited, when questions are being asked regarding the reality of the economic world and how such reality needs to be modeled, when morality and values are included in economic modeling and decision-making, and when questions on the aesthetics of economic thinking and practice are asked. Such philosophical themes open several resources to engage with the shift in economic thinking.

Speculating on how to engage fruitfully with the shifting economic mind, it is suggested to move beyond the denial that economics is changing, towards developing robust typologies of change and response, to organize processes that support a critical, but realistic engagement with shifts in thinking and to open up a philosophically astute discussion on economic methodology. These suggestions presuppose a posture that is open to learning and a commitment to an objective reality that is subjectively approached. The financial and ecological crises are good times for economists to open their minds and get their hands dirty in changing their discipline.

Notes

1 This is an observation that calls into question Soros's sharp distinction between the natural and economic sciences in his motivation for a new economics.

2 In a related approach from the field of philosophy and religion, ideals (extending beyond individual preference) are connected to desirable ends (Wolterstorff, 1984).

References

Andrews, E. L. (2008, October 23). Greenspan concedes error on regulation. *New York Times*. Retrieved from http://www.nytimes.com/2008/10/24/business/economy/24panel.html

Arthur, W. B. (1994). Inductive reasoning and bounded rationality (The El Farol Problem). *American Economic Review, 84*(2), 406–411.

Arthur, W. B., Durlauf, S. N., & Lane, D. A. (Eds.). (1997). *The economy as an evolving complex system II.* Redwood City: Addison-Wesley.

Backhouse, R. E. (2008). Methodology of economics. In S. N. Durlauf & L. E. Blume (Eds.), *The new Palgrave dictionary of economics* (2nd ed.). Retrieved from http://www.dictionaryofeconomics.com

Carpintero, O. (2013). When heterodoxy becomes orthodoxy: Ecological economics in the New Palgrave dictionary of economics. *American Journal of Economics and Sociology, 72*(5), 1287–1314.

Chandrasekhar, S. (1987). *Truth and beauty: Aesthetics and motivations in science.* Chicago, IL: University of Chicago Press.

Cobb, J., & Daly, H. E. (1989). *For the common good, redirecting the economy toward community, the environment and a sustainable future*. Boston, MA: Beacon Press.

Colander, D., Holt, R., & Rosser Jr., B. (2004). The changing face of mainstream economics. *Review of Political Economy, 16*(4), 485–499.

Davis, J. B. (2008). The turn in recent economics and return of orthodoxy. *Cambridge Journal of Economics, 32*, 349–366.

De Wit, M. P. (2011). Reflecting on how to respond to the economic and ecological crises. In L. A. Andrianos et al. (Eds.), *Ecological theology and environmental ethics* (Vol. 2). Kolymbari: Orthodox Academy of Crete.

De Wit, M. P. (2013, May 27–31). *Economics, beauty and harmony: A reflection*. Paper presented at "Building the Scientific Mind" Colloquium, Lembang, Indonesia.

De Wit, M. P. (2016). Another look at economic approaches to environmental management and policy with reference to developments in South Africa. *South African Journal of Economic and Management Sciences, 19*(5), 690–701.

Dolphin, T., & Nash, D. (2012). *Complex new world. Translating new economic thinking into public policy*. London: Institute for Public Policy Research.

Donohue-White, P., Grabill, S. J., Westley, C., & Zúñiga, G. (2002). *Human nature and the discipline of economics. Personalist anthropology and economic methodology*. Lanham, MD: Lexington Books.

Dorfman, J. (1946/1949/1950). *The economic mind in American civilization* (Vols. I, II, III). New York, NY: Viking Press.

Duncan, D. E. (1998). *The calendar. The 5000-year struggle to align the clock and the heavens—and what happened to the missing ten days*. London: Fourth Estate.

Foster, J. (2006). Why is economics not a complex systems science? *Journal of Economic Issues, 40*(4), 1069–1091.

Furnham, A., & Lewis, A. (1986). *The economic mind. The social psychology of economic behaviour*. Brighton: Wheatsheaf.

Galbraith, J. K. (1987). *A history of economics. The past as the present*. London: Penguin.

Gore, C. (2000). The rise and fall of the Washington consensus as a paradigm for developing countries. *World Development, 28*(5), 789–804.

Haferkamp, H., & Smelser, N. J. (Eds.). (1991). *Social change and modernity*. Berkeley, CA: University of California Press.

Hamermesh, D. S. (2011). *Beauty pays. Why attractive people are more successful*. Princeton, NJ: Princeton University Press.

Hands, D. W. (2008). Philosophy and economics. In S. N. Durlauf & L. E. Blume (Eds.), *The new Palgrave dictionary of economics* (2nd ed.). Retrieved from http://www.dictionaryofeconomics.com

Heise, A. (2014). The future of economics in a Lakatos–Bourdieu framework. *International Journal of Political Economy, 43*(3), 70–93.

Hodgson, G. M. (2012). Toward an evolutionary and moral science. *Journal of Economic Issues, 46*(2), 265–275.

Huy, Q. N. (2001). Time, temporal capability, and planned change. *The Academy of Management Review, 26*(4), 601–623.

Institute of New Economic Thinking (INET). (2009). *Our purpose*. Retrieved from https://www.ineteconomics.org/about/our-purpose

Janssen, M. C. W. (2008). Microfoundations. In S. N. Durlauf & L. E. Blume (Eds.), *The new Palgrave dictionary of economics* (2nd ed.). Retrieved from http://www.dictionaryofeconomics.com

Kay, J. (2012). The map is not the territory: Models, scientists, and the state of modern macroeconomics. *Critical Review, 24*(1), 87–99.

Krugman, P. (2009, September). How did economists get it so wrong? *New York Times*. Retrieved from http://www.nytimes.com/2009/09/06/magazine/06Economic-t.html

Lawson, T. (2003). *Reorienting economics*. London: Routledge.

Lewens, T. (2015). *The meaning of science* (A Pelican introduction). London: Pelican.

Mäki, U. (2008). Scientific realism and ontology. In S. N. Durlauf & L. E. Blume (Eds.), *The new Palgrave dictionary of economics* (2nd ed.). Retrieved from http://www.dictionaryofeconomics.com

Nussbaum, M. (2011). *Creating capabilities: The human development approach*. Cambridge, MA: Harvard University Press.

Orrell, D. (2012). *Truth of beauty. Science and the quest for order*. New Haven, CT & London: Yale University Press.

Pelenc, J., & Ballet, J. (2015). Strong sustainability, critical natural capital and the capability approach. *Ecological Economics, 112*, 36–44.

Rutherford, M. (1998). *The economic mind in America: Essays in the history of American economics*. London: Routledge.

Sagoff, M. (1986). Values and preferences. *Ethics, 96*(2), 301–316.

Sen, A. (1985). *Commodities and capabilities*. Amsterdam & New York, NY: North-Holland.

Sen, A. (2009). Capitalism beyond the crises. *New York Review of Books, 56*(5). Retrieved from http://www.nybooks.com/articles/2009/03/26/capitalism-beyond-the-crisis/

Singer, H. G. (1950). Book review: The economic mind in American civilization, Vol. III: 1865–1918, by Joseph Dorfman. *American Journal of Sociology, 56*(1), 121–122.

Soros, G. (2010, April 8–11). *Anatomy of crisis—the living history of the last 30 years: Economic theory, politics and policy*. Paper presented at the INET Conference at King's College.

Thomson, H. (1965). Adam Smith's philosophy of science. *Quarterly Journal of Economics, 79*(2), 212–233.

Walker, T. (2017). Approaches to critical realism: Bhaskar and Lonergan. *Journal of Critical Realism, 16*(2), 111–127.

Wolterstorff, N. (1984). *Reason within the bounds of religion* (2nd ed.). Grand Rapids, MI: William Eerdmans.

Worland, S. T. (1972). Radical political economy as a 'scientific revolution.' *Southern Economic Journal, 39*(2), 274–284.

INTERMEZZO 13

Looking Back with a View to Looking Forward

Jan Visser

> I have known rivers ancient as the world
> and older than the flow of human blood in veins.
>
> LANGSTON HUGHES

∵

Reflections inspired by Federico Mayor: 'Invent the future'

I assume that the large majority of my readers are of an age that allows them to look back over a considerable stretch of time at a world, long ago, in which they once lived. This allows appreciating, critically, the differences between the world that was and the one that is. Most of us have known our parents and grandparents, and some may have known one or more of their great-grandparents as well. We have listened to their stories. Their experiences and worldviews became interwoven with our own. They provide us with a window, letting us look into our more remote past. We appreciate artistic expressions of our time and are able to relate them, while visiting museums and concert halls, to art work and music from centuries ago. Access to humanity's historic and archeological heritage stretches our memory even further into the past, to thousands, even tens of thousands of years ago. Living amidst a rich variety of life forms different from our own, from the micro to the macro, and aware that we are ourselves host to a microbiota of a great many organisms on which the nature and quality of our life depends, we become aware how we 'dance to the tune of life' (Noble, 2017), in harmony with all other forms of life. Life on our planet originated long ago. It took the universe even longer to evolve to the extent that planets could be born. We look back at and into 13.82 billion years of history, a 'big history' (Visser, 2016). This is our life! But what is next?

During my 80 years of life I have seen dramatic change. Compared with the stories I heard from my parents and grandparents, change has greatly accelerated since the time when they lived, more so, probably, than when compared with the change that happened between their lifespan and that of their parents

 | DOI: 10.1163/9789004416802_026

and grandparents. The world population has increased more than threefold since I was born. Science and technology have developed to levels that have given us, not only collectively, but more and more also individually, power that we must still learn to master. But learning takes time and often the pace of innovation is faster than the time we give ourselves to learn and adjust. An only partly controlled chaos ensues.

This is the world in which we find ourselves. It is the starting point for learning to invent a future worthy to live in, adequately conceived for rekindling harmony among ourselves and with our planetary environment. Federico Mayor's chapter on 'Inventing the future' aims at getting us on the road.

References

Noble, D. (2017). *Dance to the tune of life: Biological relativity.* Cambridge: Cambridge University Press.

Visser, J. (2016). Human learning and the development of mind in the Anthropocene: Reflections against the backdrop of big history. *Origins, 6*(9), 3–12. Retrieved from https://www.researchgate.net/publication/308612239

CHAPTER 14

Invent the Future

Federico Mayor

> Humanity is facing new threats, unlike anything in History, which require unprecedented global solutions.
>
> AMIN MAALOUF (2009)[1]

∵

1 Introduction

"No problem of human destiny is beyond human beings. Man's reason and spirit have often solved the seemingly unsolvable—and we believe they can do it again," said President John F. Kennedy in June 1963.[2] This is a very important statement: the impossible today can be possible tomorrow and there is no place in the scientific mind for fanaticism, dogmatism, and fatalism. The past cannot be re-written. The future is to be invented.

2 Science and Conscience for Radical Changes

2.1 *Who Are We, What Are We Like, Where Are We?*

We generally avoid the essential questions—"I find myself fleeing from myself precisely when I find myself," wrote José Bergamín[3]—when we tread the difficult paths of our innermost selves, searching for answers that we know will never be reached.

The thin line between certainty and uncertainty is where freedom, the great gift to the human species, enables us at any time, without pressure, to tip the balance in one direction or the other. Each human being is unique, unrepeatable, and capable of that immense and unexplainable excess that is creativity. Wrapped in perfect, temporal and ephemeral biological structures—I realized this many years ago and I repeat this frequently—human beings are endowed, in contrast to their limitations, with an infinite spiritual space.

 | DOI: 10.1163/9789004416802_027

And herein lies the dilemma, the great puzzle or perhaps miracle of human existence, but a mystery nonetheless: we have the capacity to think, to imagine, to invent, to foresee … I believe it is appropriate here to recall a poem that I wrote in September 1981 in Teotihuacan:

I know nothing,
other than that I am,
that I am here,
astonished.
Other than that I see, think and tremble.

I am nothing, other than that I know,
perplexed and confused,
that when man was born
—conscience of the earth,
and the sea,
and the wind and the clouds—
the universe became the dwelling place of light,
of creators.

And I know that I can rebel,
and that I can do this or that,
that I can be a source of light or darkness,
that I can rise or fall …

Our conscience allows us to progressively discover what we are like, what we are composed of, and how that is reflected in physiopathology. Researching, researching incessantly to uncover the realities of all living beings. And of the planet we inhabit. And of the whole universe. Contemplating and reflecting on the cosmos, on the galaxies, on their colossal dimensions … and doing so as well with the smallest particles, the elements that comprise the minimum parts of matter. From the most distant stars to quarks, our curiosity, our desire for greater knowledge, knows no bounds. And all of this enables us to discover *what we are like* and *where we are*, but it adds little to *who we are*.

From the onset, we must ensure that science and conscience interact, each in its own realm, so that we always act upon our own reflections, that is, following what our acquired knowledge and conscience dictate at any given moment. Conscience is the ability of the human spirit to recognize itself, in its essence and in its changes … It is the sum of present or past experiences that enable men to perceive an image of their physical and moral personality.

Science and conscience—in order to firmly and boldly rectify present tendencies—form the basis for a different future of equal human dignity, providing all people access to the means for freely developing their individual abilities.

Each human being is unique and able to create. This is the great hope for humanity. Until recently this was the only glimmer in the trajectory of humankind dominated by absolute masculine power, in which people were invisible, anonymous, silent, fearful, and submissive ... But now, for the first time in history, mankind has a global conscience, is able to contemplate Planet Earth in all of its dimensions as a whole, and has realized that the future has yet to be invented.

Human beings, the 'eyes of the universe.'

We live in an increasingly more complex world. It is a world that confronts us with uncertainty, forces us to recognize that we never fully understand, and thus compelling us to 'seek to understand' an integral element of our lifelong pursuit. Authors of this book stress the importance of 'Building the Scientific Mind' as a vital condition for understanding our place in that complex world and in the universe and for making appropriate choices as to how we interact with all that surrounds us. The quality of our own earthly habitat depends on such choices. Reimagining our way of being in the world has become crucial for our own survival as a species and for the survival of countless other species. We, therefore, want the scientific mind to be a defining attribute of humans of the 21st century and beyond. This requires an essential education process that enables to fully exercise the distinctive capacities of the human species.

In the early 1990s, I commissioned, in my role of Director-General of UNESCO, European Community President Jacques Delors to produce a Report on Education for the Twenty-first Century.[4] The Report was drafted by a committee of prominent scholars, versed in diverse disciplines at different levels, including as well, among others, educators, sociologists, and philosophers. The Report defined four principal pillars of the educational process: learning to know; learning to do; learning to be; and learning to live together. Among the four, I would like to underscore the principle of 'learning to be.' A century ago Francisco Giner de los Ríos stated that education is the capacity "to sensibly manage one's own life."[5] Yes, learning to use the distinctive and infinite powers of the human species: thought, imagination, foresight, creativity! To the Delors Committee's pillars, I added 'learn to undertake' since—and I have mentioned this many times before—I remember that after a long stay in the Biochemistry Department at Oxford University, whose county coat of arms reads "Sapere aude!" (dare to know!), when I returned to Spain I thought that *while daring to know, one must also know how to dare*, since risk without knowledge is dangerous, and knowledge without risk is useless. Learn to dare, learn to undertake, to innovate, remembering those mountaineers who once said, "we did it because we didn't know it was impossible." That is how real change occurs.

It is time to reflect and to act accordingly. Education must be for all and throughout life. It must be permanently mindful of the human and our earthly environment.

Jan Visser, in the *International Handbook of Lifelong Learning* (2001), stated that "human learning is the disposition of human beings, and of the social entities to which they pertain, to *engage in continuous dialogue with the human, social, biological and physical environment* so as to generate intelligent behavior to interact constructively with change" (Visser, 2001). We produce change and, at the same time, we interact with such change as caused by us and others.

2.2 *In-Depth Knowledge of Reality*

Science is the compendium of knowledge about ourselves and nature as a whole, acquired in different stages that are progressively larger in number and size. When analyzing data gleaned from research on any aspect, researchers must exclude interpretations that are biased or conditioned by beliefs, ideology, etc. Thus, scientific rigor is a required and obligatory element of any description, research or working hypothesis that may be formulated.

New knowledge comes from observation and the capacity to progressively delve deeper into the subject of our research, stripping away all of the 'packaging' that often conceals unknown aspects of nature.

The first quality that a researcher must possess is curiosity. The Spanish philosopher Julián Marías used to say that it is difficult to examine in detail the things that we see every day. For that reason, scientific experimentation, based on hypotheses derived from other knowledge, requires reflection and contemplation from different angles and perspectives. I recall Professor Hans A. Krebs's observation that "research is seeing what everyone else sees while thinking what no one else ever thought."[6] In effect, often it's not a matter of having instruments that enable us to gather large quantities of data. It's a matter of meditating upon and 'mulling over' the data already obtained that enables us to perhaps make a new contribution.

It is often said that we need to urgently support 'applied research,' that is, promote the training of many scientists capable of using the knowledge acquired. And that is all very fine if one has a broad spectrum of knowledge that, under given circumstances, actually has a specific application. I will never forget my visit to the laboratory of Argentine Nobel Laureate Bernardo Houssay with my mentor Professor Angel Santos Ruiz, when I was an instructor in biochemistry. A sign over the entry to Professor Houssay's laboratory read: "There can be no applied sciences if there is no science to apply,"

It is a mistake to believe that the majority of scientists seek to profit from the knowledge that they or other scientists have contributed. No: the immense majority of scientists are simply fascinated by the mystery of existence, by the possibility of learning in greater detail how living beings function, of taking the study of the universe one more step forward and to be able to ensure that their efforts result in preventing, or at least, lessening human suffering.

"Knowledge for foresight, foresight for prevention" is the adage that the scientific community must constantly keep in mind. Foresight, the capacity to avoid points of no return in potentially irreversible processes, is undoubtedly one of the fundamental abilities that scientists must continually exercise and are in ever greater need to develop.

It is obvious that parliamentarians should not address and propose solutions to matters of a scientific nature without the advice that only scientists can provide (for example, transgenics, diseases derived from pandemics, risks from different types of energy sources, recapturing of carbon dioxide), in order to ensure that political leaders adopt the best decisions for individual, collective and environmental well-being.

3 Invent the Future

3.1 *Experience, Innovation, Prospective*

The past must be reflected accurately and with scientific rigor in each area of knowledge, to encourage those achievements that are beneficial to humanity and to hinder those that are detrimental. Scientists work tirelessly to gain additional knowledge, so that their findings, conclusions and diagnostics may become more precise and timely action may ensue so as to attain optimally beneficial effects. This means continuing practices that were already found to work in the past, while studying new means for resolving problems, particularly pathologies, for which no adequate approach has yet been found.

Looking at the progress made in this area—so well exemplified by our advancing longevity—we realize that the scientific community has made excellent advances, from vaccinations to the most modern antitumoral treatments. The same may be said of diagnostic systems, from the most elemental auscultation to modern systems of physical introspection that enable us to observe surgical procedures live. And, thinking of other areas in which progress has been made, the same may likewise be said of the incredible changes in our knowledge of aerodynamics and flight, of ultra-rapid transport, and of communication.

Additional years have been added to our lives and additional life to our years. But the new neurodegenerative diseases have now become an extraordinary challenge, requiring scientists' priority attention and care, calling for a greater effort, increased financial resources, and additional researchers.

The future has yet to be written. Nothing is unchangeable. The reins of the future must be placed individually and collectively in the hands of human beings who are capable of tracing the roads to tomorrow. This is especially relevant at this time, given the foreseeable re-founding of the United Nations, reshaping it to provide competent world governance, following the disastrous attempts at dominance on the part of groups of wealthy countries, such as the G6, G7, G8, …, and G20. These attempts have led not only to serious ethical, democratic and economic crises but have also concealed growing social inequalities and environmental damage, some of which is hardly reversible, if at all. It is now urgent to assume the responsibility that each generation must embrace to ensure that our planet is really habitable. Inventing improved human capacity to anticipate will be necessary to save the succeeding generations from inhuman living conditions.

Universities play an important role in Building the Scientific Mind, but in doing so they must be socially responsible, providing intellectual leadership to overcome inertia, facilitating evolution and avoiding revolution. They must devote themselves to "the Peoples", as the Preamble of the United Nations Charter so wisely insists. It doesn't mention states or governments, but rather the "Peoples," the Peoples of the United Nations who have decided to "save succeeding generations from the scourge of war."[7] This means that they assume that supreme commitment—so often evoked by President Nelson Mandela—to the future generations. *It is the duty of each generation to take the next one into account*, and, for once and for all, *to enable peace to prevail over war, and words to prevail over force*. From the beginning of time, the powers that be have always followed that perverse adage: 'If you want peace, prepare for war.' And thus, at the end of World War I the Republican Party prevented the United States from joining the League of Nations that President Wilson had created to implement his 1918 "Covenants of Peace" program.[8] One sees similar forces at work exactly a century later.

President Roosevelt's magnificent plan at the end of World War II placed "the Peoples" at the forefront of history. And the Constitution of UNESCO,[9] as the United Nations intellectual institution, proclaimed that educated people must be *free and responsible*, that they must be guided by the *democratic principles of justice, equality and intellectual and moral solidarity*. But all these, as well as the *free flow of ideas by word and image* were all too soon ignored and forgotten by the great powers that invariably had security as their supreme concern.

3.2 *Priorities*

The priorities that scientists must urgently address in close cooperation at the global level include:

- *Food*: Food production using agriculture, aquaculture and biotechnology; preservation and suitable distribution of food, promoting as far as possible the local raising of crops and livestock.
- *Water*: Appropriate use and management of water resources, adequate agriculture techniques, water production through desalination, etc. are essential aspects of ensuring the welfare of all of the earth's inhabitants.
- *Health*: This will undoubtedly become the most important field of scientific research in the next few years, given its increasing 'personification.' Much progress has been made but, given that each life is a wonder that must be nurtured with the utmost care, in-depth studies are required in such areas as genetics, epigenetics, autoimmune symptomatology, neurological deterioration due to age. All of these are areas that deserve special attention. Prevention is undoubtedly the 'top priority.' However, it is most difficult to find popular support for these areas since they are 'invisible.'
- *Environment*: For the first time the human species is living in an age in which human activity has a global impact. It is called the 'Anthropocene' in reference to the fact that today human beings, given the fantastic development of their creative imagination, are able to modify parameters that not long ago were beyond their reach. It is in this area where specialists on energy sources, recapturing carbon dioxide and other greenhouse gases, forecasting the effects of the partial melting of polar icecaps, particularly in the Arctic, should focus their research to adopt measures capable of at least containing or lessening the present rate of environmental deterioration.
- This is a requirement of intergenerational solidarity. A rapid and coordinated action to reduce the impact of natural disasters (wind, water, fire, ...) is another of the measures that citizens of the world demand, alarmed by the immense amounts of money devoted to military spending, while the aid needed to rehabilitate areas devastated by earthquakes or tsunamis, and to return the victims to normalcy, is always too little and comes too late.
- *Education*: As is the case with health, water, and food, education is a social component of a life of dignity—to which we all have a right—and there should thus be no limitations to access to education at any age, to learning and studying to provide all human beings with an education, that is, to

enable them to act upon their own reflections and not under the influence of ideological or religious dogmas or the dictates of others.
- *Peace*: In order to transition from a secular culture of war, imposition and violence to a culture of peace, the perverse adage "if you want peace, prepare for war," must be replaced by "if you want peace, work to build it each day in your daily lives." Here again, to achieve a re-founded United Nations and competent world governance, scientists must endeavor to contribute to the development of mechanisms capable of rapidly resolving inevitable conflicts as well as addressing global threats.

In fact, it is not a matter of simply preserving human life, whatever its quality. Rather, we must ensure that all human beings can live a *dignified life*. At the time of writing, an Oxfam report has just come out asserting that 82% of the wealth created in 2017 went to the richest 1% of the world population, while the poorest half of humanity got nothing.[10] Such insanely unequal wealth distribution is found year after year. It amounts to a *daily genocide of hunger and neglect*. This can't continue. Such intolerable inequalities in the distribution of all types of resources just can't exist. And it doesn't require sophisticated scientific thinking to come to that conclusion! Instead, it requires good thinking—a solidly formed scientific mindset—to start inventing a future without such abhorrent injustices.

Now that the voice of the people can be heard, we must ensure that everyone understands that *implementing human rights benefits all of us equally*. Everyone must understand that accumulating immense fortunes adds nothing to an already comfortable life. I like to repeat a simple observation that I read one day in a small chapel in the south of France: "Les linceuls n'ont pas de poches" ("funeral shrouds have no pockets"). Whether we are born to wealth or poverty, death is the great equalizer. Thus, it is essential that the *scientific, academic, intellectual, and artistic communities become the true protagonists of this change* and achieve now what in 1945 was impossible: cooperation, working together, adequately sharing in a context of democratic principles that subject other dimensions of coexistence to social justice and the conditions inherent in a life of dignity for all.

In view of the foregoing, these are the great objectives that higher education should pursue, using all of its influence and capacity for mobilization, both in the classroom and in cyberspace:

1. A re-founding of the United Nations System.
2. Nuclear disarmament.
3. Strengthening genuine democracy.
4. Coordination at the global level of associations of higher education institutions.

The Earth Charter,[11] one of the most lucid documents from the last decade, states:

> We stand at a critical moment in the Earth's history, *a time when humanity must choose its future*. As the world becomes increasingly interdependent and fragile, the future at once holds great peril and great promise. To move forward *we must recognize that in the midst of a magnificent diversity of cultures and life forms we are one human family and one Earth community with a common destiny. We must join together to bring forth a sustainable global society founded on respect for nature, universal human rights, economic justice, and a culture of peace* ... We must realize that when basic needs have been met, *human development is primarily about being more, not having more*. We have the knowledge and technology to provide for all and to reduce our impact on the environment. *The emergence of a global civil society is creating new opportunities to build a democratic and humane world.*

Today, for the reasons previously set forth, there is no doubt that universities, united to reach these great objectives, can have an extraordinary impact on the great changes that could *make the 21st century the century of the peoples, the century of a life of dignity for all.*

4 A 'New Beginning'

4.1 *We, the Peoples*

Until recently the world's inhabitants were confined both territorially and intellectually to extremely limited spaces. They knew very little of what transpired 100 or 200 kilometers away and, consequently, closely obeyed the standards of behavior established by their local 'lords.' From the beginning of time, men exercised absolute power in a world in which women only fleetingly appeared on the stages of power.

The immense majority of humans were undoubtedly aware that their possibilities for action were extremely limited. And they had to continue to obey in silence because their ability to communicate with others was practically nonexistent.

It was scientists who progressively enabled rapid communication over great distances, who facilitated the transport of goods and persons, who gradually provided the means to combat the pandemics that decimated the earth's inhabitants in the past.

If today we are experiencing a fascinating moment for change in these tendencies, which from time immemorial have so limited or prevented personal development, it is likewise due to the availability of communications and information technologies that enable all of us for the first time in history to have a *global conscience*. It allows us to not only know how the inhabitants of distant places live and die, but it also makes us aware of the needs of others, so that we can better appreciate the things we have. A global conscience not only enables the scientific community to react to the most urgent needs of others, it also encourages us all—we, the peoples—to decrease social inequalities and the most critical risks for the environment while, thanks to the possibility to participate in this process from wherever we are, it is now possible to work toward rebuilding and consolidating real democracy, both at the local and global levels (United Nations).

It is this global conscience that awakens a sense of solidarity and compassion, which clearly shows us the path that science should take down the road to tomorrow.

At last we see science and conscience, going hand in hand, able to ensure the sustainability of the planet. At last, science and conscience can unite to prevent the use of scientific progress for the benefit of a privileged few who have subjected the rest of the world's citizens to their will. At last, science and conscience can forge the transition from a culture of domination, imposition, and violence to a culture of dialogue, conciliation and peace. That great historical turning point from force to words will be accompanied by the inseparable pairing of science and conscience.

No longer the reason of force, but the force of reason, with every human being able to create, to develop science and art. Our hope. Science and art inventing the future together; humanity holding the reins of our common destiny in its hands.

Notes

1 In French, the quote reads: "L'humanité est confrontée à de nouveaux périls, sans équivalents dans l'Histoire, et qui exigent des solutions globales inédites."

2 John F. Kennedy's Commencement Address at American University in Washington on June 10, 1963 (see http://www.presidency.ucsb.edu/ws/?pid=9266).

3 From the poem "Voy huyendo de mi voz." In Spanish the line reads: "me encuentro huyendo de mí / cuando conmigo me encuentro."

4 The report is available at http://www.unesco.org/education/pdf/15_62.pdf

5 "Gobernar con sustantividad su propia vida" as cited in Junta de Andalucía (n.d.). *Francisco Giner de los Ríos*, p. 6. Retrieved from http://www.juntadeandalucia.es/educacion/vscripts/wginer/w/rec/3275.pdf

6 Personal communication. Sir Hans Adolf Krebs was a German-born British physician and biochemist. See also https://en.wikipedia.org/wiki/Hans_Adolf_Krebs

7 See http://www.un.org/en/sections/un-charter/preamble/index.html

8 See http://avalon.law.yale.edu/20th_century/wilson14.asp

9 See http://portal.unesco.org/en/ev.php-URL_ID=15244&URL_DO=DO_TOPIC&URL_SECTION=201.html

10 Oxfam press release 2018-01-22, retrieved from https://www.oxfam.org/en/pressroom/pressreleases/2018-01-22/richest-1-percent-bagged-82-percent-wealth-created-last-year

11 See http://earthcharter.org/virtual-library2/the-earth-charter-text/

References

Maalouf, A. (2009). *Le dérèglement du monde: Quand nos civilisations s'épuisent*. Paris, France: Éditions Grasset.

Visser, J. (2001). Integrity, completeness and comprehensiveness of the learning environment: Meeting the basic learning needs of all throughout life. In D. N. Aspin, J. D. Chapman, M. J. Hatton, & Y. Sawano (Eds.), *International handbook of lifelong learning* (Vol. 2, pp. 447–472). Dordrecht: Kluwer Academic Publishers.

INTERMEZZO 14

Culture of Differences vs. Difference of Cultures

Jan Visser

Reflections inspired by Ralf Syring: 'Seeking to find why things happen: Variations on a theme of Diallo Sampa's grandfather'

Some pieces of music occupy an almost indelible place in one's long-term memory. Why that is so may depend on both personality and circumstance. Meeting persons who are similarly impassioned by the same instantiation of human creativity establishes a bond between those who are passionate about it. For instance, I love Brahms, Not everyone does. Brahms's Deutsches Requiem cuts deep into my soul. So does his first piano concerto and his sextet for strings Opus 18, to name but a few of the works that console me and open my heart to their indescribable beauty. I feel close to those who have similar feelings.

At age 16 I was hospitalized; a first for me. Symptoms that didn't go away of their own accord accompanied the feelings of generalized discomfort and uncertainty I had at that stage of my adolescence. The symptoms worried my parents and the various medical professionals they consulted. At the advice of the latter I was admitted to the same hospital where I was born and kept under observation for six weeks while I was subjected to all kinds of examinations, some of them painful, until I was finally operated upon. An abnormality was found, and removed, that had nothing to do with the symptoms in question. Those subsided over time and were gone after another year.

Interestingly, those six weeks in hospital coincided with the Holland Festival, a yearly event of the performing arts that takes place in Amsterdam every year in June. It was founded in 1947. All concerts were broadcast on national radio. I listened to virtually all of them. One work stood out that year. It had apparently been chosen by multiple performing ensembles to be part of their program. It was Brahms's 'Variations on a theme by Haydn.' As sometimes happens with something inherently beautiful, I didn't initially appreciate its beauty but after six weeks I was hooked and have remained so since.

This particular composition makes me feel close to Ralf Syring, the author of the chapter on 'Seeking to find out why things happen: Variations on a theme of Diallo Sampa's grandfather.' Ralf wrote the chapter following the same structure as Brahms's variations, a work that touched us both. He and I were born

 | DOI: 10.1163/9789004416802_028

in countries that had been at war. One had been the occupier of the other and destroyed the city where I spent my childhood. We are both aware of our conflicting historical backgrounds. Such awareness contributes to the closeness we feel as friends and colleagues. We share the experience of having lived and worked for a considerable part of our life in parts of the world where our native cultures contrasted with those of the people who were locally born. It's an experience that helps one overcome preoccupation with the 'difference of cultures' as one embraces the 'culture of differences' to which we all pertain.

CHAPTER 15

Seeking to Find out Why Things Happen: Variations on a Theme of Diallo Sampa's Grandfather

Ralf Syring

Theme, 8 variations, finale

1. *Theme. Corale riguardo le approssimazioni alle spiegazioni.* Andante
2. *Variation I.* Poco più animato
3. *Variation II.* Più vivace
4. *Variation III.* Con moto
5. *Variation IV.* Andante con moto
6. *Variation V.* Vivace
7. *Variation VI.* Vivace
8. *Variation VII.* Grazioso
9. *Variation VIII.* Presto non troppo
10. *Finale.* Andante

1 Theme: Chorale riguardo le approssimazioni alle spiegazioni – Andante

This story takes us to the village of Prábis in Guinea-Bissau. Renate Welsh, an Austrian author of children's books, wrote it down and published it in German in 1979. Here is my translation of part of it:

Diallo Sampa was sitting next to his grandfather. In front of them sat the teacher. The teacher was young, only a couple of years older than Diallo. He himself still went to school—in the capital city.

Two times a week he came to the village in the early evening. He came in a rickety old car which then continued the trip and took his colleagues to more remote villages.

They learned to read.

The grandfather was reading aloud. Everybody was listening. He read with big effort, word by word. While reading he swayed back and forth on his heels, and he held the book very tightly. He frowned. Wrinkles seemed to creep up and down, to and fro. He read slowly, but he didn't make any mistake.

 | DOI: 10.1163/9789004416802_029

When grandfather was ready, Diallo breathed out long. Grandfather was one of the oldest in the village. He had worked on the farm, when it was still owned by an agent of the Portuguese secret police. He had fought against the Portuguese. He had led the men of the village when they took weapons to the Southern Front of the liberation army. Grandfather did not tell much about that time, but other men from the village had told how grandfather alone had defended a store house for sugar cane, in which they had hidden the wounded.

And now he learned to read. In the beginning it gave him more trouble than Diallo. But now he had read an entire page without a single mistake.

Everybody looked at grandfather.

"How do you feel now?" asked the young teacher, "now that you are so good at reading?"

Grandfather swayed more intensely on his heels. Diallo moved a bit closer to him. Grandfather cleared his throat—once, twice. Then he passed the book to Diallo and showed his hands. He looked at his palms and held them in a way that the others could see them. "This scar," he said, "is from the first hut I built. And this one here is from the big fire, when my first son was born. And this one is from the first bazooka which I set off. That was when I did not yet know how to handle it. Every scar says what happened. It does not say why. The books say why."

2 Variation I: Poco meno animato

"Every scar says what happened. It does not say why. The books say why." The grandfather impresses us. He seems to be a wise old man, telling us what the meaning of learning to read is for him. We respect and admire him for that, and it is good to do so. However, is he actually right?

There are many different books. Those of us who have read a lot of books in their lives—and those who read this book, probably have—know that there are 'good books' and 'bad books.' If the grandfather is right in his belief that "the books say why," could some books then mislead him, suggesting wrong explanations to him?

Let us imagine, how the story might continue:

Some months later the grandfather went to the market. He did not go there very often. It was the task of women to go there and sell what they could spare of the food they produced in the village. The grandfather wanted to buy new batteries for the little radio, which some time ago a person who had come from the near-by capital city Bissau, had given to him. While he was looking for the batteries, he found a small book in a garbage bin, not in good shape, but still good enough to read. He took it and it made him feel sad that people were throwing books away. Back home he started to read—slowly, but he managed.

"When we arrived at the center of Biombo region," grandfather read—Biombo is the region in which Prábis lies—"we tried to set fire to the storehouses where the village people kept food, mainly maize, with which they supported the terrorists. We were attacked by some of them and could not get close enough. In one area we managed to get hold of one of these black bastards. I hit him with my knife on one of his arms, but in the turmoil he managed to escape. May the big scar remind him of his crime during his entire life."

Grandfather did not understand all the words. However, he understood the sentence about the scar. And he knew what a crime was. He looked at the scar which remembered him of the first bazooka. It was in the time of war against the Portuguese army. No, he did not feel that he had committed a crime fighting them. And the idea made him feel sad.

The book said why, it attributed a meaning to the scar, but it was a different meaning than the one grandfather had in mind. Is one of the meanings correct and the other wrong? Is the book the grandfather had found in the garbage bin, a bad book?

3 Variation II: Piú vivace

At the time when the grandfather was hurt, he was, of course, not a grandfather.

He was a young man, strong—he smiled while thinking of that. When his shoulder was burnt from the bazooka, he went quite a long way to meet with one of the mobile clinics, which were run by comrades of the liberation movement. He might have got treatment less far away, as there was a healer nearby, but he knew that the comrades had some problems with regard to the healers. He never understood quite well the reason for that, although they tried to explain it.

He remembered one meeting in which a medical doctor, who had studied in Portugal, told them something about medical science and about progress and the need to fight superstition. It seemed to him that the doctor was talking about the healers in the villages when he was telling them about superstition, but he had apparently not quite well understood what the healers were doing and how important they were. That was not a problem. He had been in Portugal for a long time and been at the university. So how could he still understand what was important for the people in the villages? He was now helping the comrades, and they all appreciated that. It didn't matter that much that he did not understand the health services which they always had had in the villages.

The *Encyclopaedia Britannica* defines superstition as "belief, half-belief, or practice for which there appears to be no rational substance."[1] Well, that explains one difficult term with another. In fact, the authors of the relevant entry in the Encyclopaedia Britannica mention that it is an ambiguous term, which probably cannot be used except subjectively. The doctor, who had studied in Portugal, probably tried to explain that there are many beliefs among people which, if we look further into them, turn out to be wrong, which in this case means that they suggest people to do things that may be harmful for them.

For example, in some areas of Guinea-Bissau, people believe that pregnant women should not eat eggs. Yet, it is important for pregnant women, and for the children they bear, that they feed themselves well. Eggs, which contain proteins, are thus important for them.

In my own experiences as a medical doctor I have come across situations in which 'superstition' may have played a role. Once, an old woman, who had worked for decades in the village as what now is called a 'traditional birth attendant,' called me to attend a delivery. When I arrived she took me aside and said that the baby would die during delivery, because she had heard it cry in the womb. She then went away and left me alone with the parturient in the dark. Delivery was difficult, and I was unsure about the situation and afraid. However, the baby was born and was alive, and, in the end, the mother and I were happy.

This happened in a time of war, and I was not able to find the 'traditional birth attendant' again. So, I could not ask her why she had called me while she was sure that the baby would die. I could not ask her either to explain in more detail what had happened when she had heard the fetus cry in the womb. Perhaps she was not as sure as she made it appear. Maybe she called me because of another 'superstition,' namely that the white man might have power against what she had heard? Or did it mean that she just assumed that, under the conditions in which they normally lived, the baby would die?

What is called 'superstition' is frequently an attempt to give meaning to a phenomenon that is not understood. This happens in many cultures with thunder. People are often afraid of thunder. It therefore leads them to attribute a divine force to it. After they learn that thunder is a phenomenon that follows lightning, they should no longer be afraid of thunder, but rather of lightning.

What doctors 'know' may also depend on where they were trained. The following comparison may make this clear. If the doctor who appears in the story about Diallo Sampa's grandfather had studied in what was then called West Germany instead of Portugal, he might have learned a different approach to treat a sore throat in children than what he had been taught in Portugal. Still many years after West Germany ceased to exist as a geopolitical entity; the number of tonsillectomies in Germany was nearly twice as high as in Portugal. More striking, even, are the differences found between districts in Germany, which were reportedly 12-fold for the surgical treatment of 'chronic tonsillitis' and even 58-fold for 'hyperplasia of tonsils.' (Noltin, Zich, & Deckenbach, 2013).

So, what then are the reasons for decisions on how to approach disease with the intention of healing? The grandfather in Prábis as well as the medical doctor who studied in Portugal, would like to believe that it is with health and disease—and those whose profession it is to treat people against disease and for health—as it is with scars and books. The scars or the fever or the rash or the pain say what happened, but they do not say why. The doctors and the healers say why. Do they?

4 Variation III: Con moto

Health, according to the World Health Organization,[2] is "a state of complete physical, mental and social well-being and not merely the absence of disease or infirmity."[3] That's a nice definition, no doubt, but it is unhelpful to clearly demarcate levels and areas of responsibility and authority at which people and entities intervene in the domain of health. Yes, the definition does make us aware of the fact that "there is more to health care than medical care" (Airhihenbuwa, 1995, p. 5). However, there is the risk that it turns most people into health workers and that health workers feel competent about anything that may contribute to social well-being, thus attributing an authority to them which is difficult to distinguish from that of an imam, a priest, or a rabbi. Working with this statement as a guiding principle, a 'Ministry of Health' might claim to have competence also in matters of education, agriculture, employment, energy …, in fact, in nearly everything. In reality, most ministries of health limit themselves to the role of a 'Ministry of Disease,' and that is what they deal with.

The definition by WHO can be understood as part of a recognition in 'Western' or 'modern' medicine that health depends on so much more than only medical intervention. This concept has been worked out more extensively in recent years by the Commission on Social Determinants of Health, created by the World Health Organization in 2005. The commission proposed a variety of elements that influence people's health (see Figure 15.1).

The grandfather in Prábis would look amazed at such a diagram, which he might consider a work of art from a different culture. He would not discuss social determinants of health, but if somebody would try to explain to him what it means he

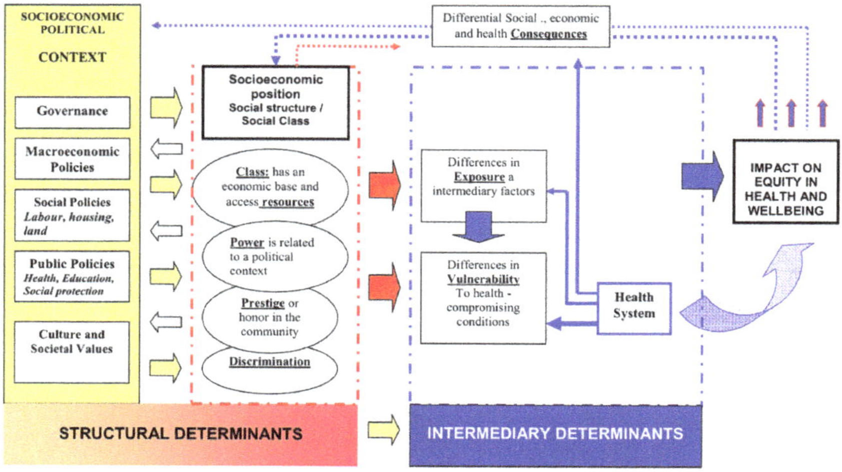

FIGURE 15.1 Social determinants of health, as proposed by the special commission of the WHO (from Commission on Social Determinants of Health, 2007)

would smile and say: So finally the people who studied at the university understood. And he would be happy with that.

The authors of the final report of the Commission on Social Determinants of Health might consider the result of their work 'innovative.' However, much of what is new in 'Western' or 'modern' medicine, has long been known and practiced in 'autochthonous medicine.' There seems to be a tendency now in 'Western' medicine to move in the direction of rediscovering and recognizing the value of long-known autochthonous concepts. Something similar happened longer ago with the idea of 'psycho-somatic' disease, which in 'Western' medicine was discovered only about a hundred years ago.[4]

However, there are also tendencies in the opposite direction, in which modern medicine moves further away from autochthonous healing concepts. A case in point is 'commercialization,' the idea that medical intervention is like replacing or adding spare parts to an object for a profit (if you do not like the shape of your nose, the doctor can fix it the way you want it to be as long as you pay what it's worth for you). The development of ever more sophisticated technological solutions to medical problems plays into this tendency.

5 Variation IV: Andante con moto

Different perspectives on 'health' are possible. For most people the idea of health is probably what WHO considers as being insufficient, i.e., "the absence of disease or infirmity." In societies in which most people are struggling for a decent livelihood, health is an issue only in case of its absence. Being concerned with the amount of cholesterol in food or any other substance is a phenomenon that arose in relatively wealthy societies because of specific imbalances that are inherent to the construction of wealth. So when most people are dealing with health, it is when they are ill or medically challenged, i.e., when they are in a weak position. But it is also possible to look at health from a positive perspective. In that case health could also be described as 'feeling well.' A problematic view of health exists when health is interpreted as the ability of a person to be productive, to contribute in some useful manner to the society, particularly its economy. Such a view is problematic because it excludes people who are not able to contribute to the society. It declares them ill by definition, although that may not be what they themselves feel.[5]

Health or disease professionals look, of course, differently at the concepts of their expertise. There are numerous types of health and disease concepts. One of the main concepts in terms of providing services to big numbers of people is the concept of what is frequently called 'traditional medicine.' However, 'tradition' is not specific for that concept. I rather call it 'autochthonous medicine,' as it is linked to cultural concepts that are geographically limited and does not claim universal validity.

Most of these concepts have three characteristics in common: Health is a state of harmony with the surrounding nature, with the surrounding society, i.e. the community people live in, and with the spirits of the ancestors. Disease[6] is the expression of some kind of disruption of that harmony. This concept is considered adequate in societies in which community or collectivity is valued higher than individuality. "In the African traditional healing context, an understanding of social conflict is pivotal to curing a person who has become ill. This is where the healer deals with the *ultimate* cause of an illness: who or what caused it, and why. The *proximate* cause, or specific etiology, tells only how it happened, not why it happened. For people who believe in this concept of disease nosology, knowledge of germ theory is useful only for explaining the medium through which health problems manifest themselves" (Airhihenbuwa, 1995, p. 51).

There are within this general framework multiple specificities and specializations. The concept is linked to its own particular way of training its professionals. Knowledge acquisition occurs by apprenticeship in the form of experiential learning, not by didactic transfer of knowledge following a "donor-deficit-model" (Ibid., p. 57).

The latter is the main approach in training medical doctors in a 'Western' cultural context.[7] Knowledge is being transferred from somebody who knows to somebody who does not know. The contents of teaching claim universal validity. They can do so as long as they limit themselves to biological and physiological processes which are the same for all humans. (I hesitate writing this sentence. While I find it necessary to stress that there is no biological or physiological difference between human beings, knowing that the assertion of such differences has supported theories of superiority and inferiority among human beings, I also know that we know about biological and physiological processes only by describing them, and in doing so, cultural differences come into the picture.) However, taking the WHO definition of health seriously, universal validity quickly transforms itself into universal hegemony. 'Well-being' is not a category which is felt in the same way by human beings across the board.

In 'Western' medicine more and more concepts have been developed that go beyond biological and physiological processes: psychosomatic diseases, behavioral models, salutogenesis, mental health concepts, etc. In these areas 'Western' medicine needs to understand that it is culturally determined like any other concept which deals with health as well-being.

We may assume that the grandfather in Prábis felt well when he was able to read. Yet his main way of getting to know something continues to be listening. He would feel intimidated if he would be invited to a library. He liked the young teacher who came twice a week to his village, but it also happened that he wondered why these young people were so hectic. And they forgot—oh yes, they quickly forgot what they were told. It was as if words only were important to them if they were written. And that made the grandfather sometimes wonder what would happen to all the memories in the village, which never had been written down anywhere. Sometimes he thought that it might be a good idea to write all these memories down. He thought of that when a couple in the village was getting very ill, first the father and then his wife. They were still young, and their two children were small. The father had been away for two years, and people told that he went to Angola in order to find diamonds. However, when he came back, he was not rich, but ill. People were saying that he had a disease which they called AIDS, and that he would die. A nurse from the nearby hospital in Cumura had suggested that the parents should write some kind of diary to leave something to their children after their death, a 'memory book.' The grandfather thought that it might be a good idea, and for the first time it occurred to him that his lessons had not only taught him to read, but that—with some effort—he might also try to write.

6 Variation V: Vivace

Independence of Guinea-Bissau was declared on 24th September 1973. It was a big event in Madina do Boé. The grandfather had not been able to go there, because at that time, Portuguese military still controlled Bissau, the capital city, and Bafatá, the second city of the country. Portugal continued to consider the country as their 'province' at that time, although they did no longer have access to the biggest part of the country. Sometimes today, when the grandfather is sitting under the big tree together with other men of his village, they remember that time and they tell each other their stories. These stories are very different. Some tell how proud they were that they had been able to win the war, to force a big and technically well-equipped army out of their country. Others tell about the suffering during the war. You cannot forget if you have seen a comrade and friend die just beside you. There are so many invisible scars. And they are not only different from the visible ones because of visibility, but because they continue to hurt—some of them even after a long time.

The old men under the tree also remember the prisoners of war of that time, these young Portuguese boys, who were so terribly afraid. The grandfather always thought that they suffered more than his own comrades, because they were forced to fight this war without having any other reason for it. During the time of General Antonio de Spínola as a governor of the country, those boys, when arriving from Portugal, were forced to go out of the barracks and bring back an African man—alive or dead.[8] *They were ordered to commit crimes, and many of them did. It produced so many scars on all sides ...*

When not long after these events and in a different part of the world soldiers of the army of the United States of America came back to their country after they had lost the 'American War' in Vietnam, many of them were found to be no longer able to 'function' well in their society. Psychiatrists were not sufficiently prepared to deal with this. A 'post-Vietnam syndrome' was diagnosed[9] and its symptoms were added, in 1980, to the third edition of the Diagnostical and Statistical Manual of Mental Disorders (DSM III) of the American Psychiatric Association as 'Post-Traumatic Stress Disorder' or PTSD.

The concept was offered to the entire world. It became part of the 'Western' idea of medicine as science and therefore universally correct and valid. However that did not work. In 1995, a group of psychologists and social workers in Luanda tried to support demobilized child soldiers in an attempt of reintegration into their communities. They had received some funding from an organization in the USA, together with a manual which explained the PTSD-concept to them. Yet, they were not able to work with it. They were not traditional healers. They had been academically trained according to 'Western' standards, but they were working in a context which was different from what was the theoretical framework of PTSD.

Leslie Swartz (1998), South African Professor of Psychology at the University of Cape Town, wrote about the difference in interest between DSM IV–based diagnosis and African diagnosis:

> Introducing her book on healing amongst Zulu people, Ngubane (...) notes the distinction between *umkhuhlane* and *ukufa kwabantu*. The former means illness that 'just happens' (...), such as smallpox or influenza. These illnesses are considered universal and able to be treated biomedically, in a non-ritualized way. *Ukufa kwabantu* ('diseases of the African people'), on the other hand, are specifically African ... The interrelationship of this difference in diagnostic practice is profound. First, there is considerable interest in local variation in phenomena, and it is regarded as a diagnostic achievement rather than a diagnostic failure that a phenomenon like *amafufunyana* (a form of spirit possession or hysteria occurring amongst Zulu and Xhosa-speakers ...), for example, may have different diagnostic features in different contexts. Second, where Western psychiatry tries to incorporate new phenomena into existing categories, and is cautious about developing new categories, there is not necessarily the same caution in African systems .... The heavy reliance on contextual information in making sense of these phenomena—a hermeneutic approach—contrasts with the attempt to make the DSM-IV as context-free as possible. (p. 66)

Psychologists of the Instituto Latinoamericano de Salud Mental y Derechos Humanos (Latin American Institute of Mental Health and Human Rights – ILAS) found in Guatemala's Petén province after the war in that country that young people were talking to them about their experiences in the first person singular: "I have seen ...." It took them some time to understand that this was a symptom of a severe traumatic process, because talking like this was unusual in these communities, where people would speak in the first person plural: "We ..." (see also Becker et al., 1994).

The point here is not that 'Western' healing methods should be replaced by autochthonous methods in Africa or Latin America. Both may have their strengths and weaknesses in a healing process. However, the inception of the healing process needs an adequate approach, and that approach needs to address the specific context.

Another conceptual problem has been highlighted in the PTSD-concept. The term itself considers 'trauma' as an event and the resulting illness as something that occurs after the event ('post'). It has, however, been shown that it is rather a process, which is continuing, does not have an end.

Protagonists of 'Western medicine' might say that this may be true for mental health issues, but certainly not for somatic diseases which are the same in any human being. This distinction, however, is a specific culturally determined statement. From a different perspective it can be argued that it is always the entire person we are talking about, if we consider health or disease. So it cannot be expected that cultural and social determinants would not apply in what we call somatic diseases.

7 Variation VI: Vivace

The grandfather learned to read, because he was aware that he knew what happened, but he did not know why it happened. He expected the books to tell him why. Later he learned that not all books told him why. So he needed to be careful when reading.

It is not only his scars and the things that happened in the past that make him wonder why things happened and happen. When the men are sitting under the tree to talk, one of the most important subjects is health: What to do to make ill people healthy again? And, of course, in order to do that you need to know why they are ill. The grandfather knows now that there are also books which tell why a person is ill. However, as with the other books, the difficulty is to know which of the books tell what the reason really is and which would rather mislead you.

In the village they are lucky with respect to that problem. They do not need the books, because they have people among themselves who know. There is also a health center which is run by the government's ministry of health and has recently been extended, and in one of the nearby 'tabancas' (that is how they call the small villages) there is even a hospital, run by Christian people, but that is mainly for people with leprosy. Yet they are also treating others in an attached health center.

So there are quite some possibilities to get healthy. The Prábis health center, however, does not have a real medical doctor. There are nurses. The problem with them is that they examine you and then give you a sheet of paper with some medicines written on it, which you are supposed to buy and to take. However, they do not tell you why you feel ill, and that is in fact what people in the village are looking for. So they prefer to ask the village healers.

'Western' medical concepts do give reasons why a person is ill. Trained doctors within that concept do not always do that. Quite often they do not say anything and only hand out a prescription, a list of drugs the patient then is expected to buy and consume. However, if they tell something it is in many cases about bacteria or about a virus. I remember how I tried to convince people in a village in Eastern Angola's Moxico Province to accept that their children be vaccinated against poliomyelitis. We were explaining about the virus and the ways of transmission. Finally mothers came with their children and we vaccinated them. Later, when I understood better what people did in that village, I learned that they had come because they trusted us. We were nice people and so it might well be an advantage to do what we suggested. However, when we explained about the virus, we thought that we answered their questions. We assumed that they would want to know what we were explaining, but we answered questions that nobody had asked and we did in no way respond to the real questions,[10] which were about more profound reasons for why a person would be ill and another one not.

An article in the Guinean newspaper *Nô Pintcha* of July 7, 2017, about the situation in Prábis, mentions that "the utilization of the hospital by the community has been weak, because many are animists and rely more on traditional treatments by local healers and sorcerers."[11] As a solution to that 'problem,' readers are told that the population is being sensitized.[12] This is the most frequent reaction to people's inclination to consult the healers in their villages. They are considered as not being sufficiently sensitive with regard to their own health and therefore not doing the right thing, which is to go to a health center or a hospital.

In many projects which are then planned for this type of 'target population,'[13] the concept of 'behavior change communication' is applied. Much of that communication can be summarized by saying: The people are behaving wrongly, and that is why they have problems. They need to be told the correct behavior, and then the problem will be solved. In much of the literature on 'behavior change,' the word 'culture' only appears by way of an adjective to the word 'barrier.' This has changed more recently and 'behavior change communication' has been renamed in some contexts 'social and behavior change communication.'[14,15] Consequently, those who go to the villages to 'sensitize' their 'targets,' should take time to get sensitized themselves about the context. The concept of the population as a 'target' does not leave any space to first create a relationship in which mutual learning might be possible, which then could probably open different and maybe more effective ways of communication. In order to solve problems, the 'Western' or 'modern' approaches needed, and need, much more attention to the complexity of context factors—people behave in a way that makes sense to them with regard to certain conditions—whereas 'traditional' concepts often underestimate the importance of the individual person.

During the demobilization of child soldiers in Angola, for example, young people were accepted back in their communities after the war on the condition that they passed through a purification ritual. Many of them had been forced to attack their own village. However, after the ritual their relationship with the community was considered as equal to any other community member. That did improve the individual situation of the young people. However, many of them experienced nevertheless results of their traumatic experiences during the war, often in the form of what from a 'Western' medical point of view would be called psycho-somatic diseases. In the community, these could no longer be related to the experiences of the war, as the purification ritual was considered to have wiped out these experiences from the memory of the young individuals as well as that of the community.

Behavior change communication has been and continues to be a main approach to people living with HIV and in HIV prevention.

7.1 *Aspects of the History of HIV-Related Interventions in Sub-Saharan Africa*

After more than 25 years of work to address the HIV and AIDS challenge in Sub-Saharan Africa in the areas of prevention and care, we still face a serious situation with regard to the impact of HIV in African societies. There is no doubt that there is considerable success in treatment. Where available, antiretroviral drugs have transformed infection with HIV into a mode of a chronic disease. Availability, however, is still limited to only a small proportion of infected people. At the same time, new infections continue to contribute to an increasing number of people who will need treatment in the future. It looks as if prevention efforts have not had a significant impact in most of the societies, if compared to the employed resources.

This raises a question about what has been missing in the attempts to contain HIV. One of the weaknesses that can be identified in the approaches to prevention is the basic idea that sexual practices—which represent the dominant way of transmission in Sub-Saharan Africa—follow universal concepts and that they can be influenced everywhere by the same messages. So messages have been created which aim at behavior change without being much concerned with the specific determinants of existing practice.

Even the publication of a *Communications Framework for HIV/AIDS* by UNAIDS in 1999 with the subtitle *A New Direction* has not significantly influenced the dominant approach. This is particularly striking, because the "new direction" contains a chapter on "contextual domains," which are divided into five paragraphs: government policy, socioeconomic status, culture, gender relations, and spirituality (UNAIDS, 1999).[16]

7.2 *Areas Which Are Touched If We Address HIV and AIDS*

HIV is particularly difficult to deal with as it is connected to several sensitive areas of life at the same time: sexuality, death, and blood or body fluids. These are areas for which all cultures have specific rigorous rules, and often special rituals. Those diseases and occurrences which are linked to death, are considered socially dangerous, and require the respect of diverse rules which are profoundly anchored in African societies, rural as well as urban. These practices and rules, and more so their violation, may have influence on the proliferation of HIV insofar as they are performed according to paradigms and concepts of social contagion which differ from the biologic contagion and so have an effect on how people view HIV and behave towards it.

Even in the so-called Western cultures, where the commercialization of sexuality may create the impression that its special character is no longer acknowledged, rituals related to marriage and death continue to be important. The number of people in Sub-Saharan Africa who take part in monotheistic religious practices may lead others to assume that these people do no longer live according to the paradigms that are connected to the respect towards the ancestors. However, reality is often different. The monotheistic religions—whether Islam, Christianity, or Judaism—are practiced in Africa often in a way that is complementary to their traditional beliefs and do not eliminate people's own culture, which is profoundly anchored socially and emotionally. Looking more closely, it becomes clear that the rituals around sexuality, death and blood continue to be practiced even in countries that are up to 95% Christianized (like Rwanda) or Islamic (like francophone West-African countries).

A subject which is related to sexuality has necessarily as well to do with reproduction and fertility and is linked to the relationships between men and women (which does not mean that it is only between men and women that sexuality is practiced) and the differences between them. This refers to the biological, and most importantly to the social, differences, i.e., gender relations. These are socially and culturally constructed, and we need to know and understand them very well, if we want to address communities about HIV.

The impact of HIV, and particularly of AIDS, causes also economical and juridical problems, the last ones especially with regard to heritage rules.

As HIV infection is related to so many sensitive problems and as it meant that infected people were doomed to death when it first appeared, mechanisms of social protection have been pursued, of which exclusion and stigmatization of infected people are part. This situation has created a need for care beyond mere medical treatment. Infected people also need spiritual and psycho-social care, areas that only recently have been recognized by 'Western' medicine and in which autochthonous medicine often has a special strength.

In light of this situation it seems naïve to believe that precisely in these highly sensitive areas of concern messages that claim universal validity might contribute to slowdown the spread of HIV, or that interventions, in which no relation with the specific social and cultural context of each society is established, can be effective. Statistics, which continue to be published, suggest a different reality. If they are true, if new infections with HIV—as UNAIDS wants to make us believe—can be stopped by 2030 using messages and methods that claim universal validity, it still remains to be seen, at what cost that might possibly happen.

7.3 *Gender*

Among the areas which are connected to the spread of HIV, gender relations are particularly important.

The most frequent way of biological transmission of HIV is sexual intercourse with penetration of the penis into the vagina or the anus (of a man or a woman). We know that biological differences between men and women put women into a disadvantageous position compared to men as far as HIV transmission is concerned. The probability of transmission of the virus from a man to a woman during penetrative sexual intercourse between the two is much higher than that from a woman to a man during the same act. However, consistent use of protection mechanisms could make that difference insignificant. The reasons which hinder consistent use of known protection mechanisms cannot be found in biological differences, but have to be looked for in social relationships.

The inferior position of women in many societies is often attributed to culture, and it is men and women alike who justify the lack of equity between themselves with culture.

If we compare societies, we may find that equity is much more related to the socio-economic situation. Progress in equity of gender has been achieved in countries that have known a better economic situation for the majority of their population. In history, European culture, for example, has not been more favorable for women's rights than Bantu culture. A whole range of different factors like increased income and more general access to education have created a situation in which women were enabled to fight for their rights and men to think about their roles.

The socio-economic situation of a society is certainly one of the sources of cultural dynamism and creativity. Absence of poverty turns relations of superiority and inferiority between men and women socially dysfunctional. However, this is not a predetermined sequence. Improved economic conditions do not automatically lead to equity between men and women, and poverty does not necessarily create an inferior position of women.

It is striking that within cultures the justification for the absence of equity between women and men is often not based on cultural arguments. On the contrary, apologists of inequality evoke nature, i.e. the biological constitution of men and women, from which they try to deduce a distribution of tasks attributed to women and men, as well as power differences. These justifications eliminate the difference between nature and culture. Whereas humankind certainly is part of nature and depends on it to a large extent, it differs at the same time from animals precisely in its capability of cultural creation. Humans can, for example, protect the weak and hamper the natural process of selection of the strongest for life. In the same way humankind can also—and, in fact, does—construct masculinity and femininity. These are social constructions, certainly linked to cultural expressions, into which natural characteristics are integrated, which determine the roles of women and men in a society. As such, these relations are not predetermined, and they can be changed by the members of a society.

If roles and relations are to be changed, they first need to be understood. It is not enough to describe them, but they need to be analyzed. Existing social relations are not understood without regarding their histories and their cultural interpretations. It is certainly necessary to protect women against violence perpetrated against them by men in most societies, and a man who rapes a woman is guilty and needs to be punished. However, in social and cultural analysis of relations, we do not look for culprits, but we try to understand phenomena. That allows us to change relations whenever we agree on the need for such a change.

As gender relations are relations of power, it is not easy to change them, especially not by intervention from outside. If we intend to work against the spread of HIV, it is important to emphasize HIV and to deal in the first place with those elements of gender relations involved in the spread of HIV. It could be that through a variety of such contributions from multiple thematic areas, more profound change can be achieved.

8 Variation VII: Grazioso

We may well assume that the grandfather has 12 grandchildren and three great-grandchildren. More than half of them are girls—and he smiles by thinking of them as girls. Even his youngest grandchild is now a real woman and when he saw her the last time, she was angry when his neighbor was calling him from outside: "Look, your girl is coming." He remembers one of the long evening talks under the tree with the other old men of the village. They were talking about school and whether girls should be sent there. Well, it was in a way a useless talk, because they went anyway, unless someone would prevent them by force. One of the families in the village had done that during some time, but it created a lot of trouble and they did not like that. Finally it was the old men, who still remembered the struggle against Portuguese rule who convinced some of the others, the younger ones among the old: It is better, if also girls learn to read and write—you never know what may happen. To learn to calculate, you do not have to go to school. The girls learn it anyway, only look at the women at the market. However, they should also read and write and learn history and to drive a car—remember our women comrades in the war.

The grandfather thought of the nurses in the health center of Prábis. Three of them were women, and of those two came from the district. Recently the grandfather had had several days with an awful headache. He went to the village healer, who looked at him for a long time and finally said: I can make you better, but it will cost you a bottle of whiskey. The grandfather returned in a sad mood. He could not afford such a price. A chicken—yes, that could have been possible, but a bottle of whiskey—no, that was too expensive and not appropriate. So he went to the health center. One of the nurses from Prábis examined him. She said that he needed to rest and gave him a prescription for some pills, which were cheap. She was kind, but she did not tell him, where his headache came from. The grandfather was thinking that it might be a good idea, if the nurses and the healers worked together. However that seemed a faraway dream. He knew that they did not trust each other, and there were many good reasons for that.

Many attempts have been made to overcome the normally wide gap between the different paradigms of health and disease from which people choose. One example related to HIV in which I participated, has been tried in 2010/11 in the district of Kwango in the Democratic Republic of Congo. The combination of research, intended to find out about the connotations of HIV and AIDS in communities and the terms and metaphors used by chiefs and healers to describe it and to deal with it, and information/training of biomedical personnel and healers has initiated a process of structured dialogue between followers of the different medical paradigms in the district.

We started together with district health authorities in Kwango district (a 90,000 km^2 area with two million inhabitants approximately 300 km east of Kinshasa) a process to integrate HIV related work into the cultural context. Kwango district is rural and has a long border with Angola's diamond rich areas. It has a history of violent confrontations with Belgian colonial rule during the first part of the 20th century. The population is predominantly Yaka and Suku, both speaking Bantu languages and having a matrilineal structure.

Under scientific guidance of the anthropologist Sophie Kotanyi, meetings were held and in-depth interviews conducted with chiefs and different types of healers as well as biomedical personnel in five of 14 health zones in the district. This was an initial step to learn about existing connotations with AIDS in the area, and about the terms, metaphors etc. used to describe and to deal with the phenomenon of HIV and AIDS. The approach is building further on experiences from Mozambique. The main underlying hypotheses are:

- Both, the autochthonous and the 'Western' medical paradigm can be used complementarily to combat the spread of HIV and to care for HIV-infected people.
- Procedures of treatment in autochthonous medicine can be a risk for spreading HIV, but contain as well elements that are favorable for prevention.
- A basis of dialogue between practitioners of autochthonous medicine and 'Western' medicine can be created, if both recognize their own cultural specificity and abstain from considering their approaches as being universally valid.
- A dialogue based on mutual respect creates conditions for negotiating specific procedures and promoting testing and treatment adherence through referral and counter-referral of patients.

The intervention in Kwango district had to end in 2011 because funding was no longer available. Much research is still required. Several meetings, however, had already taken place between healers and biomedical personnel in the health zones that were visited.

Further research will have to investigate the existence of initiation rites for young women and men in the district. The possibility of integrating HIV-related messages into these rites needs to be explored.

The approach faces a number of problems, one of them being gender relations. While the promotion of 'traditional' customs like "girls need to be virgins until marriage" might contribute to a delay in onset of sexual activity and so protect young women from being infected with HIV, it also stabilizes existing gender relations, which are one of the main factors for high HIV prevalence. It is not yet clear whether this type of dilemma can be part of the intended dialogue.

9 Variation VIII: Presto non troppo

The grandfather remembered that a short time after Portugal withdrew its military from the country and recognized its independence in 1974, other people from European countries came to Guinea-Bissau. They came from Sweden, from the Netherlands, from two different countries both called Germany, and other countries. They were no soldiers, and they said that they came to help. To help to rebuild what had been destroyed during the war.

It was difficult for the grandfather to think about all the time that had passed since then. Once, still during the war, he had been in a meeting where their leader, Amílcar Cabral, had spoken to them. They all were proud of him. He talked to them in a way everybody could understand. And they knew already at that time that he spoke as well to well-known leaders of the world, in the United Nations, to the Pope ... It was so sad that he was killed in early 1973 in neighboring Conakry, even before Guinea-Bissau declared independence. Amílcar Cabral[17] *always had told them that what needed to be done was not only to rebuild what the war had destroyed. Destruction had started long before that. They did not want to re-build, they wanted to build. The grandfather still remembered very well that it took him and other comrades some time to understand the difference. However, yes, that was what they wanted, even if they would not have been able to explain it so well. They wanted something new.*

The people who came after the war, were also talking about something new, but in the end ...? Yes, there were new things, but that were not the ones they had dreamed about.

Why is it like that? The grandfather has not yet found the book that would explain it to him. And if he only chooses one of the big number of problems that sometimes cause him a headache for many days, only the one to find out, which of the healers are the good ones, who of them would really tell him why ...

If international aid—now called 'cooperation'—is more and more recognizing context, it is in most cases because of tactical reasons, not because of a real interest, which would imply that the international donors' own concepts would be questioned and have to be considered as culturally specific. That is particularly true for everything having to do with health and disease. The cases in which I have met colleague medical doctors with a genuine interest in autochthonous medical approaches are rare. In most cases where interest was expressed it was for tactical reasons in order to better impose what was considered universally valid.

Those many people who came and come to Guinea-Bissau and to many other places in the world, do not have much time. They come with a 'project,' which has a start and an end and a specific funding. Many of these projects do not last longer than three years. Sometimes they are extended, but not for a very long time. They are evaluated and accounted for with regard to those who give the money, not to those whose situation they claim to improve.

After thinking and discussing a lot, it still seems that we cannot propose a convincing answer to the question how to distinguish between good and bad books and how to distinguish between good and bad healers. There may be, however, some criteria that may take us in the direction of an answer.

9.1 *The Recognition of Cultural Diversity and Specificity*

We are not in a position to choose whether we want or do not want to include culture in whatever we do. We are doing it anyway, because culture is like our own shadow which can be short or long, but we cannot escape from it.[18] So it should just be recognized and not be denied. This can help us to be modest and eager to find out about different cultural approaches. Yet, it is in early 2018 more difficult to speak about culture and cultural identity than it has been before, as the terms 'cultural identity' and 'identity' alone have been hijacked by people who use it as a justification for exclusion of those whom they consider different and inferior.

There are many differences between cultures, and there are many similarities. A culture is not just good or bad. It has different and contradictory aspects. If we start learning about a culture, it is helpful to start with its strengths: "As I have recommended and now insist, all programs developed for a community must begin with the positive. It is the responsibility of the interventionist to find out what these positive values are. If none can be found, it may indicate that the interventionist is incapable of working in such a community and he or she must leave before he or she becomes a part of the problem" (Airhihenbuwa, 2007).

9.2 *Curiosity*

A health worker who is not curious, cannot be a reliable health worker: curious about the patient, curious about the patient's condition, curious about science and research findings, curious about meaning ...

9.3 *Sympathy*

The Hungarian psychoanalyst Sándor Ferenczi reflects in his clinical diary of 1932 (Ferenczi, 2013) on his relationship with his patients. He goes as far as stating "Only sympathy heals," and "Understanding is necessary in order to apply the sympathy at the right moment (analysis). No healing without sympathy (at most insight into the genesis of the suffering)" (Ibid., p. 265).[19]

This is similar to what Collins O. Airhihenbuwa (2007) has been telling us, when he wrote: "To have a depth of understanding of a culture requires a certain emotional investment" (2007, p. 63).

This reminds us that a person feeling ill, does not only ask why, but is also looking for healing. Both are, however, closely linked.

This is the minimum we may expect from those who write books as well as from those who practice healing. We still need to come further and know and understand more. That is why we need a scientific mind. We need research. "I seem to be saying two things that contradict each other. On the one hand, we trust scientific knowledge, on the other hand, we are always ready to modify in-depth part of our conceptual structure about the world. But there is no contradiction, because the idea of a contradiction comes from what I see as the deepest misunderstanding about science: the idea that science is about certainty."[20] Remaining with the uncertainty, with lots of open questions, is not a mistake; it is just the human condition. If we want to contribute to a more human world, it is with such uncertainty we will have to do it.

So, at the end of these variations we come back to the original story, the theme. The theme has not changed. We come back to the same story. However, after going through the variations, we may have gained a more profound understanding about the potential and the complexity which is hidden in the theme, in the story, and so enjoy listening to it again.

10 Finale: Andante

This story takes us to the village of Prábis in Guinea-Bissau. Renate Welsh, an Austrian author of children's books, wrote it down and published it in German in 1979. Here is my translation of part of it:

Diallo Sampa was sitting next to his grandfather. In front of them sat the teacher. The teacher was young, only a couple of years older than Diallo. He himself still went to school—in the capital city.

Two times a week he came to the village in the early evening. He came in a rickety old car which then continued the trip and took his colleagues to more remote villages.

They learned to read.

The grandfather was reading aloud. Everybody was listening. He read with big effort, word by word. While reading he swayed back and forth on his heels, and he held the book very tightly. He frowned. Wrinkles seemed to creep up and down, to and fro. He read slowly, but he didn't make any mistake.

When grandfather was ready, Diallo breathed out long. Grandfather was one of the oldest in the village. He had worked on the farm, when it was still owned by an agent of the Portuguese secret police. He had fought against the Portuguese. He had led the men of the village when they took weapons to the Southern Front of the liberation army. Grandfather did not tell much about that time, but other men from the village had told how grandfather alone had defended a store house for sugar cane, in which they had hidden the wounded.

And now he learned to read. In the beginning it gave him more trouble than Diallo. But now he had read an entire page without a single mistake.

Everybody looked at grandfather.

"How do you feel now?" asked the young teacher, "now that you are so good at reading?"

Grandfather swayed more intensely on his heels. Diallo moved a bit closer to him. Grandfather cleared his throat—once, twice. Then he passed the book to Diallo and showed his hands. He looked at his palms and held them in a way that the others could see them. "This scar," he said, "is from the first hut I built. And this one here is from the big fire, when my first son was born. And this one is from the first bazooka which I set off. That was when I did not yet know how to handle it. Every scar says what happened. It does not say why. The books say why" (Welsh, 1979).

Notes

1 https://www.britannica.com/topic/superstition
2 http://www.who.int/about/mission/en/
3 The German physician and politician Rudolf Virchow seems to have used a similar concept earlier: "Die Medicin ist eine sociale Wissenschaft, und die Politik ist weiter nichts, als Medicin im Grossen" [Medicine is a social science, and politics is nothing but large-scale medicine] (Virchow, 1848, p. 125).
4 An important milestone is Georg Groddeck (1923).
5 It is sad that some organizations of people living with disability claim rights with the argument that they can do useful things and contribute to society. While they have to struggle against exclusion, they themselves exclude some of the people who should rather be protected by them. Human rights are not based on any useful contribution to society, but just on the mere existence of a human being or any being born from a human being.
6 I leave out here the distinction of the terms disease, illness and sickness in the English language.
7 While this is still true, different approaches are being practiced like 'Problem Based Learning,' which has been pioneered by the Canadian McMaster University Medical School since the 1960s and subsequently been taken up by select institutions around the world.
8 This part of the story is based on personal communication with Portuguese soldiers who had come back from Guinea-Bissau, in Coimbra in 1971.
9 https://historyofptsd.wordpress.com/recent-history/
10 See note 4 above!
11 *Nô Pintcha* (http://www.jornalnopintcha.gw/administracao-de-Prábis-com-enormes-dificuldades-para-projetar-o-desenvolvimento/) – original text: "A adesão da comunidade ao hospital tem sido fraca porque muitos são animistas e apostam mais em tratamentos tradicionais feita pelos curandeiros e feiticeiros locais" (translation by the author).

12 "A este respeito, o secretário da AFAP, Marvana Nanque, garantiu que a comunidade, através da associação local e com a orientação dos responsáveis da área, atua na sensibilização sobre as boas práticas da saúde e na importância de frequentar o hospital."

13 This term is frightening. It may well be meant in a peaceful way, but its original meaning is "a person, object, or place selected as the aim of an attack" (https://en.oxforddictionaries.com/definition/target). Even if it may in this context have lost its specific meaning of attack, it is still significant that it is used in connection with 'development' interventions. These are coming to a population from outside itself and do not consider that the population itself might have its own ideas and ways of dealing with a particular subject.

14 One example is the University Research Company (URC), a public health organization in the USA, which describes one of its methods of intervention like this: "URC uses social and behavior change communication (SBCC) to enable individuals, families, communities, and societies to improve their quality of life. Our approach recognizes that human behavior is complex—the result of a combination of context-specific influences, including environmental factors, social norms, health policies, the quality of health and social services, and individual knowledge" (http://www.urc-chs.com/social-behavior-change-communication).

15 Interestingly the British government's forestry commission has done research on behaviour change and among other interesting reports published a summary of underlying theories: Jake Morris, Mariella Marzano, Norman Dandy, Liz O'Brien: Theories and models of behaviour and behaviour change (https://www.forestry.gov.uk/pdf/behaviour_review_theory.pdf/$FILE/behaviour_review_theory.pdf).

16 A review and reformulation of this document seems to have been done in 2007. However, in January 2018 it could no longer be found on the UNAIDS website, whereas the version of 1999 is still accessible.

17 Selected works of Amílcar Cabral have been re-edited in 2013 in Praia/Cape Verde (Cabral, 2013). Translations of parts of his work exist in different languages.

18 Claude Gumbucha Mararike, Professor at the Department of Sociology of the University of Zimbabwe in Harare, remarked this at the regional conference of SAfAIDS at Johannesburg on 12th and 13th April 2010.

19 Original text: "Nur Sympathie heilt (…). Verständnis ist notwendig, uns die Sympathie an der richtigen Stelle (Analyse), in der richtigen Art anzuwenden. Ohne Sympathie keine Heilung. (Höchstens Einsichten in die Genese des Leidens)." Translated by the author.

20 Carlo Rovelli, quoted from https://edge.org/conversation/carlo_rovelli-science-is-not-about-certainty-a-philosophy-of-physics

References

Airhihenbuwa, C. O. (1995). *Health and culture: Beyond the Western paradigm.* Thousand Oaks, CA: Sage Publications.

Airhihenbuwa, C. O. (2007). *Healing our differences. The crisis of global health and the politics of identity.* Lanham, MD: Rowman & Littlefield Publishers.

Becker, D., Morales, G., & Aguilar, M. I. (Eds.). (1994). *Trauma Psicosocial y adolescentes Latinoamericanos: Formas de acción grupal.* Santiago de Chile: Ediciones Chile América CESOC.

Cabral, A. (2013). *Obras escolhidas. Unidade e Luta, Vol. 1 A arma da teoria, vol. 2 A prática revolucionaria.* Praia, Cape Verde: Publisher missing!

Commission on Social Determinants of Health. (2007, April). *A conceptual framework for action on the social determinants of health.* Discussion Paper for the Commission on Social Determinants of Health. Draft.

Ferenczi, S. (2013). *Das klinische Tagebuch.* Gießen: Psychosozial-Verlag.

Groddeck, G. (1923). *Das Buch vom Es. Psychoanalytische Briefe an eine Freundin.* Leipzig/Wien/Zürich: Internationaler Psychoanalytischer Verlag.

Nolting, H.-D., Zich, K., & Deckenbach, B. (2013). *Faktencheck Gesundheit: Entfernung der Gaumenmandeln bei Kindern und Jugendlichen.* Im Auftrag der Bertelsmann Stiftung, Gütersloh, Germany. Retrieved from https://www.bertelsmann-stiftung.de/de/publikationen/publikation/did/faktencheck-gesundheit-gaumenmandeln/

Swartz, L. (1998). *Culture and mental health. A Southern African view.* Cape Town: Oxford University Press.

UNAIDS. (1999). *Pennsylvania State University: Communications framework for HIV/AIDS: A new direction.* Geneva: UNAIDS.

Virchow, R. (1848, November 3). Die Medicin ist eine sociale Wissenschaft, und die Politik ist weiter nichts, als Medicin im Grossen [Medicine is a social science, and politics is nothing but large-scale medicine]. *Der Armenarzt, Die medicinische Reform, 18*, 125.

Welsh, R. (1979). *Ich verstehe die Trommel nicht mehr. Erzählungen aus Afrika.* Wien/München: Jungbrunnen.

INTERMEZZO 15

Reverence for Life, Whatever Its Manifestations

Jan Visser

Reflections inspired by Stephen Hinshaw: 'Nontraditional pathways to the development of a scientific mind: Examples from the domain of psychopathology'

It was the fall of 1998. I forget the exact month. It was cold, which is not uncommon in a city on the shore of Lake Michigan. The next day I drove from Chicago to Batavia, IL, a 45-mile distance. "Take the Eisenhower," I had been advised by my host. He had kindly invited me to stay over at his place, but we would first meet at his lab. He had been its director until 1989, when he retired. Since then he had dedicated himself entirely to the teaching of science; making science accessible, loved and understood by the general public; and promoting science literacy. I had invited him to join us at the planned AERA symposium in April 1999 on 'Overcoming the Underdevelopment of Learning,' mentioned in Chapter 1. This was what we would be talking about. We talked while we sat in his office, while he showed me around at the lab, while we drove from place to place and finally ended up at his home. It was the beginning of an inspiring and enduring relationship that greatly contributed to the various key programs we implemented at the Learning Development Institute.

He was known to be a great scientist. Over the years I have grown to know him as a great human being, someone willing to apply his formidable intellect to making the world better. He would no doubt have been a contributor to this book had he lived and retained the power of that intellect. Leon Lederman, Nobel Laureate in Physics, died 20 years after that first meeting we had, on October 3, 2018. This Intermezzo is dedicated to his memory. Like Leo Vroman and Harry Kroto he didn't make it to be part of this book, but he is there in spirit. It is appropriate to mention his name in connection with Stephen Hinshaw's chapter on 'Nontraditional pathways to the development of a scientific mind.' Leon died of the complications of dementia.

We finally understand that people, whose physical abilities are different from ours, deserve our respect and recognition as valued members of our communities and societies. They are just differently positioned along the spectrum of possible expressions of human life. But what about those whose mental functions work differently than ours? Their different abilities are often

 | DOI: 10.1163/9789004416802_030

still stigmatized. Once, while we were discussing the scientific mind, Leon remarked that it could be convenient if, on occasion, we could forget what we hold in memory, as it hinders our ability to develop fresh ideas. He reached that stage during the final years of his life and may have had much to tell us. But were we ready to listen? Are we ready to listen to the 9.9 million dementia patients that are added, every year, across the world? Albert Schweitzer coined the phrase 'reverence for life' as the "principle in which affirmation of the world and ethics are joined together" (Schweitzer, 2009, p. 155). Doesn't that refer to all life, however it expresses itself?

Reference

Schweitzer, A. (2009). *Out of my life and thought: An autobiography* (A. Bultmann Lemke, Trans.). Baltimore, MD: Johns Hopkins University Press. (Original title: *Aus meinem Leben und Denken*)

CHAPTER 16

Nontraditional Pathways to the Development of a Scientific Mind: Examples from the Domain of Psychopathology

Stephen P. Hinshaw

1 Introduction

This chapter will, in all likelihood, have a different flavor from other contributions to this volume. Specifically, as opposed to discussing scientific and philosophical issues related to the development of scientific minds across the majority of individuals, I focus on psychopathology as a 'lens' for comprehending the ways in which some minds do not, at least initially, appear headed in a scientific direction but may actually emerge with real advantages in terms of scientific development. The first instance relates to children with attentional and learning disorders, whose genetic predispositions, life experiences, and combinations of the two push them toward ways of perceiving, thinking, and behaving that are in many ways antithetical to the development of literacy, scholastic competence, and scientific mindsets. Indeed, longitudinal research on the lives of children with such neurodevelopmental disorders reveals core compromises to learning and achievement. Yet closer examination uncovers some surprising ways in which such atypically developing youth may be prone to espouse several core aspects of scientific values and principles, especially innovation, divergent thinking, tenacity, creativity, and (under the right circumstances) strong motivation. In this discussion, I necessarily comment on different phases of the scientific process, including the crucial distinction between its discovery and confirmatory stages.

Second, I use a more qualitative angle to discuss how family experiences of severe mental disturbance may lead not only to trying emotional consequences—including processes of internalization and self-blame when adequate discussion of the family's difficulties is prohibited—but also to life directions aimed toward providing support to others in similar situations and especially toward intense drive to understand the roots of such disturbance. Many people who engage in the so-called helping professions do so, in fact, because of difficult experiences early in their own development. Less discussed, however, is how, under the right circumstances, such experiences

 | DOI: 10.1163/9789004416802_031

may forge a strong motivation to engage in scientific inquiry related to psychological, psychiatric, and neural underpinnings of both typical and atypical development. My own developmental course will serve as a key example of how family and societal silence related to the denial of psychotic-level mental disturbance, followed by revelations of the truth, prompted me to get to the bottom of key scientific issues associated with mental disorder, its stigmatization, and its treatment.

Given space limitations, each portion of this chapter will embrace a more illustrative than definitive stance. I hope to convey that in psychopathology, multiple causal pathways may converge into life paths marked by similar symptoms and problems—embodying the construct of equifinality (Cicchetti & Rogosch, 1996; see also Beauchaine & McNulty, 2013, for an explication of ontogenic process models). In parallel, several differing developmental routes may also lead to the development of scientific minds and lifestyles. In particular, exploration, inquisitiveness, ingenuity, and creativity may well have their origins, in at least some cases, alongside developmental, even heritable precursors (e.g., cognitive styles; behavioral tendencies) associated, at their extremes, with pathological variants of behavior.

Overall, it is a mistake to assume that the scientific process is rigid, buttoned-down, and formulaic. Core scientific discoveries have been made and will continue to be made in the context of personal and familial psychopathology and by minds that have not necessarily traversed a straight and narrow path. So long as we do not fall prey to the tantalizing myth that all forms of mental disorder are inevitably linked to creativity and scientific prowess, there is much to be gained by weighing the evidence and considering the multiple precursors of and the pathways to scientific advances.

2 Behavioral and Learning Disorders in Youth

My clinical and research work, over the years, has involved large numbers of youth and families experiencing the conditions known as attention deficit hyperactivity disorder (ADHD) and so-called 'specific learning disorders' of various kinds, as detailed in the Fifth Edition of the Diagnostic and Statistical Manual of Mental Disorders (American Psychiatric Association, 2013), the official U.S. guidebook to psychiatric diagnoses known as *DSM-5*. It would require a textbook-length discussion to begin to catalogue the symptom patterns, underlying mechanisms, causal pathways, and developmental trajectories of youth who experience the relevant behavior and cognitive patterns (for details, see Barkley, 2015; Beauchaine & Hinshaw, 2017; Cicchetti & Cohen,

2016). I therefore illuminate a few key points of information about these conditions, along with developmental principles and etiologic mechanisms associated with their emergence. Then, despite the inherent associations these conditions have with academic and learning deficits, in the subsequent section I present information regarding the potential for at least some subsets of such individuals to show unexpected proclivities toward scientific ways of thinking and behaving.

2.1 *ADHD and Specific Learning Disorders*

The neurodevelopmental conditions called ADHD and learning disorders are present at with remarkably consistent rates around the world—at least in societies with compulsory education. I begin with ADHD, a topic of intensive study throughout my career, and then proceed to learning disorders.

2.1.1 ADHD

ADHD involves developmentally extreme levels of the symptom clusters of inattention, disorganization, hyperactivity and impulsivity, expressed at frequencies and intensities that interfere with learning and socialization (e.g., Barkley, 2015). Across the past century this syndrome has been variously termed minimal brain dysfunction, hyperkinesis, and hyperactivity, before the terms attention deficit disorder and then ADHD took over in the 1980s and beyond (the term *hyperkinetic impulse disorder* is still used in the United Kingdom and some other nations). Often stereotyped as a psychiatric excuse for ineffective parenting or insufficiently enriching classrooms—or a ploy by pharmaceutical firms to dispense strong stimulant medications to unsuspecting youth and their families—ADHD is decidedly real (Hinshaw & Scheffler, 2014). Indeed, it has deep psychobiological roots (see below) and is associated with rampant academic failure, a high likelihood of rejection from one's peer group because of social skills deficits, and alarming rates of accidental injury, as well as huge cost to society regarding the consequent need for special education, substance abuse treatment, or eventual underemployment or unemployment.

Much of the current controversy surrounding ADHD stems from its soaring rates of diagnosis across the last decade, both in the U.S. and around the world. These have been fueled by (a) recognition of this long-ridiculed psychiatric condition, (b) the common provision of diagnostic evaluations in all-too-brief 10 to 15-minute sessions by non-specialists, and (c) the major push for academic and vocational performance in today's competitive climate, along with the growing recognition that a diagnosis of ADHD may well provide accommodations and special services (e.g., extra time on standardized tests; special education placement; etc.). With thorough, evidence-based assessments, ADHD can be distinguished from trauma-related reactions, neurological conditions (e.g., seizure disorders), and/or other psychiatric disorders; but considerable time, training, and effort are needed for differential diagnosis. Moreover, many psychiatric conditions can and do occur alongside ADHD, exemplifying what is called "comorbidity." It is essential to document such co-occurrence, as different treatment strategies may be required for the separable conditions.

The symptom clusters linked to ADHD are highly heritable, meaning that genes rather than environments produce the major source of variation across the population. Yet, the advent of compulsory education in the 19th century provided the context in which extremes of inattention and impulsivity were made salient in a subset of children. Moreover, parenting practices and school contexts are still highly relevant for shaping the developmental course of symptoms and impairments across the lifespan (see Harold et al., 2013, who showed the importance of parenting for behavioral outcomes in an adopted sample of youth with ADHD, thereby removing the influence of genetic relatedness between generations). Genes and environments work together, not separately, in the genesis of such behavior patterns.

A large number of brain regions are undoubtedly implicated in ADHD-related symptomatology (Castellanos & Proal, 2012), but the frontal lobes and their interconnections with the striatum and other subcortical structures appear particularly salient. Beyond genes, prenatal and early-life exposures to neurotoxins, as well as interactions between difficult temperament and suboptimal childrearing practices, are implicated in developmental trajectories (see Nigg, 2006, 2017).

ADHD does not stop, as was formerly contended, during the transition to adolescence. Indeed, adults with ADHD are now the fastest-growing segment of the population receiving diagnosis and treatment (Hinshaw & Scheffler, 2014; Hinshaw & Ellison, 2016). Symptoms may change form with time—for example, motoric hyperactivity may transform into mental restlessness—but underlying problems with impulse control, regulation of attention, and self-control persist in a majority of cases.

One model of ADHD underlying cognitive dysfunction centers on problems with the brain's executive functions (or "cognitive control"), a set of higher-order skills related to planning, working memory, maintenance of set during task performance, and error correction (Brown, 2013). Another model zeroes in on inhibitory control, signifying the ability to withhold a previously rewarded response in the service of long-term planning (Barkley, 1997). Still another implicates deficient development of intrinsic motivation as the core deficit, related to low activation of brain pathways associated with the neurotransmitter dopamine (Volkow et al., 2009; see also Sonuga-Barke et al., 2010, for a synthesis). Genetic variation in dopaminergic and noradrenergic systems, interacting with child rearing practices and school environments, shape the ultimate display of the constituent behavior patterns.

Overall, ADHD is a neurodevelopmental disorder with strong likelihood of producing school-related difficulties, problems in navigating complex social environments, and risk for personal and family harm (see Hinshaw et al., 2012). It has strong heritability and clear neurobiological roots, while at the same time its life trajectories are shaped through interactions and transactions with family and school settings. On the face of it, individuals with ADHD would not appear to be geared for developing strongly scientific minds—indeed, rates of school and vocational failure are far above normative levels (Barkley, 2015). Yet, some of the relevant symptoms and behaviors, if 'met' by optimal contexts and settings, might yield important aspects of scientific minds and pursuits.

2.1.2 Specific Learning Disorders

Here, parallel to the situation regarding ADHD, a number of labels have been used to define problems with academic achievement over the past century and a half (e.g., word-blindness or dyslexia for poor reading, dyscalculia for poor mathematics, and dysgraphia for poor writing; learning disabilities, as general term for all such conditions). With the advent of *DSM-5*, these learning problems are now termed 'specific learning disorders,' with the acronym LD still often invoked. The underlying concept is that a subset of children, who lack obvious perceptual or sensory conditions or overall intellectual disabilities, fall markedly behind their peers with respect to reading, mathematics, or writing skills (American Psychiatric Association, 2013). A diagnosis mandates persistent difficulties in these academic subject areas, typically indicated by poor 'decoding' skills early in the reading process (i.e., difficulties linking sounds with symbols) and poor comprehension in later development; written expression that lacks clarity; and problems remembering number facts or inaccurate mathematical reasoning. Moreover, in order to receive a diagnosis for the specific learning disorder, it must not be possible to explain these problems by developmental, neurological, sensory (vision or hearing), or motor disorders—and they must prominently interfere with academic achievement, occupational performance, or activities of daily living.

As might be imagined, if a child reveals overall cognitive functioning within the wide bounds of the normative range but is unable to utilize the typical school curriculum to master reading, math, and/or writing skills, major adverse consequences with respect to success in school, developing a sense of self-worth, adjustment to classroom behavioral norms, and the long-term attainment of literacy and job-related skills are highly likely. Data reveal wide variability in adult outcomes for individuals with LD, but persistent issues regarding literacy, vocational adjustment, and even interpersonal relations are salient (e.g., Maughan, 1995; Shaywitz & Shaywitz, 2005). Nonetheless, exposure to systematic, evidence-based instruction can and does lead to significant improvements in academic functioning.

For many years the predominant view was that LDs were largely related to problems in visual processing. For example, regarding the most prevalent of such conditions—specific reading disorder—it was believed that children who failed to learn to read normatively were subject to reversals of letters (or whole words), impairing their development of fluent decoding skills. Whereas this may be the case in a small subset of deficient readers, and whereas young children (particularly boys) are prone to such reversal errors early in the development of visual processing, the view has radically switched to the perspective that the underlying problem lies in the domains of (a) phonological coding, defined as "the ability to use speech codes to represent information in the form of words and word parts" (Vellutino, Fletcher, Snowling, & Scanlon, 2004, p. 4) and (b) higher-order skills related to phonological awareness, orthographic awareness, and syntactic awareness (for a thorough overview, see Vellutino et al., 2004). Mathematical disorders are potentially even more complex, involving issues of working memory, sequencing, or comprehension (with the latter related to word problems). Space does not permit elaboration of writing disorders at a mechanistic level, but integration across differing brain modules, and interactions with motoric output, are salient.

As is the case with ADHD, the heritability of learning disorders is at least moderate, with brain regions linked to linguistic processing (in the case of reading disorders) strongly implicated. An incisive review of Shaywitz and Shaywitz (2005) reveals that two posterior, left-hemisphere brain pathways are highly involved in the genesis of specific reading disorders. Moreover, two major subgroups of reading disordered individuals appear to exist, one marked by strong heritability and the other reflecting environmental inputs. It is essential that such information be well publicized, to counter the prominent belief that children who fail to master the basics of academic work are lazy, stupid, or incompetent. These kinds of self-perceptions are devastating. They inhibit attempts to pursue important life goals. Still, it took the advent of compulsory education, a century and a half ago, to place into sharp relief the inherent difficulties in academic performance experienced by some children. Indeed, human brains did not evolve to read (a cultural 'task' that has existed for only several thousand years) nor did human bodies evolve to sit still in classrooms for multiple hours per day. As I have emphasized, even though biologically and genetically mediated individual differences in inhibitory control (ADHD) and language-related processing (specific reading disorder) are known to be involved in etiologic pathways, important environmental and cultural shifts have made such differences in performance salient. Finally, these conditions are not 'either/or,' as both ADHD-related symptoms and the cognitive problems underlying specific learning disorders each exist on a spectrum, distributed continuously throughout the population.

For much of the 20th century, it was believed that the diagnosis of a learning disorder involved the assessment of a significant gap between the individual's intellectual potential, as indexed by an overall IQ score, and his or her proficiency in the academic domain under consideration (Vellutino et al., 2004; Tannock, 2013). The clear psychometric corollary was that only individuals with relatively high intelligence quotients could qualify for an LD diagnosis; those youth with average or subaverage IQ scores (often those of lower socioeconomic status) could not, by definition, show such a discrepancy. Even more, until the child had failed academically in school for several years, there could not be a sufficiently large IQ-achievement gap upon which to base an LD diagnosis (see, e.g., Tannock, 2013). Thus, children had to wait until their academic failures were truly egregious before they could qualify for any special services, which disallowed important efforts toward preventive intervention and often led to strong demoralization. Indeed, the bulk of relevant research reveals that the fundamental processing deficits are in place regardless of the presence of a major discrepancy between IQ and achievement. In all likelihood, more children with learning disorders exist than traditional assessment practices reveal.

In short, specific learning disorders involve both heritable and environmentally induced problems in the cognitive constituents of literacy and mathematical abilities—with instructional climate and support extremely important for shaping eventual outcome. Given the utter importance of attentional focus, self-regulation, and literacy development for success in our post-modern world, it would certainly appear, on the face of it, that specific learning disorders (as well as ADHD) would predict poor development of the potential for competent scientific minds.

3 Alternative Pathways

In order to address whether and how attention and learning problems might actually promote the development of scientific minds in certain cases, I first delve into Visser's (Chapter 1, this volume) explication of the many properties of the development of such minds. From this list, I highlight several: *exploration, inquisitiveness, the ability to think metacognitively, the desire to understand (and do so profoundly), ingenuity, readiness to transcend,* and *creativity.*

This is an intriguing list, indeed. Nowhere does it mention precision, orderliness, dispassion, or linearity of thinking, which might be high on listings of stereotypes of the practices of scientists. Of course, other terms and concepts from Visser (Chapter 1, this volume) cover these more tried-and-true aspects of scientific thought and process, but the subset above actively connotes action, motivation, curiosity, and strong affect.

Lest one think that such characteristics are overly soft—unnecessary parts of scientific endeavors or mere frills—I hasten to highlight systematic research on the scientific thinking and output of major scientists. Rothenberg (1996) performed structured interviews with 22 Nobel laureates in chemistry, physics, physiology, and medicine, emerging with a four-phase 'map' related to major scientific breakthroughs. He labeled these complex stages of the development of scientific endeavors as follows: (1) motivation to create, (2) deviation or separation (from established canons), (3) simultaneous opposition or antithesis (of established ideas), and (4) construction of the theory, discovery, or experiment related to weighing new evidence. In other words, truly groundbreaking science is not a simple, linear process of the rigorous, objective testing of canned hypotheses. Rather, it involves considerable cognitive (and even emotion-laden) work that deconstructs and reconstructs prior thinking and paradigms in a given field, requiring many of the traits noted above.

Indeed, the participants in Rothenberg's research, along with earlier giants in the sciences, were often possessed by strong emotions regarding their subject matter, speaking of 'suffering' in light of unresolved findings, being 'haunted' by conflicting theories, and working through 'unbearable' periods during the research process. They also valued the 'beauty' of their models and directly showed character traits of fortitude and even resistance, above all when separating their models from those more established paradigms. Particularly in this initial—motivational and emotion-laden—phase, such personal characteristics of the scientists in question were highly relevant to major discoveries. Without them, the science would never have been done at all, nor would established paradigms in the fields of inquiry have been challenged.

In short, to dichotomize the situation, the practice of science can be demarcated as having an initial 'discovery' phase, marked by a scientist's intensive familiarization with the theories and methods in a particular field and paired with emotion-laden confrontation with the apparently unresolvable conflicts and inconsistencies within that field, along with intensive motivation and drive to yield new insights and solutions (see Kuhn, 1962; see also Pylyshyn, 1972, who argues that rigidly codifying such discovery-related procedures is not possible). In contrast, the 'verification' phase—often the subject of stereotypes regarding objective science—places a premium on objectivity, dispassion, and value-free testing of hypotheses and subsequent presentation of findings. These latter traits and practices are of course essential to any science. Yet without the motivational and affective force of the discovery phase, science would proceed in piecemeal fashion, full of drudgery, with an absence of major breakthroughs.

Back to learning and attention disorders: are there characteristics of people grappling with such problems that might facilitate scientific discoveries (see, for example, Andreason, 2005; West, 1991)? At a qualitative and anecdotal level, considerable lore exists that several luminaries in the world of science contended, early in their lives, with learning or attention problems. Commonly listed examples are Albert Einstein, Alexander Graham Bell, and Thomas Edison. In addition, expanding into entrepreneurial and artistic realms, this list grows appreciably (e.g., Will Rogers, Isadore Duncan, and Robert Rauschenberg; more modern business successes like Paul Orfalea and David Neeleman). In these accounts, the individual typically recalls feeling out of place in school and emphasizes the considerable energy and motivation they developed to pursue their calling, along with nontraditional ways of thinking and approaching problems.

A provocative instance from the world of art emanates from Mattison's (2003) book-length examination of the creative process of Robert Rauschenberg, who freely admits to lifelong dyslexia (specific reading disorder). Mattison presents a fascinating analysis of the artist's creative process, which was inspired, in part, by Rauschenberg's unique way of seeing the world—which he described as seeing it in fragments. In terms of his work as a printmaker, he was quoted as saying that he had a clear advantage given that he already saw things backwards! Finally, in the realm of science, Collin Diedrich, a molecular virologist, writes in clear and moving prose about the ways in which his serious reading disorder has not deterred him from pursuing a scientific career. Among many other traits, he cites his goal orientation, perseverance, and passion as central (PLoS Pathogens, 2014).

Whereas such accounts may well contain considerable truth, we need far more than an assemblage of case studies in order to bolster the argument. Indeed, case studies cannot account for additional variables and factors and may be highly subject to selection bias. Turning to empirical research, in an extremely important investigation from several decades ago, Tarver, Ellsworth, and Rounds (1980) began by noting an association between unfocused attention, incidental learning, and creativity. That is, in the realm of what is termed *selective attention*—involving the choice to focus on central versus peripheral material in the environment—prior research had revealed that youth with LD were typically deficient in recalling material directly related to task performance but actually performed *better* than the control group in remembering 'incidental' information. In other words, their seemingly poor selective attention may have actually allowed them access to a host of supposedly irrelevant information, which might fuel innovative thinking.

With this formulation as a guidepost, Tarver et al. (1980) contrasted LD vs. non-LD children in first, third, fifth, and seventh grades on tests of figural creativity (completing a squiggle to make a picture) and verbal creativity (discussing alternate uses of everyday objects). Intriguingly, the LD group scored higher than the comparison group on measures of figural creativity, displaying a penchant for originality and uniqueness, but not verbal creativity. Also of note was that for several indices of verbal creativity, the LD group scored particularly high in first grade but not during later grades, suggesting that traditional schooling (and negative feedback from teachers and peers) may have served to squelch the initially creative utterances of such youth.

Indeed, during the 1990s there was a call for a different perspective on what were then termed learning disabilities, based in part on the seminal work of Howard Gardner on multiple intelligences (Hearn & Stone, 1995; see Gardner, 1983, 1993). In short, as opposed to century-old convictions that what we call intelligence comprises core abilities in language and logical deduction, Gardner has contended that each person possesses seven or more basic forms of intelligence, spanning domains of *linguistic* and *logico-mathematical* (measured by most IQ tests) as well as *musical-rhythmic, visual-spatial, bodily-kinesthetic, interpersonal*, and *intrapersonal* skills (usually measured sparingly if at all on such tests). It could be the case that individuals with some forms of LD, who are strong in non-traditional forms of intelligence, comprise learners whose skills are not well-matched to the predominant values of most schools.

It is a mistake, however, to think—as some cultural constructivists do—that LD is entirely a socially created entity, the instance of a mismatch between certain children's cognitive and learning tendencies, on the one hand, and traditional schooling on the other. Certainly, societies need to create schools that far transcend the factory-like models of public education that predominated at the onset of compulsory schooling and that still appear in many venues today (Rose, 2013). Children who lack phonological awareness and other building blocks of reading or math, as well as those with inherent difficulties in impulse control and attentional focus, can and should have options for education that are far more active and engaging than the passive model still too often apparent. It may also be the case that some of their lack of apparent focus can serve them well when incidental learning of seemingly irrelevant information that transcends the subject matter at hand, would be valuable. Yet placing the entire blame on inadequate schooling is as reductionist on the sociocultural side as the biomedical/psychiatric perspective that the entire problem is based inside one's nervous system.

Back to ADHD: Beliefs here are in many ways parallel. That is, the lack of inhibitory control that marks a primary symptom domain of individuals with this condition may yield an exploratory style, a talent for 'out of the box' thinking, and creativity. Healey and Rucklidge (2006) performed one of the few careful empirical investigations in this domain, showing that a group of well-defined creative children often showed many signs of ADHD (though none met formal diagnostic criteria; see Healey, 2014, for a review). Moreover, by contrasting subgroups of children who crossed ADHD vs. non-ADHD and creative vs. non-creative dimensions, they showed that even creative children with high ADHD-related symptomatology suffered deficits in naming and processing speed, though they outperformed the ADHD group without creativity with respect to "higher order" cognitive functions. In short, the fear of mistaking normal-range creative youth, bored by typical academic curricula, for those with bona fide ADHD is probably unfounded. Moreover, some aspects of creativity may well be present in subgroups of children with ADHD (see also Abraham et al., 2007; Fugate, Zentall, & Gentry, 2013, for additional information on some limited but important areas of strength and creativity in youth with ADHD).

As for adults with ADHD, the careful research of White and Shah (2011) yielded supportive findings. That is, compared to adults without ADHD, those with the disorder not only showed higher levels of creative thinking on a standard verbal task but also displayed higher levels of real-world creative achievement. In addition, and not surprisingly, these diagnosed adults had a preference for *idea generation*, whereas the comparison group showed greater affinity for *problem clarification*. Both of these processes are important for scientific inquiry, as noted above, though at different phases of the scientific process.

I hasten to point out that many advocacy groups contend that ADHD is invariably a hidden gift (for an egregious example, see Honos-Webb, 2005; for a more evidence-based perspective, see S. Kaufman, 2013). The extreme perspective here is that individuals with ADHD are uniquely talented but pulled down by rigid social and academic conventions, which emphasize caution and restraint rather than exploration and boldness. The high rates of accidental injury, school and social failure, and social costs linked to ADHD argue strongly against such a simple (or, more accurately, simplistic) conclusion (Hinshaw & Scheffler, 2014). Moreover, as reviewed by Antshel (2008), the presence of ADHD symptoms in individuals with extremely high IQ is often impairing rather than facilitating of greater achievement and productivity (see also Antshel et al.,

2007). Thus, whereas some traits associated with ADHD may well predict the kinds of exploratory, disinhibited traits linked to scientific curiosity and even scientific vision, one must be extremely careful lest such ADHD-related features are glorified.

3.1 *Larger Issues of Mental Illness and Creativity*

There has been vast speculation—and a large number of empirical investigations—related to the more general topic of mental illness and creativity. In fact, creativity has been the subject of voluminous speculation and research for nearly the entire history of psychology (see Sternberg, 1999, for a major compendium). Indeed, this entire chapter could be devoted to debating the many ways of operationalizing this elusive yet crucial construct. Specifically, in a variety of forms, but focused mainly on schizophrenia, bipolar disorder, and other conditions featuring psychotic behavior, investigators have examined potential linkages to creative processes. This literature spans case studies, case-control designs, behavioral and molecular genetics investigations, and evolutionary psychology perspectives and models. Many strong assertions have been made that several forms of divergent and creative thinking are correlated with the kinds of primary process thinking involved in psychosis. Even more, considerable evidence exists that the first-degree relatives of individuals with such mental disorders—assumed to have a partial but not full genetic loading—are particularly marked by creative thought and accomplishment (see Kyaga et al., 2013, for positive evidence in a population-wide, 40-year investigation in Sweden.; see also Johnson, Tharp, & Holmes, 2015, who did not find such an association for bipolar disorder per se in a smaller yet extremely careful study).

Recent syntheses cast real doubt on the overall assertion that mental illness is inevitably linked to creative process (for a masterful compilation of relevant research, see J. Kaufman, 2014). Still, questions remain unanswered and debate continues. To take just one example, Carson (2010) has built a program of research examining the construct of latent inhibition from models of learning in psychology. Latent inhibition is defined as "repeated pre-exposure to a stimulus that is not reinforced retards future associability to that stimulus" (Carson, 2010, p. 183). Yet for some individuals, who may be prone to psychotic thinking (characterized by delusions and hallucinations), a core deficit in latent inhibition may place a number of seemingly 'irrelevant' mental products in conscious awareness, predicting disorganized thought. Yet if such individuals simultaneously have high overall intellectual ability, they may be able to utilize such information for creative thought and action. In the words of Carson (2010, p. 183), "... psychotic vulnerability (including the LI [latent inhibition] deficit) is associated with creativity only in the presence of high IQ or other cognitive strengths."

Overall, tantalizing—though inconsistent—evidence suggests that incidental learning, bold cognitive style, and strong motivation to overcome impairments in the basics of academic functioning and self-regulation can foster scientific mindsets in a subset of individuals with LD and ADHD. I reiterate, however, that glorifying or idealizing LD or ADHD as magic pathways to creativity, vision, and artistic/entrepreneurial/scientific breakthroughs is overstated and simplistic. Yet particularly in the discovery phases of scientific endeavors, some of the constituent traits could be advantageous.

4 Family Experiences of Psychopathology

My second, and extremely brief, example emanates from personal and family experience. I hesitate to overgeneralize from any case example, much less my own, yet I believe that the presence of serious family psychopathology provides a backdrop for the emergence of motivation for not only clinical interest in mental disorder but also intensive focus on research and scientific initiatives.

As described in Hinshaw (2002, 2010, 2017), I grew up in a loving, intellectual home in the U.S. Midwest, yet I was mystified by the sudden disappearances of my father, a philosopher, who vanished from time to time as though silently abducted in the middle of the night. I didn't know at the time that he was being held in mental hospitals or that his first major episode had occurred at the age of 16, when he came to believe that his newly discovered ability to fly—a clear delusion—would send a message to the leaders of the free world to stop the oncoming Nazi threat. He attempted such a flight, leaping from the roof of his family home. He was institutionalized for half a year, nearly starving to death from the delusional belief that the Fascists had poisoned the food of the Southern California back-ward where he was held. After recovering spontaneously (his sole 'treatment' was being tied to his bed each night to prevent wandering), he earned highest honors at Stanford and Princeton, studying with Bertrand Russell and Albert Einstein at the latter. Yet he again became floridly psychotic in the spring of 1945, just after completing his dissertation. He was incarcerated at Philadelphia State Hospital, later known to be the worst mental facility in the U.S., where he was beaten by fellow inmates and staff, carrying the belief that he was imprisoned in a concentration camp in Europe. His life of high attainment and utter madness was now firmly in place.

After receiving a professorship at Ohio State, he met my mother, a stellar graduate student in history. They fell in love and got engaged, but the only thing my father mentioned about his past was some 'trouble' during high school and at Princeton. Shame and stigma pervaded his life. It was not until my mother was pregnant with me and then with my sister that she understood the full force of his wildly psychotic episodes, which had been diagnosed as schizophrenia for many years.

During Dad's many hospitalizations during the 1950s, when he received high doses of Thorazine and many electroconvulsive treatments, he believed that he must have somehow been responsible for his confusing, disruptive, and devastating periods of psychosis, doubtless related to the maltreatment he had received at the hands of his stepmother, decades before. Serious mental disorder, as noted earlier, is nearly always the product of genetic vulnerability acting in concert with trauma or major negative life events.

When my father asked his doctors what my sister and I should be told about his mysterious absences, the answer was unambiguous: we were never to learn, lest this information permanently destroy us. As a result, nothing about his disappearances was ever discussed, and my parents did everything possible to hide the worst of the episodes from us. Many children who know that something is wrong in their lives but are not told any information about the situation tend to blame themselves—that is, internalize. I was no exception. After all, it's probably better at some level to believe that devastating events have *some* cause, rather than emanate from a random, cruel world, even if one's self-image takes a hit when the blame is self-directed. I tried to keep emotions at bay through my rigid routines of school and sports. The ultimate paradox was that when Dad returned, he provided care and support. My own life, then, was as inconsistent as his.

I didn't know at the time that family silence about a parent's mental illness actually intensifies the child's risk for experiencing mental disorder. The noted, extremely perceptive psychiatrist William Beardslee has devised a form of family therapy in which much of the work is devoted to the parents' coming up with a narrative about their emotional struggles, in language understandable to the children (see Beardslee, 2002). Such treatment reduces long-term risk for offspring mood disorders, revealing the power of communication and understanding to replace internalization and shame.

It was only when I returned home from my first spring break from college that my father decided, finally, to defy the doctor's orders and disclose some of his key life events to me. Thus began a series of talks, several times per year, that continued for the next 25 years. Initially, I was terrified: my father was disclosing his lifelong, severe mental illness—in the form of schizophrenia—and I would undoubtedly be next in line to experience this kind of madness.

Throughout college and for several years afterwards, I lived in fear, each night, that I was about to lose control of my mind. I resorted to several forms of self-torture to keep the demons at bay and stay sane, often forcing myself to vomit in order to exhaust myself enough to sleep. Only gradually did I open up to others and come to the realization that I was not destined to repeat my father's fate, which was unfortunately shared by other family members as well.

At the same time, however, from our first talk I understood that I had a mission in life—to understand my father's mental illness and perhaps all forms of mental disorder as well. I undertook intensive volunteer experiences as an undergraduate, working as a Big Brother, teaching in a prison, serving as an in-home therapist's aide on a multidisciplinary mental health team. I tried to learn as much as I could about psychology, social sciences, and biological aspects of the field. After spending several years, following graduation, working at two nearly simultaneous jobs with children and families, I entered the doctoral program in clinical psychology at UCLA. There, I studied feverishly to learn about developmental principles, clinical theories, and statistical methods, all to formulate a means of classifying and understanding the multiple forces that conspire to place vulnerable youth on trajectories toward psychopathology. As a clinical psychologist, I also immersed myself in psychopharmacology, assessment, and psychological therapies, in order to develop, deliver, and evaluate multi-modal interventions.

Around this time I worked with other family members to convince my father's doctors that he actually experienced episodes of bipolar disorder (manic-depressive illness), rather than schizophrenia. Family members often get placed in roles that far transcend their own training and experience, another legacy of living in a family experiencing such problems.

Throughout my career I've blended scientific work on causal factors related to child psychopathology with active clinical trials designed to treat—and further understand—long-term outcomes. I've gradually come to understand that a grand diagnostic scheme, a master framework for understanding linkages within and across the many forms of mental disorder, is premature until the field learns far more about basic biological, psychological, social, and cultural factors and forces. Even more, any objective science related to mental illness must be leavened with first-hand accounts of the *experience* of psychopathology in individuals and family members. Indeed, unless there is true humanization, the pervasive stigma that still clings to mental illness will remain entrenched (Hinshaw & Stier, 2008; Martinez & Hinshaw, 2016).

Thus, at least in my own case (but doubtless in many others as well), the early loneliness, confusion, and isolation I experienced, related to the utter silence surrounding the entire topic in my family, prompted intense interest in both clinical and scientific efforts designed to tackle the entire domain of mental disorder. As is always the case, such outcomes are overdetermined. That is, I grew up in a household revering learning; I pushed myself to work hard academically; and I inherited a partial loading of the genes from my father's side of the family, doubtless fueling my high energy and drive (though also the mercurial swings of mood and perspective I still experience to this day). But without the 'burning question' of my father's unfortunate plight, which robbed him of many productive years of functioning and left him perennially feeling that he was somehow to blame for his episodes and experience, I would not have had the motivation to pursue, with the passion I have, the multiple aspects of psychology and mental health that have comprised my life's work.

5 Conclusion

I remain convinced that multiple pathways toward the development of scientific minds exist. Learning and attention disorders clearly compromise the language- and self-regulation-related skills that typically serve as the underpinnings of academic success and, presumably, a scientific orientation toward the world. Yet some characteristics of these conditions, such as a propensity for exploration, incidental learning, and creativity, may well be conducive to scientific ways of perceiving and acting in the world, especially in the presence of additional traits (e.g., intelligence, diligence) and specifically if learning environments embrace multiple means of active learning. Indeed, lest we forget, extremely young children—who have clearly not developed the intellectual structures and abilities to act like 'real' scientists—show mental flexibility, openness to experience, and utter curiosity, which has prompted my colleague Alison Gopnik of Berkeley to contend that they are natural scientists and experimenters (Gopnik, 2009; Gopnik, Meltzoff, & Kuhl, 1999). In addition, it may sometimes be the case that painful experiences in childhood can serve as strong motivation toward gaining scientific understanding, as portrayed in the example of my own inspiration to become a clinical scientist and developmental psychopathologist after experiencing the silence around and subsequent revelation about my father's serious case of bipolar illness. The hope is that we do not view science as an enterprise subject to pervasive stereotypes of ultra-objectivity and even drudgery. Discovery phases of the scientific process mandate high motivation and an emotion-driven urge toward exploration and theory testing; perseverance and passion are required for confirmatory phases as well.

References

Abraham, A., Windmann, S., Siefen, R., Daum, I., & Gunturkun, O. (2006). Creative thinking in adolescents with Attention Deficit Hyperactivity Disorder (ADHD). *Child Neuropsychology, 12*, 111–123.

American Psychiatric Association. (2013). *Diagnostic and statistical manual of mental disorders* (5th ed.). Washington, DC: American Psychiatric Press.

Andreasen, N. C. (2005). *The creating brain*. New York, NY: Dana Press.

Antshel, K. M. (2008). Attention-deficit hyperactivity disorder in the context of a high intellectual quotient/giftedness. *Developmental Disabilities Research Reviews, 14*(4), 293–299.

Antshel, K. M., Faraone, S. V., Stallone, K., Nave, A., Kaufmann, F. A., Doyle, A., ... Biederman, J. (2007). Is attention deficit hyperactivity disorder a valid diagnosis in the presence of high IQ? Results from the MGH longitudinal family studies of ADHD. *Journal of Child Psychology and Psychiatry, 48*(7), 687–694.

Barkley, R. A. (1997). *ADHD and the nature of self-control*. New York, NY: Guilford Press.

Barkley, R. A. (Ed.). (2015). *Attention deficit hyperactivity disorder: A handbook for diagnosis and treatment* (4th ed.). New York, NY: Guilford Press.

Beardslee, W. R. (2002). *Out of a darkened room: When a parent is depressed, protecting the children and strengthening the family*. Boston, MA: Little, Brown.

Beauchaine, T. P., & Hinshaw, S. P. (Eds.). (2017). *Child and adolescent psychopathology* (3rd ed.). Hoboken, NJ: Wiley.

Beauchaine, T. P., & McNulty, T. (2013). Comorbidities and continuities as ontogenic processes: Toward a developmental spectrum model of externalizing psychopathology. *Development and Psychopathology, 25*, 1505–1528.

Brown, T. L. (2013). *A new understanding of ADHD in children and adults: Executive function impairments*. New York, NY: Routledge.

Carson, S. (2010). Latent inhibition and creativity. In R. E. Lubow & I. Weiner (Eds.), *Latent inhibition: Cognition, neuroscience and applications to schizophrenia* (pp. 183–198). New York, NY: Cambridge University Press.

Castellanos, F. X., & Proal, E. (2012). Large-scale brain systems in ADHD: Beyond the prefrontal–striatal model. *Trends in Cognitive Sciences, 16*, 17–26.

Cicchetti, D., & Cohen, D. J. (Eds.). (2016). *Developmental psychopathology* (Vols. 1–4, 3rd ed.). Hoboken, NJ: Wiley.

Cicchetti, D., & Rogosch, F. (1996). Equifinality and multifinality in developmental psychopathology. *Development and Psychopathology, 8*, 597–600.

Fugate, C. M., Zentall, S. S., & Gentry, M. (2013). Creativity and working memory in gifted students with and without characteristics of attention deficit hyperactive disorder: Lifting the mask. *Gifted Child Quarterly, 57*(4), 234–246.

Gardner, H. (1983). *Frames of mind*. New York, NY: Basic Books.

Gardner, H. (1993). *Multiple intelligences: The theory in practice.* New York, NY: Basic Books.

Gopnik, A. (2009). *The philosophical baby: What children's minds tell us about truth, love, and the meaning of life.* New York, NY: Farrar, Straus and Giroux.

Gopnik, A., Meltzoff, A. N., & Kuhl, P. (1999). *The scientist in the crib: What early learning tells us about the mind.* New York, NY: William Morrow & Company.

Harold, G. T., Leve, L. D., Barrett, D., Elam, K., Neiderheiser, J. M., Natsuaki, N. M., … Thapar, A. (2013). Biological and rearing mother influences on child ADHD symptoms: Revisiting the developmental interface between nature and nurture. *Journal of Child Psychology and Psychiatry, 54*, 1038–1046.

Healey, D. (2014). Attention-deficit/hyperactivity disorder and creativity: Ever the twain shall meet? In J. C. Kaufman (Ed.), *Creativity and mental illness* (pp. 236–249). New York, NY: Cambridge University Press.

Healey, D., & Rucklidge, J. J. (2006). An investigation into the relationship among ADHD symptomatology, creativity, and neuropsychological functioning in children. *Child Neuropsychology, 12*(6), 421–438.

Hearn, D., & Stone, S. (1995). Multiple intelligences and underachievement: Lessons from individuals with learning disabilities. *Journal of Learning Disabilities, 28*(7), 439–448.

Hinshaw, S. P. (2002). *The years of silence are past: My father's life with bipolar disorder.* New York, NY: Cambridge University Press.

Hinshaw, S. P. (2010). Growing up in a family with bipolar disorder: Personal experience, developmental lessons, and overcoming stigma. In D. J. Miklowitz & D. Cicchetti (Eds.), *Bipolar disorder: A developmental psychopathology approach* (pp. 525–556). New York, NY: Guilford Press.

Hinshaw, S. P. (2017). *Another kind of madness: A journey through the stigma and hope of mental illness*. New York, NY: St. Martin's Press.

Hinshaw, S. P., & Ellison, K. (2016). *ADHD: What everyone needs to know.* New York, NY: Oxford University Press.

Hinshaw, S. P., Owens, E. B., Zalecki, C., Huggins, S. P., Montenegro-Nevado, A. J., Schrodek, E., & Swanson, E. N. (2012). Prospective follow-up of girls with attention-deficit/hyperactivity disorder into early adulthood: Continuing impairment includes elevated risk for suicide attempts and self-injury. *Journal of Consulting and Clinical Psychology, 80*, 1041–1051.

Hinshaw, S. P., & Scheffler, R. M. (2014). *The ADHD explosion: Myths, medication, money, and today's push for performance*. New York, NY: Oxford University Press.

Hinshaw, S. P., & Stier, A. (2008). Stigma as related to mental disorders. *Annual Review of Clinical Psychology, 4*, 269–293.

Honos-Webb, L. (2010). *The gift of ADHD: How to transform your child's problems into strengths* (2nd ed.). Oakland, CA: New Harbinger.

Johnson, S. L., Tharp, J. A., & Holmes, M. K. (2015). Understanding creativity in bipolar I disorder. *Psychology of Aesthetics, Creativity, and the Arts, 9*(3), 319–327.

Kaufman, J. C. (Ed.). (2014). *Creativity and mental illness.* Cambridge: Cambridge University Press.

Kaufman, S. B. (2013). *Ungifted: Intelligence redefined.* New York, NY: Basic Books.

Kuhn, T. S. (1962). *The structure of scientific revolutions* (2nd ed.). Chicago, IL: University of Chicago Press.

Kyaga, S., Landén, M., Boman, M., Hultman, C. M., Långström, N., & Lichtenstein, P. (2013). Mental illness, suicide and creativity: 40-year prospective total population study. *Journal of Psychiatric Research, 47*(1), 83–90.

Martinez, A., & Hinshaw, S. P. (2016). Mental health stigma: Theory, developmental issues, and research priorities. In D. Cicchetti & D. J. Cohen (Ed.), *Developmental psychopathology* (4th ed.). Hoboken, NJ: Wiley.

Mattison, R. S. (2003). *Robert Rauschenberg: Breaking boundaries.* New Haven, CT: Yale University Press.

Maughan, B. (1995). Long-term outcomes of developmental reading problems. *Journal of Child Psychology & Psychiatry, 36*(3), 357–371.

Nigg, J. T. (2006). *What causes ADHD? Understanding what goes wrong and why.* New York, NY: Guilford Press.

Nigg, J. T. (2017). Attention deficit hyperactivity disorder. In T. P. Beauchaine & S. P. Hinshaw (Eds.), *Child and adolescent psychopathology* (3rd ed., pp. 407–448). Hoboken, NJ: Wiley.

PLoS Pathogens. (2014, June 10). *I'm a scientist with learning disabilities and that's OK!* Retrieved from http://blogs.plos.org/speakingofmedicine/2014/06/10/im-scientist-learning-disabilities-thats-okay/

Pylyshyn, Z. W. (1972). Are there discovery procedures? *Bulletin of the British Psychological Society, 25*(88), 189–192.

Rose, L. T., & Ellison, K. (2013). *Square peg: My story and what it means for raising innovators, visionaries, and out-of-the-box thinkers.* New York, NY: Hyperion.

Rothenberg, A. (1996). The Janusian process in scientific creativity. *Creativity Research Journal, 9*(2–3), 207–231.

Shaywitz, S. E., & Shaywitz, B. A. (2005). Dyslexia (specific reading disability). *Biological Psychiatry, 57*(11), 1301–1309.

Sonuga-Barke, E., Bitsakou, P., & Thompson, M. (2010). Beyond the dual pathway model: Evidence for the dissociation of timing, inhibitory, and delay-related impairments in attention-deficit/hyperactivity disorder. *Journal of the American Academy of Child and Adolescent Psychiatry, 49*, 345–355.

Sternberg, R. J. (Ed.). (1999). *The nature of creativity: Contemporary psychological perspectives.* New York, NY: Cambridge University Press.

Tannock, R. (2013, May). Rethinking learning disorders. *Scientific American*. Retrieived from http://blogs.scientificamerican.com/mind-guest-blog/rethinking-learning-disorders/

Tarver, S. G., Ellsworth, P. S., & Rounds, D. J. (1980). Figural and verbal creativity in learning disabled and nondisabled children. *Learning Disability Quarterly, 3*(3), 11–18.

Vellutino, F. R., Fletcher, J. M., Snowling, M. J., & Scanlon, D. M. (2004). Specific reading disability (Dyslexia): What have we learned in the past four decades? *Journal of Child Psychology and Psychiatry, 45*, 2–40.

Volkow, N. D., Wang, G., Kollins, S. H., Wigal, T. L., Newcorn, J. H., Telang, F., ... Swanson, J. M. (2009). Evaluating dopamine reward pathway in ADHD: Clinical implications. *Journal of the American Medical Association, 302*, 1084–1091.

West, T. G. (1991). *In the mind's eye: Visual thinkers, gifted people with learning difficulties, computer images, and the ironies of creativity*. Amherst, MA: Prometheus Books.

White, H. A., & Shah, P. (2011). Creative style and achievement in adults with attention-deficit/hyperactivity disorder. *Personality and Individual Differences, 50*(5), 673–677.

Homo Ludens

Jan Visser

Reflections inspired by Carlo Fabricatore and Ximena López: 'Education in a complex world: Nurturing chaordic agency through game design'

We are a playful species. We play games all the time for different purposes and with different means. The Dutch historian and cultural theorist Johan Huizinga considers play an element of culture, not something that occurs within a culture but that instead is an integral part of it. There is no culture without play. In addition to the designations 'Homo sapiens' and 'Homo faber,' he therefore suggests that we should also call ourselves 'Homo ludens.' Writing, in 1938, about the nature and significance of play as a cultural phenomenon, he states that play

> goes beyond the confines of purely physical or purely biological activity. ... [T]here is some sense to it. In play there is something 'at play' which transcends the immediate needs of life and imparts meaning to the action. All play means something. (Huizinga, 1980, p. 1)

Just as children are born scientists, they are also born actors and playwrights. From a very early age on, children demonstrate a remarkable natural urge to explore, inquire, discover things, and try things out in play. The question 'What if...?' seems to be continually on their mind. Play is also an important coping mechanism as one can see when observing children in circumstances of war and distress. It helps children to live with the reality at hand. I remember well the day in November 1944 when there were soldiers in the street in front of our house and my father was taken away.[1] As it happened, my sister and I started playing, making fun of the soldiers, modifying and singing a nursery rhyme to suit the situation. The incident has remained stuck in my memory because I have long wondered why we played. The answer came when, in the early 1970s, I was working on a documentary film about the plight of Palestinian refugee children. One child explained how her house had been destroyed by bombs and how she and her family had fled. Upon completing her story, she told the researcher: "It was a dream." There you are. When you can't live with the reality

 | DOI: 10.1163/9789004416802_032

at hand, you better do something about it. Play around with thoughts and songs to build a different world in the mind.

Playing games while experimenting with different scenarios helps us to prepare, imagine, and experiment with different futures. This is crucial at the current juncture in time as the world is facing problems of a size and complexity that cannot be adequately addressed using reductionist thinking. Effective interaction with such problems requires, as Carlo Fabricatore and Ximena López argue in their chapter on 'Education in a complex world,' different approaches. Change that works in our complex world comes from 'chaordic agency,' agency that is able to understand and embrace complexity and is ready to navigate confidently in a world that is full of uncertainty and ambiguity.

Note

1 He eventually returned safe and sound after the war was over a year later.

Reference

Huizinga, J. (1980). *Homo ludens: A study of the play element in culture*. London: Routledge & Kegan Paul [Translation from the original Dutch: *Homo ludens: Proeve eener bepaling van het spel-element der cultuur.*]

CHAPTER 17

Education in a Complex World: Nurturing Chaordic Agency through Game Design

Carlo Fabricatore and Ximena López

1 Introduction: Actors in the Wicked Drama of a Complex World

We live in a world driven by dynamics that challenge and very often defeat traditional reductionist approaches to ethics and problem solving (Rittel & Webber, 1973). This is a world energized by global phenomena whose consequences are massive, unpredictable and uncontrollable even though they may emerge from the interplay of apparently small, limited events (McDaniel & Driebe, 2005).

This world is the realm of wicked problems, a special category which includes phenomena such as environmental emergencies, geopolitical crises, cultural decline and pandemics. Wicked problems are not fully knowable, have no definitive description, no stopping rule, and no template solution (Rittel & Webber, 1973). Ordinary problems can be conceptualized as discrepancies between a current and a desirable state of affairs. Instead, wicked problems cannot be reduced to a 'what is' vs. 'what ought to be' dichotomy. A wicked problem involves multiple interacting and often conflicting dynamic systems, stakeholders, rules and constraints. All these change over time, continuously altering the circumstances and characteristics of the problem. The complicacy and changing nature of the elements that define a wicked problem situation makes it impossible to fully and definitively know the current state of affairs. In addition, the very same attempts to address a wicked problem may generate unexpected consequences that further increase the unknowability and uncontrollability of the problem situation (cf. Rittel & Webber, 1973; Buchanan, 1992). Likewise, it is very hard, if at all possible, to unequivocally define a desirable state of affairs. Unknown factors—argue Rittel and Webber (1973)—changing circumstances and uncertain ramifications make it difficult to identify possible futures. Furthermore, they argue, what might seem desirable from one perspective could engender consequences undesirable from other perspectives: there is no 'right-or-wrong,' 'true-or-false' solution to wicked problems. Thus, they say, wicked problems defy traditional problem-solving logics because of

 | DOI: 10.1163/9789004416802_033

interdependencies, conflicting stakeholders and uncertainties, and because they are often symptoms of, or cause for other problems.

The inability to cope with wicked problems is at the origin of dire consequences for human development, of which migration crises, global terrorism and financial collapses are notable examples. Wicked problems define the drama of our complex world. This is a drama in which we are the key actors, and which becomes stark tragedy whenever we fail to timely identify dangers and act upon their origins. Even though wicked problems cannot be definitively solved, stopped or removed in their entirety (Rittel & Webber, 1973), the plot of this drama is not fixed. We can co-author it, through influencing wicked problem situations so that collectively beneficial equilibria may emerge from seemingly chaotic and detrimental dynamics originated by the interplay of complex adaptive systems.

Influencing wicked problem situations therefore means influencing the involved complex adaptive systems (henceforth referred to as CASs), first among them human social systems. CASs are compounded of entities which interact with one another and their environment so that a whole system can preserve its existence, evolving and adapting to changing environmental conditions in a self-regulated manner. The adaptive and self-regulated nature of CASs makes it impossible to fully know, predict or externally control them (cf. Fabricatore & López, 2012, 2014a). Nonetheless, the structures and behaviors of CASs can be indirectly influenced through altering their environmental conditions and mediating their interactions. This allows transforming wicked problem situations, if one has the ability to interpret them in order to continuously identify and promote possibilities for positive change (cf. Beinhocker, 1997). Developing such ability is a lifelong process that entails continuous and purposeful engagement with wicked problem scenarios and the involved CASs (cf. Fabricatore & López, 2012, 2015). This requires foundational capabilities and sensibilities that should be promoted in all educational settings. Formal educational settings play a pivotal role in this sense, given that they should be designed to prepare individuals to shape our society through developing, leading, influencing and working in its institutions.

How do CASs define wicked problem situations? What makes CASs difficult to treat? Which capabilities and sensibilities are required to influence wicked problem situations through influencing the involved CASs? What can be done to nurture these capabilities and sensibilities within formal educational settings? In order to address these questions, this chapter is articulated in four parts. The first part explores key characteristics of complex adaptive systems, explaining their key role in defining wicked problem situations and their implications for our global society. The second part outlines core capabilities and sensibilities required to deal with wicked problem situations through influencing the involved complex adaptive systems. The third part introduces Engage-Adapt-Learn (EAL), an educational framework that we have developed to foster learners' engagement with real-world wicked problem situations and the development of related capabilities and sensibilities. The fourth part presents an application case of EAL in game design higher education. This chapter concludes outlining key impacts that EAL has produced in the presented case, and proposing related reflections to promote further research and educational practice.

2 CASs as the Core of Wicked Problem Situations

2.1 *CASs as Self-Organized, Adaptive Systems*

In order to understand how wicked problem situations can be influenced and transformed, it is firstly important to comprehend key characteristics that distinguish CASs from other types of systems, and at the same time make them very difficult to treat. In everyday language, "complex" and "complicated" tend to be synonymized. This is not appropriate when dealing with CASs. CASs are aggregates of large numbers of heterogeneous elements interrelated and interoperating in manifold ways. As such, they are intrinsically complicated. However, the complexity of a CAS transcends the mere complicacy of other types of systems, because of how elements in a CAS relate to one other and interact to influence the state of the whole system.

A car engine is a good example of a complicated system. It contains numerous interacting parts and is organized according to a design. This defines the mechanics of individual parts and their interactions, so that the engine can function as a whole. Because of the high number of interacting parts, and how difficult it can be to understand their functions and relationships, a car engine appears complicated. However, since every aspect of a car engine is defined by an immutable design, even an untrained eye can unravel its complicacy, provided the availability of adequate time, tools, information and cognitive abilities required to explore the engine system. Each part in a complicated system can be understood in isolation, and properties and behaviors of the system as a whole can be understood as a composition of properties and behaviors of individual parts.

The design underpinning a complicated system allows to explain how and why the system works, and to predict how it will function over time. Complicated systems are fully knowable and predictable, and hence they are controllable. Conversely, CASs are more than aggregates of components that harmoniously interact as governed by a central design. CASs elude knowability, predictability and controllability because of their scale, self-organization and evolutionary adaptation capabilities, and their sensitivity to environmental conditions.

Within a CAS large numbers of entities interact in ways that may change over time, originating systemic features that define the CAS as a whole and cannot be fully understood nor predicted through examining behaviors and relationships of individual parts (cf. McDaniel & Driebe, 2005; Mitchell, 2009). Agents are the key type of entities that define the global structure and behavior of a CAS. Agents are highly interdependent entities that interact with one another and their environment, subject to often conflicting constraints. Influenced by these constraints, agents collaborate or compete in order to achieve and maintain states suitable for the preservation of the whole system, and their own preservation within the system (McDaniel & Driebe, 2005). Ant colonies are a good example of all this. The colony as a whole represents a CAS, and ants are the agents. While ants have individual survival needs, their behaviors are collectively aimed at preserving the colony's gene pool (cf. Johnson, 2002; Camazine, Deneubourg, Franks, Sneyd, Theraulaz, & Bonabeau, 2003). This defines the organization of the entire colony: the queen is the breeder, builders construct and maintain the nest, soldiers protect the nest, scouts and foragers find and retrieve food, and so on and so forth. Ants collaborate for the wellbeing of the colony, as in the case of foragers that connect to one another to build bridges in order to reach food otherwise unreachable, or collaborate to transport food too heavy for a single ant.

Interactions among agents do not require any form of centralized control. Agents interact in a self-organized manner, based on simple rules and limited knowledge of their proximate environment (cf. Camazine et al., 2003; McDaniel & Driebe, 2005; Mitchell, 2009). Agents 'learn' about their environment through sensing it and receiving signals sent by other entities. They then use this information as feedback loops to iteratively define their behaviors and relationships, in order to adapt to one another and to the rest of their environment. Depending on feedback loops, agents may change ongoing behaviors and relationships or preserve them, originating in the latter case patterns of global organization beneficial for the perpetuation of the CAS as a whole (cf. Johnson, 2002; Camazine et al., 2003; McDaniel & Driebe, 2005). Even though these patterns are originated by local interactions and relationships among agents, they cannot be fully understood as a linear superposition of properties and behaviors of individual agents, nor can they be reliably predicted by examining isolated local interactions. Ants are once again a good example of this. In order to find and retrieve food, scouts and foragers self-organize expeditions through sensing their surrounding environment and communicating with one another via chemical signals (pheromone trails) (cf. Johnson, 2002; Camazine et al., 2003). Scouts discover food sources through exploring and sensing the environment. After discovering a food source, they return to the nest depositing a pheromone trail. Foragers leaving the nest then sense and use this trail as feedback to follow the route to the food source. As the number of foragers along the trail increases, more pheromone is deposited on the route, the pheromone trail is strengthened, and the feedback is amplified. The stronger the feedback, the easier it is for even more foragers to identify and follow the route to the food source. Thus, in a brief time, a proper foraging expedition is organized, leading to a global order that permits the efficient retrieval of food from a source initially discovered by a solitary scout. No need for centralized control. No queen in sight. To accomplish their mission ants only need basic perception and reaction abilities, and simple rules defining how to scout for food, retrieve it, communicate with neighbors, and select and follow pheromone trails. Furthermore, the global organization of the expedition cannot be fully explained nor predicted through the observation of specific ants at specific points in time.

Agents' adaptive learning processes and the consequent emergence of patterns of global order define CASs as super-organisms that co-evolve with their environment, changing in order to enhance their fitness to the environment, while at the same time changing the environment in order to improve their fitness landscape (cf. McDaniel & Driebe, 2005). Equilibria resulting from co-evolutionary processes are not permanent. CASs are open systems, susceptible to external perturbations that can alter their equilibria and elicit new adaptation processes requiring agents to change patterns of organization. Perturbations may be originated by changes in environmental conditions. For example, some ant species cluster their own bodies to build dome-like temporary nests, and react to external temperature changes by opening or closing ventilation ducts in order to maintain an acceptable temperature inside the nest (Franks, 1989). Perturbations may also be generated by other CASs. For example, when swarms of army ants deplete vast areas of food sources, other colonies abandon these areas and build nests elsewhere, to avoid starvation (Camazine et al., 2003).

Finally, CASs are highly sensitive to small variations in the conditions defining their evolutionary adaptation process. Analogous systems (e.g. similar ant colonies of the same species) operating in apparently similar conditions might develop in time significantly different patterns of organization, simply because of the effects of minimal differences in their environmental conditions. This is what is often portrayed as the 'butterfly effect,' analogizing the influence of marginal environmental conditions on large scale complex phenomena with the flap of a butterfly's wings spurring a tornado in a distant place.

2.2 *The Core of Wicked Problems: Unknowability, Unpredictability, Uncontrollability and Self-Preserving Nature of CASs*

All accounted for, CASs can be regarded as organic wholes that adapt and evolve responding to internal needs and external perturbations through self-organized interactions with their environments, including other CASs. Through evolutionary processes CASs transition from one temporary state of equilibrium to another, traversing phases of disorganization and generating perturbations and environmental changes that can affect other CASs. Throughout these processes CASs can only be explored relying on limited information gathered from observation of groups of agents, their local interactions, and their proximate environment. This is insufficient to fully account for CASs' scale (high number of interconnected agents), properties that change due to adaptive evolutionary behaviors, and global features that cannot be reduced to superposition of local behaviors and other properties of individual agents. This makes CASs not fully knowable. By extension, CASs cannot be fully predicted, because of their unknowability, and because it is not possible to ascertain or predict all the environmental conditions that might affect their behavior and evolution over time. Unknowability and unpredictability make CASs not controllable. Furthermore, because CASs interact and co-evolve with their environments driven by self-preservation needs, and because different CASs may have conflicting needs and constraints, the evolution of a CAS may generate consequences detrimental (if not lethal) for other CASs, and even for itself, on a longer time scale.

Our whole world can be seen as a large-scale CAS, a super-organism composed of CASs of different types and scales that interact in often conflictive ways. Due to CASs' characteristics it is extremely difficult to forecast, prevent or mitigate undesirable consequences of their evolutionary processes and conflictive interactions. This is especially true when dealing with human social systems, a pivotal type of CAS in our world (cf. Johnson, 2002; Mitchell, 2009). Their importance is crucial because human beings can generate impacts on our whole world that are unattainable by other species, influencing other systems at all levels and scales, for good or for bad (cf. Third Nobel Laureate Symposium, 2011; Jones, 2014). In comparison to other CASs, human social systems present an increased level of complexity. Human beings have multiple individual needs that often conflict with the global needs of the systems they belong to. At the same time, different social systems often clash because they are driven by conflicting needs. All this generates over time systems of co-occurring and increasingly interrelated problems that defy universally optimal and definitive solutions: wicked problem systems. These are situations in which multiple social systems and other constituents interact, originating dynamics that are beneficial to some and detrimental to others (cf. Rittel & Webber, 1973; Jackson, 2003; Jones, 2014). In these situations, global equilibria can only be achieved by embracing collaboration and compromise, striving to harmonize local with global needs and constraints, in the pursuit of conditions that can satisfy all affected parties to an acceptable level (cf. Ackoff, 1994; Checkland & Scholes, 1990; McDaniel & Driebe 2005; Third Nobel Laureate Symposium, 2011; Jones, 2014). This requires a collective process of definition and agreement on the problems to address, collectively beneficial conditions to attain, and strategies to pursue. Such process may prove a challenging endeavor for multiple reasons. Firstly, social systems tend to give priority to self-preservation and this may compromise the willingness to prioritize collective interests over own interests. Secondly, incomplete and uncertain information in changing conditions may limit the ability to identify and understand all the stakeholders involved in a problem situation, their needs, constraints, and relationships with their environment. Thirdly, a lack of care for others' needs may act as a barrier to engage in fair and transparent agreement. All this may lead to negative consequences for vulnerable stakeholders and for the environments in which social systems interact.

Social systems are a central cause of wicked problem situations but they are also pivotal to their management. Through critical reflection and collective action we, as human beings, can promote and shepherd change towards better futures. This requires the ability to continuously interpret complex systems, their interactions and driving needs, in order to identify and promote conditions that can satisfy all the parties affected by the wicked problem situation. It is therefore crucial to reflect on what defines such ability and how this can be developed through formal education.

3 The Need for Chaordic Agency

Since wicked problems cannot be stopped, removed nor definitively solved, they can only be addressed through transforming conditions and dynamics that originate and govern them. For this, wicked problem situations must be continuously managed, influencing environmental conditions and mediating the interactions of the involved CASs so that these can co-evolve through chaotic phases of transformation towards collectively beneficial equilibria. We label 'Chaordic agency' the ability required for this. We use 'chaordic' as a portmanteau word to ally the ideas of chaos and order into one. Chaordic agency is not an isolated skill. Rather, it is the resultant of diverse and yet complementary capabilities and sensibilities that human beings must develop, integrate and apply to tackle wicked problem situations.

3.1 *Attitudes*

At the heart of chaordic agency lie fundamental attitudes required to motivate and orient engagement with wicked problem situations. Building upon our past research (Fabricatore & López, 2012, 2014a, 2015), these attitudes can be conceptualized as:

1. Global belongingness, i.e. a sense of belonging to a global social community that forms an organic whole with its environment.
2. Compassionate altruism, i.e. genuine concern for the well-being of others, along with proclivity to seek ways to support others.
3. Emotional commitment to the sustainable development of our global community and its environment.
4. Adaptivity, i.e. the acknowledgement that change, uncertainty and uncontrollability are ineluctable conditions that drive our world, and that as things change we can as well change, adapt and develop.
5. Confidence in individual and collective capacity of action.
6. Hope in impacts that individual and collective action can generate.

Global belongingness, compassionate altruism and emotional commitment to the sustainable development of our global world are essential to feed and orient a deep sense of responsibility for life and diversity, present and future. Adaptivity is crucial to build resilience to external changes, and to co-evolve with our environment as this changes unexpectedly and uncontrollably. Confidence in capacity of action and hope in its effects are essential to overcome potentially paralyzing sensations of powerlessness originated by the scale, unknowability, unpredictability and uncontrollability of wicked problem situations.

Altogether, these attitudes are essential to motivate and guide the exploration of wicked problem scenarios and the pursuit of changes beneficial for our global world. This pursuit also requires capabilities involving domain knowledge pertinent to the problem areas of concern, and diverse skillsets for dealing with wicked problem situations (Fabricatore & López, 2014a, 2015). We explain each of these elements below.

3.2 *Trans-Domain Knowledge*

Addressing wicked problem scenarios requires the integration of knowledge from multiple domains (cf. Buchanan, 1992; Dunne & Martin, 2006). Although important, trans-domain knowledge alone is not sufficient to ensure success. Because of the unknowability that characterizes CASs and their changing nature, no amount of information can ever be sufficient to fully define a wicked problem situation. Wicked problems cannot be formulated in a definitive and unique way (Rittel & Webber, 1973). Information will always be incomplete and changing, and a specific problem will always be susceptible to different formulations, depending on the perspective that drives its analysis. All possible formulations will engender some degree of uncertainty, and lead to different, likewise uncertain strategies (Rittel & Webber, 1973; Buchanan, 1992; Jones, 2014). This situation will not change, however much information may be available. Knowledge from multiple domains is therefore an important tool, but its importance is secondary to the skillsets required to interpret, analyze, monitor and act upon the complex networks of CASs involved in wicked problem situations.

3.3 *Skillsets*

Wicked problem situations can only be influenced through iterative and adaptive processes integrating exploration, planning and action, informed by environmental feedback from provisional outcomes of action and other ambient events (cf. Rittel & Webber, 1973; Argyris, 1977; Buchanan, 1992; Jackson, 2003). Environmental changes should inform the adaptive (re)definition of objectives, strategies and techniques involved in a plan, and the revision of the assumptions underpinning existing plans and ongoing actions (cf. Argyris, 1977; Beinhocker, 1997). Building on our past research (Fabricatore & López, 2012, 2014a, 2014b, 2015) and on key perspectives on the management of complex problems in socially-implicated contexts (e.g. Buchanan, 1992; Ackoff, 1994; Checkland & Scholes, 1990; Jackson, 2003; Jones, 2014), we posit that such processes require the integration of key operational, social and higher-order thinking skillsets. The application of these skillsets should be oriented by the above-described attitudes and informed by trans-domain knowledge.

3.4 *Operational Skillsets*

These are key to be able to interact with reality. Requirements for operational skills may vary depending on the domains and disciplines involved in a problem situation of interest. Furthermore, because of the mutability of wicked problems, required operational skillsets may also vary for a specific problem situation, as the situation evolves over time.

Operational skillsets are essential to implement planned action. More is however required for the iterative exploration of problem scenarios and the consequent planning of strategies to pursue. For this, social and higher-order thinking skillsets are paramount, as they drive the investigation and interpretation of a problem situation of concern; the collective definition of shared goals and agreed strategies to pursue; and the interpretation of results of action (cf. Dunne & Martin, 2006; Dorst, 2011).

3.5 *Social Skillsets*

These are essential to enable and drive social negotiation processes (cf. Rittel & Webber, 1973; Checkland & Scholes, 1990; Buchanan, 1992; Jackson, 2003; Jones, 2014). As wicked problems involve multiple and diverse stakeholders, dialogue and social negotiation are key to construct pluralistic perspectives that accommodate diverging and often conflictive needs and worldviews. Pluralistic perspectives are then crucial to drive the collective interpretation of reality, the definition of commonly agreed objectives and strategies, and the shared evaluation of results and implications of action. Social skills are therefore central to enable collective processes and ensure a fair participation of all the parties affected by a wicked problem situation.

3.6 *Higher-Order Thinking Skillsets*

Tackling wicked problems requires systematic processes of action-driven inquiry into reality, in order to iteratively interpret wicked problem situations and consequently plan and pursue desirable changes. Such processes rely on the thinking skillsets involved in problem solving, systems thinking, creative thinking and analogical reasoning (cf. Buchanan, 1992; Jackson, 2003; Jones, 2014).

3.7 *System Thinking and Problem Solving*

In its simplest form, a problem can be viewed as discrepancy between a current, stable and knowable state of matters and a new, desirable goal state, whose attainment requires a process of transformations to transition from initial to goal state (Ackoff, 1978). Addressing well-defined, stable problems requires a linear process consisting in a problem definition followed by a problem solution process (cf. Ackoff, 1978; Dorst, 2011). The problem definition process entails investigating the problem situation in order to identify the elements and laws that define and govern the problem's present state, and the properties that define the goal state. The problem solution process then leverages this knowledge to plan and implement a strategy to achieve the goal state. Such linear process entails the ability to integrate analytic and synthetic thinking.

Analytic and synthetic thinking skills are key but not sufficient to address wicked problem situations. Because of their indeterminacy, ill-definition and mutability, wicked problem situations have to be continuously 'managed' through an iterative process of design and implementation of strategies, based on emergent results of action (cf. Ackoff, 1978; Buchanan, 1992; Dorst, 2011). For this, systems thinking is essential. Systems thinking can be broadly regarded as a thinking paradigm aimed at conceptualizing and interpreting real-world entities as systems, i.e. wholes which are composed of interacting parts and present properties that emerge from the interplay of parts but cannot be explained considering parts in isolation (Ackoff, 1994).

Wicked problem scenarios should be modelled in terms of interacting systems, adopting a holistic and pluralistic approach (Checkland & Scholes, 1990; Buchanan, 1992; Jackson, 2003). Holism requires modeling a situation starting from global phenomena, and consequently modeling parts and interactions that might be originating such phenomena (cf. Checkland & Scholes, 1990; Ackoff, 1994). Pluralism requires acknowledging that in situations involving social or socially-implicated systems there is no single, objective and universally acceptable representation of reality: different people have different worldviews and tend to interpret and represent the same situations in different ways (cf. Checkland & Scholes, 1990; Buchanan, 1992; Jackson, 2003). Altogether, holism, pluralism, and the mutability and unknowability of wicked problem situations require embracing systems thinking as a continuous learning process driven by systematic inquiry. Through this process the stakeholders involved in a problem situation should learn about one another and the environment they are dealing with, and consequently act collectively in order to effect mutually beneficial changes (cf. Checkland & Scholes, 1990; Jackson, 2003). In this process system thinking is fundamental to allow stakeholders to formulate systemic models in order to achieve three key purposes (cf. Checkland & Scholes, 1990; Buchanan, 1992; Jackson, 2003; Jones, 2014):

1. Represent, reflect upon, and debate diverse and possibly conflictive perspectives on the problem situation.
2. Collectively identify shared objectives and suitable strategies to pursue.
3. Implement strategies and evaluate their outcomes based on commonly agreed criteria.

3.8 *Creative Thinking*

Creativity is fundamental for the interpretation of a situation and for the formulation of strategies. In relation to wicked problems, creativity can be regarded as the ability to purposefully generate novel ideas in order to model and act upon a situation of concern (cf. Ackoff, 1978; Buchanan, 1992; Runco, 2004). Creativity is key to face unknown situations and adapt to unpredictable environmental changes that may invalidate past knowledge, beliefs and experience (Runco, 2004). Creativity is also key to tackle the ill-definition of wicked problem situations, as it allows reasoning and acting in conditions of incomplete and uncertain information through devising novel systemic configurations, elements, relationships, and possibilities to intervene in a problem scenario (cf. Ackoff, 1978; Buchanan, 1992; Dunne & Martin, 2006). Finally, creativity is beneficial for the evaluation of the nature, value and implications of actions, as it supports the exploration and integration of diverse perspectives and frames of reference (cf. Ackoff, 1978; Buchanan, 1992).

3.9 *Analogical Reasoning*

This is a fundamental complement of problem solving, systems thinking and creative thinking. Analogical reasoning is the ability to identify current similarities between two situations, and use these to hypothesize other current or future similarities (Gentner & Smith, 2012). Analogical reasoning is important to identify recurrences in systemic relationships and behaviors for a given situation, establishing patterns, and predicting future situations accordingly (cf. Dorst, 2011; Gentner & Smith, 2012). Analogies are also useful to establish relationships between two different situations, and consequently interpret and predict the state, behavior and implications of one situation based on knowledge regarding the other (cf. Buchanan, 1992; Gentner & Smith, 2012).

Because of the importance of wicked problems, formal education should facilitate the development of chaordic agency capabilities and sensibilities, and promote learners' engagement with authentic wicked problem situations. Higher education should play a paramount role in this context, given that universities have the responsibility to enable learners to become agents of sustainable development for our society, through shaping, leading and working in its institutions (Fabricatore & López, 2015). For this purpose, we developed the Engage-Adapt-Learn framework (henceforth referred to as EAL), which will be presented hereafter.

4 Strategizing Education to Foster Chaordic Agency: The EAL Framework

EAL is a system of tenets and heuristics that aims at fostering the development of chaordic agency sensibilities and capabilities, and promoting learners' awareness, comprehension of and affective engagement with complex wicked problem situations that affect our society. EAL tenets conceptualize learners, the process of learning and its focus through the lenses of complexity science, human activity and motivation, accounting for the conditions that characterize real-world wicked problem scenarios. The heuristics operationalize the tenets, presenting guidelines to support the design of educational strategies.

EAL embraces the idea that in real-world circumstances learning happens as an ongoing process of adaptive transformations involving learners and their environments, triggered by changes in the personal state and/or environmental conditions that perturb the learner (cf. Jonassen, 2002; Bedny & Karwowski, 2006; Davis & Sumara, 2006; Visser, 2008, 2012). Perturbations can be regarded as disequilibrating stimuli that disrupt what the learner might perceive as a state of balance (e.g. conflicts; cognitive misalignments, slips and dissonances; threats; and puzzlements). As perturbations affect the learner cognitively, emotionally and/or physically, they generate a need for change. The learner reacts to this through engaging in a process of interactions with the environment, construing meaning in order to explore and pursue possibilities to attain desired changes, in the self and/or the environment. Such process evolves until the learner reaches a new, provisional equilibrium, which lasts until new meaningful perturbations arise. In this process the learner's actions occasion and are occasioned by changes in the environment, so that the learner transforms the environment as the environment transforms the learner (Davis & Sumara, 2006). Through this transformative process the learner develops continuously at cognitive, affective and behavioral levels, purposefully developing capacities relevant to interact with the environment (cf. Jonassen, 2002; Visser, 2008). This process unfolds as learners engage in complex dynamics that they consider real-world-relevant and personally significant, and which require them to deal with change, unpredictability, unknowability and uncertainty (cf. Davis & Sumara, 2006; Dunne & Martin, 2006; Dorst, 2011). From this stem the first two EAL tenets:

EAL Tenet #1: In natural conditions learning unfolds as an iterative process of reciprocal adaptive transformations of the learner and the environment, which integrates and develops the learners' cognition, affection and behavior.

EAL Tenet #2: Purposeful learning is an activity-entailed process of development of capabilities and sensibilities required to meaningfully interact with the world. It is intrinsic to all conscious and purposeful human activity, it unfolds through activity, and cannot be separated from activity.

The structure, dynamics and motivation that define human activity and the involved learning can be modeled and understood through the conceptual framework of the Activity Theory (cf. Jonassen, 2002; Bedny & Karwowski, 2006), henceforth referred to as AT. According to AT, activity can be conceived as a purposeful and contextualized process through which a subject interacts with the world in order to transform an object into a desired outcome (cf. Leont'ev, 1978; Bedny & Karwowski, 2006; Engeström, 2014). The object should be regarded as a system in itself. It may be more or less tangible, as long as it can be shared with other subjects participating in the activity (e.g. a person; a compound physical object; a social system; a body of knowledge; etc.) The subject engages in a process of material or conceptual transformations of the object in order to pursue meaningful goals, which are mental representations of a desired future state of the activity's object. A goal may require the transformation of existing properties of the activity's object, of relationships between the object and other environmental elements, or the generation of an entirely new object. Transformations may be direct or mediated by tools (cf. Kaptelinin, 1996; Bedny & Karwowski, 2006; Engeström, 2014). These may be material (e.g. instruments) or conceptual (e.g. theories), pre-existing or newly-created through intermediate stages of activity. The process of activity may be individual or collective, when several subjects interact to attain a shared goal (cf. Bedny & Karwowski, 2006; Engeström, 2014). Collective activity is mediated by mechanics that define and regulate interpersonal interactions and organize collective efforts, such as roles, responsibilities, and rules of communication and collaboration.

Goal-directed transformations mediated by external tools entail multiple types of learning. These are necessary to develop competencies (knowledge and skills) required to plan and implement action, leverage tools, and interact with other activity's participants. Exploratory learning may be required to investigate objects as the activity unfolds, and to iteratively define goals accordingly (Bedny & Karwowski, 2006). Learning may also be required to develop operational and procedural competencies (Kaptelinin, 1996). Operational competencies integrate knowledge about the functioning of tools and the skills required to operate them. Procedural competencies define the ability to plan and carry out action using tools in context. They integrate knowledge on types of higher-level goals attainable through tools, and skills to identify problems that can be addressed through the use of tools and methods to do so. Procedural competencies also involve the ability to coordinate and prioritize goals and problem-solving processes, accounting for broader contexts of activity and limitations imposed by the available tools. Interpersonal activity also entails learning intended as development of competencies required to plan and perform collective endeavors (cf. Checkland & Poulter, 1990; Buchanan, 1992; Jackson, 2003; Bedny & Karwowski, 2006; Dunne & Martin, 2006; Engeström, 2014). Interpersonal competencies integrate: knowledge of formal and informal mechanisms that regulate intersubjective interactions; knowledge of personal conditions of other subjects which may influence intersubjective activity; and skills required to explore and acquire the above knowledge, and use it to collectively define activity goals, strategies, and evaluate outcomes.

The context in which an activity takes place also originates learning. An activity context consists of interrelated conditions that afford or constrain goal definition and attainment, and influence the meaningfulness and personal significance that activity's participants may attribute to the activity (cf. Leont'ev, 1978; Nardi, 1996; Ryan & Deci, 2000; Bedny & Karwowski, 2006). The context fully defines an abstract activity as a concrete task. The context involves broad environmental conditions that relate to the historical, cultural, societal, and physical setting in which the activity takes place, and to the personal characteristics and circumstances of the activity's participants (cf. Leont'ev, 1978; Radomsky, 2013). Consequently, we posit that activity engagement entails the development of contextualization competencies. These competencies are defined by knowledge of broad environmental conditions which influence task meaning, organization and performance; the skills required to investigate the environment in order to attain such knowledge; and the ability to integrate this knowledge with knowledge pertaining to the object of the activity, goal and other elements that directly mediate activity. From this stems the third EAL tenet:

> *EAL Tenet #3*: Activity-entailed learning involves comprehending the meaningfulness of an activity, which is defined by the context in which the activity unfolds.

Intersubjective dynamics are also important for their impact on the scope and outcomes of learning processes. In conditions of complexity, learning is a trans-level process, happening both at the level of individuals and whole collectives (Visser, 2001a, 2001b; Davis & Sumara, 2006). Through intersubjective dynamics the interplay of individual understandings and knowledge produces collective learning. This, in turn, feeds back into individual learning, enhancing it beyond what would be attainable by individuals alone. The nature and relevance of collective learning lead to the fourth EAL tenet:

> *EAL Tenet #4:* Learning is an interpersonal process that originates outcomes exceeding what individuals alone could attain.

Finally, the way activity unfolds in complex scenarios also plays an important role in defining how learning develops in wicked problem situations. Activity is a dialogical process of interaction between subjects and their environments (cf. Visser, 2012): subjects change as they interact with their environment, in response to meaningful stimuli partly originated by the environment; the environment changes in response to subjects' actions, generating further stimuli; and so on. In this process, subjects learn in order to adapt, and adapt in order to learn, continuously, incrementally and iteratively, attempting to react in the best possible way to mutable and uncontrollable events. This process is self-organized and non-linear, and so is the learning involved in it (cf. McDaniel & Driebe, 2005; Davis & Sumara, 2006). Learning does not occur according to pre-defined plans. Rather, it emerges in response to the necessity to adapt to changing circumstances. Furthermore, learning does not happen because of external control. Instead, subjects learn individually and collectively depending on how they self-organize their adaptation processes, and what they need to know to define their goals, and to plan, implement and evaluate strategies. From this stems the fifth EAL tenet:

EAL Tenet #5: Learning is a self-organized, non-linear process.

The EAL tenets have key implications for educational practice. Hence, we have formulated four heuristics—delineated hereafter—which embody the EAL tenets, and are aimed at informing educational strategies to foster chaordic agency and learners' engagement with wicked problem scenarios.

EAL heuristic #1: Definition of the object of learning and scope of learning-entailing activities.

a *Trans-domain activity.* Learning should be required for the sake of developing activities involving different domains (cf. Davis & Sumara, 2006; Fabricatore & López, 2014b).

b *Focus on systems and complex problems.* Systems should be the object of learning-entailing activities. These activities should require learners to critically analyze complex problem situations in order to identify meaningful objectives to achieve, and consequently create and/or modify systems to attain these objectives (cf. Jonassen, 2002; Dunne & Martin, 2006).

EAL heuristic #2: Contextualization of learning-entailing activities.

a *Authenticity of learning scenario.* Learning-entailing activities should be contextualized in real-world wicked problem scenarios (cf. Dunne & Martin, 2006). Hence, activities should require learners to tackle complex problem situations mirroring salient aspects of real world wicked problem scenarios.

b *Ill-definition of scenario.* The problem situations tackled by learners should be ill-defined, not fully knowable nor predictable, and subjectable to different interpretations depending on the perspective embraced (cf. Ackoff, 1978; Dunne & Martin, 2006).

EAL *heuristic #3:* Promotion of collective participation and self-organization of learning-entailing activities.

a *Motivation of collective activity.* Learning-entailing activities should be meaningful, individually as much as collectively. The object and purpose of learning-entailing activities should require interaction between learners relevant to collectively define shared objectives, plan and implement strategies, and evaluate results (Fabricatore & López, 2014b).

b *Action-driven self-organization of collective activity.* The organization and goal of collective activity should not be externally defined (cf. Davis & Sumara, 2006; Fabricatore & López, 2014b). Learners should autonomously define objectives, strategies, and their roles and responsibilities. They should be enabled to organize activity based on the iterative exploration of meaningful scenarios, in response to provisional outcomes, and accounting for the personal and environmental contexts in which they operate.

EAL *heuristic #4:* Adaptive support of learning-entailing activities.

a *Support based on continuous monitoring.* Support should be generally provided in response to learners' progression in their activities, based on needs and opportunities emerging from provisional results of their strategies. Hence, support should be based on continuous monitoring of the learners' activities (Fabricatore & López, 2014b).

b *Provision of core knowledge to build shared understandings.* In order to promote intersubjective dynamics, learning instances centered on the reception of information (such as listening to lectures, watching videos, or reading texts) should be leveraged to promptly provide core knowledge. This should be the minimum knowledge required by learners to build an initial shared understanding of the scenarios to be tackled and the key methods to investigate and act upon them. This knowledge should serve as a shared language to allow learners to collectively engage in learning-entailing activities as early as possible (cf. Davis & Sumara, 2006; Fabricatore & López, 2014b).

c *Support for heterogeneity and non-linearity in learning, and adaptive provision of information.* To facilitate non-linear learning, learners should be enabled to access knowledge iteratively and incrementally, promoting a continual deepening of their understandings, and their ability to use the acquired knowledge in increasingly complex ways (Bruner, 1977). Furthermore, self-organization requires knowledge to be accessible as and when it is needed, in a non-linear fashion. Hence, learners should be enabled to autonomously identify and access knowledge relevant to the specificities and states of their activities. Finally, heterogeneous materials should be used to support different learning styles (Fabricatore & López, 2014b).

d *Enabling feedback.* Feedback should be continuously provided to learners, based on iterative monitoring of provisional outcomes of their activities (Fabricatore & López, 2014b). In order to support self-organization and non-linear progression of learning, feedback should not be prescriptive. Rather, it should enable learners to decide what to pursue, how to do so, and to ultimately achieve their purposes (Fabricatore & López, 2014b). Feedback should generally foster learners' collective inquiry into the scenarios that they are tackling (working scenario), promoting the exploration of the 'unknown' and the critical analysis, planning and evaluation of the 'ongoing' (cf. Davis & Sumara, 2006; Dunne & Martin, 2006). Hence, feedback should enable learners to investigate their working scenarios through alternative perspectives, stimulating the exploration of unforeseen possibilities of action and their possible implications. The adoption of alternative perspectives should also be promoted to facilitate the critical evaluation of ongoing strategies and their implications. Analogies between learners' working scenarios and other situations (real or hypothetical) can be leveraged to promote the use of alternative perspectives as 'lenses of inquiry.'

e *Adaptive support to intersubjective activity.* Support mechanisms should be implemented to promote the emergence of individual worldviews, and facilitate their integration and accommodation in case of conflict (Dunne & Martin, 2006). To promote self-organization, support should be provided to allow learners to define the nature and purpose of their relationships and collective action (Fabricatore & López, 2014b).

f *Enabling constraints.* Constraints are an ineluctable reality in formal educational environments (e.g. coursework requirements, deadlines, etc.), and may have negative impacts on self-organization possibilities. To prevent this, constraints should be underpinned by openness to unanticipated possibilities, orienting learners in the exploration of viable options while at the same time allowing varied and flexible responses to their circumstances (Davis & Sumara, 2006).

5 GD-EAL: An Application of EAL to Foster Chaordic Agency through Game Design

EAL is an evolving framework. We have developed it since 2013 with a specific focus on wicked problem scenarios through implementing it in video game design courses. For this, we designed an EAL-based educational strategy which leverages video game design (GD) as a core learning-entailing activity: GD-EAL. In this section we describe key aspects GD-EAL to demonstrate a case of application of the EAL framework.

5.1 *Design Thinking, Chaordic Agency and Video Game Design*

We selected video game design for its potentialities to foster the development of chaordic agency and promote learners' engagement with real-world wicked problem scenarios. Design activities can be a powerful means to develop core chaordic agency capabilities and sensibilities, if they are planned to foster design thinking. Design thinking is a pragmatic reasoning strategy, conceived to deal with wicked problem situations involving multiple interacting stakeholders driven by conflicting needs and constraints (cf. Buchanan, 1992; Dorst, 2011). Design thinking tackles ill-defined problem situations through a systematic and iterative core process which integrates the exploration of problem situations from multiple perspectives with the planning, implementation and evaluation of interventions (i.e. creation or modification of systems) (cf. Buchanan, 1992; Dunne & Martin, 2006; Dorst, 2011). This process requires interventionists (designers) to integrate abductive, inductive and deductive reasoning. Abductive reasoning is needed to formulate working hypotheses regarding how and why an ill-defined problem system functions, and what solution systems could serve to attain desired objectives. Inductive and deductive reasoning are then used in order to iteratively plan, implement, test and evaluate a solution system based on the accepted working hypotheses. Emergent results are used to review working hypotheses, possibly leading to their re-formulation and new iterations.

Design thinking involves systems thinking skillsets in order to explore and model a scenario of interest, as well as problem-solving skillsets in order to identify problematizing conditions, feasible improvements, and strategies to pursue these improvements through an intervention (cf. Buchanan, 1992; Dunne & Martin, 2006; Dorst, 2011). Design thinking leverages analogical reasoning to define interventions building upon similarities between a problem situation of concern and other contexts (Buchanan, 1992). Design thinking also relies on creativity to conceive unprecedented possibilities of improvement, and implement interventions accordingly (Dunne & Martin, 2006). Thus, design thinking requires the thinking skillsets that are core to chaordic agency. Furthermore, the design thinking core process must seek the harmonization and participation of all the stakeholders involved in the formulation of working hypotheses, plans, and the consequent implementation and evaluation of interventions (cf. Buchanan, 1992; Dorst, 2011). Hence, design thinking also elicits chaordic agency social skillsets required to accommodate potentially conflictive perspectives, and formulate pluralistic perspectives to address wicked problem situations.

The literature on design thinking suggests that formal education can promote the development of the alluded thinking and social skillsets through collective design activities, provided that these embed the design thinking core process and are focused on ill-defined problem systems (cf. Dunne & Martin, 2006; Scheer, Noweski, & Meinel, 2012). Accordingly, the GD-EAL strategy is centered on a collaborative game design project articulated mirroring the design thinking core process. The prototypical structure of a GD-EAL project consists in iterations that integrate research, design, implementation and evaluation activities, and are separated by project review milestones. Through this process GD-EAL requires designers-learners to design a game contextualized in a real-world wicked problem situation. The systemic nature of video games and the ill-definition of the wicked problem scenarios involved in the project ensure that designers-learners focus their work on ill-defined problem systems, and actually engage with real-world wicked problems.

Video games can be regarded as systems in which players engage in interrelated problem-solving activities, interacting with game elements (e.g. other human players, digital objects and virtual agents) in order to pursue desired goals (cf. Fabricatore, 2007; Fabricatore & López, 2014a; Fabricatore, 2018). Gameplay activities can be set in fictional contexts involving storylines and settings which can mirror real-world scenarios (e.g. geopolitical, cultural and historical circumstances). Game contexts allow players to interpret meanings and relationships of game events and all the involved game elements (cf. Fabricatore & López, 2014a; Fabricatore, 2018). Based on meaning-making processes, as players progress they act in response to challenges and opportunities presented by the game, transforming the game environment through their actions, and transforming themselves as they develop gameplay mastery. In this process, players experience a holistic engagement integrating the cognitive, affective and operational dimensions: players think in order to act, and feel as they think and act (cf. Fabricatore & López, 2012, 2014a; Fabricatore, 2018).

Contextualizing gameplay activities in wicked problem scenarios requires designers-learners to deeply explore such scenarios, applying systems thinking and problem-solving skills in order to understand or hypothesize the involved elements, interactions, dynamics and mechanics. Based on this exploration, systems thinking and problem-solving are also needed to model the artificial systems that players have to engage in, defining the problem situations that players should address, and the possibilities of action and viable strategies available to them. Furthermore, in order to foster players' affective engagement a game should promote meaningful aims and motivate their pursuit. Thus, contextualizing a game in a wicked problem scenario requires designers-learners to explore the real-world scenarios from an affective perspective, considering the involved stakeholders, and exploring who might feel what, and why. Through this exploration designers-learners have an important opportunity to develop their own affective engagement, sensibilities and dispositions towards the scenarios being tackled.

5.2 *Promotion of Self-Organized, Collective and Non-Linear Learning*

In addition to leveraging game design to foster design thinking and engagement with wicked problem scenarios, GD-EAL involves other aspects focused on promoting self-organization, intersubjective dynamics and non-linear learning.

GD-EAL defines project requirements to foster trans-personal learning and the development of social skillsets. The design project is collaborative, and this is intrinsically justified by the scope, amount of work and range of skillsets required to complete a project. Collaborative design stimulates meaningful social negotiation among designers having diverse views regarding how a system should be designed, and the approaches to implement and evaluate the design (cf. Davis & Sumara, 2006; Dunne & Martin, 2006). To broaden the scope and variety of social negotiations, designers-learners are asked to elaborate a player engagement strategy and conduct player testing processes. This requires studying the profile of target users, in order to hypothesize and debate when, how and why players would engage in the game, cognitively, operationally and affectively. Player testing is also leveraged to promote direct interaction with the player, thus further broadening the spectrum of social interactions among stakeholders involved in the project. Furthermore, each designer-learner has to develop a project journal, reflecting on individual and collective experiences in relation to purposes pursued, strategies adopted and outcomes attained. Designers-learners are encouraged to use their journals as an input for team discussions, in order to promote the emergence and interplay of diverse perspectives, and the positive impact that this can have on individual and collective learning (cf. Davis & Sumara, 2006; Dunne & Martin, 2006). Finally, a project process embeds continuous interactions with tutors acting as advisors, in order to implement a further type of social negotiation relevant for the exploration of working scenarios and the design of game systems.

In order to foster self-organization GD-EAL promotes flexibility in the definition of game contexts, contents, and work approaches. Designers-learners are allowed to freely define game contexts based on real-world wicked problem situations and related themes. For example, in the latest editions of our courses, designers-learners defined their working scenarios based on a cover story in TIME magazine and a specific theme. The cover story could be freely chosen, provided that it concerned a contemporary, high social impact complex problem. The theme could be freely selected from a list representing alternative perspectives on wicked problems, in order to stimulate designers-learners to explore different perspectives.

With respect to project contents, GD-EAL allows designers-learners to design any type of game they wish, as long as this meets two key requirements. Firstly, the game has to foster learning processes aimed at enhancing players' awareness, comprehension, affective involvement and sensation of agency regarding the wicked problem scenario underpinning the game. In this manner we sought to promote deep exploration of the scenario, by requiring designers-learners to hypothesize what players could learn because of the game, and how this could affect players' attitudes towards the scenario involved in the game. Secondly, the game has to be story-driven. This was decided so as to stimulate the analysis of key stakeholders involved in the wicked problem scenario underpinning the game, which would be essential to defining the key actors in the game narrative. Both requirements have been planned as enabling constraints, to orient and focus designers-learners' work without curtailing their ability to decide what to do, how and why.

Concerning work approaches, GD-EAL allows designers-learners to self-organize roles and responsibilities, work strategies and plans, and adjust them throughout the project in response to emergent outcomes. To support this, summative assessment is implemented only at the end of the academic year, in order to minimize the impact that this could have on iterative design processes.

GD-EAL relies on knowledge provision strategies planned based on the EAL heuristics. Designers-learners are provided with core knowledge through lectures delivered in the first few weeks of the academic year, supplemented by materials made available online (e.g. lecture notes, presentations, videos, etc.) This aims at facilitating the formulation of shared understandings and promoting purposeful intersubjective interactions and collaborative work as early as possible.

GD-EAL facilitates non-linear learning and on-demand access to information through the use of a bespoke type of concept maps: knowledge maps. Concept maps are generally useful tools to enable the teaching of relevant subject knowledge and to support the thinking processes in problem-solving and design activities (cf. Oxman, 2004; Kokotovich, 2008). We designed knowledge maps specifically to support game design as an iterative problem-solving process. For this, we articulated knowledge maps in a hierarchy of nodes. Each node contains a topic and related guiding questions. The topic represents a key type of game feature, and the questions support creative inquiry to decide how to design that feature and why. Knowledge maps associate specific knowledge with each node. This knowledge includes: 'big ideas' to be considered when designing a specific game feature; strategies useful to design that feature; and additional resources (e.g. articles and podcasts) to deepen the understanding of that feature and related design issues. This structure has been conceived to support on-demand access to knowledge depending on the state of the project. As designers-learners tackle a specific game feature, they can use, iteratively and incrementally, the knowledge map contents associated with that feature. Following a 'what, how and why' logic, guiding questions naturally lead to big ideas, which in turn lead to strategies and supporting resources. Guiding questions, big ideas and strategies were also planned to be non-prescriptive, in order to promote designers-learners' inventiveness and autonomous search for knowledge.

5.3 *Adaptive Support to Learning*

GD-EAL promotes collective work and non-linear, incremental learning through adaptive support provided to designers-learners. This consists in multiple forms of tutor mediation based on the project progress and emerging outcomes of designers-learners' work.

Self-organized team project revision sessions are mediated by tutors in order to stimulate and support discussion on current project state, future plans and individual responsibilities. As part of these revision sessions, designers-learners are encouraged to collectively discuss reflections that they recorded in their project journals.

Non-linear and incremental learning is fostered through project workshops. These are organized so that tutors can provide support to teams and individual designers-learners, based on emerging project outcomes rather than plans predefined by tutors. Intermediate project revision milestones are also planned for this purpose. On these occasions designers-learners receive formative feedback from tutors and peers, without summative assessment pressures that could compromise self-organization processes.

The feedback provided by tutors throughout the project is advisory and exploratory. Advisory feedback aims at supporting designers-learners to improve their current strategies. Exploratory feedback aims at helping them to identify and explore possibilities and implications that might have not been sufficiently considered. Case studies are used by tutors to promote analogical reasoning, helping designers-learners to reflect on their projects based on analogies with other cases.

Multiple tutors with different disciplinary backgrounds are involved in the project process. This is done to stimulate designers-learners' engagement with multiple perspectives, and strengthen the transdisciplinary nature of their work.

In order to adapt their mediations, tutors monitor project progress and emerging outcomes continuously. Provisional outcomes are ascertained based on direct observation of work and interaction with teams. Project journals are also used as key feedback to systematically monitor progress and collective dynamics for each team, and to get a sense of engagement of each team member. For this, designers-learners are required to compile a journal entry after each project review milestone, reflecting on contents, quality of their projects, issues to address, trajectories for future improvement, and key aspects of personal engagement, individual performance, and dynamics of the whole team.

6 Conclusive Reflections: Is This Working? What Next?

We live in a world of wicked problems defined by non-fully knowable, predictable or controllable dynamics, at the center of which are human social systems. These dynamics have important and possibly dramatic implications for our future. Dealing with them requires specific capabilities and sensibilities. To identify and nurture these capabilities and sensibilities, it is important for formal education to adopt perspectives based on complexity theory, and emphasize learners' collective engagement with real-world wicked problem scenarios. For this, formal education should regard these scenarios as manifestations of co-evolutionary interactions involving CASs and their environments. Consequently, it should implement educational strategies to promote the investigation of wicked problem scenarios, and the exploration of their possible origins, implications and possibilities to promote desirable changes. It is through this process that learners should be supported in the development of the capabilities and sensibilities involved in chaordic agency. This motivated and oriented our work on the EAL framework.

In our past work we have presented results indicating that strategies mirroring EAL can be beneficial to developing skills required to face complex dynamics through self-organized team work; promoting trans-personal learning; and enhancing individual and collective learning (Fabricatore & López, 2014b). We have also reported outcomes suggesting that EAL can promote the understanding of wicked problems, the appreciation of their relevance, and learners' desire to actively contribute to improving wicked problem situations (Fabricatore & López, 2015). Although it is beyond the scope of this chapter to present further analyses of quantified impacts, it is worth highlighting some notable qualitative results of the application of the EAL framework through the GD-EAL strategy.

The game systems created by designers-learners are generally highly articulated, presenting multiple interacting elements that are at the origin of phenomena which require players to face ill-defined problems, change and uncertainty. The games are generally contextualized in novel scenarios that accurately incorporate elements of real-world wicked problems. These scenarios frequently embody original perspectives, reflecting creative exploration of wicked problem situations. Altogether, this suggests that GD-EAL can promote deep exploration and understanding of wicked problems, and the development of core chaordic agency skillsets.

The rationales elaborated by designers-learners to explain and justify why players should wholly engage with their games often suggest deep reflection on ethical implications of real-world wicked problems. The projects' storylines frequently demonstrate that designers-learners are able to interpret wicked problem situations through the eyes of key stakeholders involved in them, appreciating and valuing the diversity of their needs, purposes and backgrounds. In some notable cases designers-learners' decided to continue game design work based on wicked problems even outside our courses. Altogether, this suggests that GD-EAL can promote learners' affective engagement with real-world wicked problem scenarios.

Observation of group dynamics indicated that GD-EAL fosters the development of social skillsets, self-organization, individual and collective learning, and the improvement of project outcomes. Designers-learners' ability to self-organize and collaborate purposefully tends to improve throughout the project, and they are usually aware of this. We have observed that ongoing team discussions on wicked problem situations can stimulate increasing interest and progressive engagement of individuals who are initially reticent and/or uninterested in the course. We have also observed cases in which project work has led to sustained dialogue on wicked problems outside the course, both among designers-learners, and between designers-learners and tutors. Finally, we appreciated that designers-learners' ability to dialogue with stakeholders outside the team (e.g. tutors and players) increases over time, as does their ability to extrapolate and leverage new ideas and perspectives emerging from these dialogues.

Overall, our observations so far suggest that EAL and game design activities can indeed promote significant learners' engagement with wicked problem scenarios, and the development of key chaordic agency capabilities and sensibilities. This is, in our view, highly encouraging. Looking onwards, we believe that further research is required to expand our work on GD-EAL, promote further applications of the EAL framework, and explore alternative approaches to foster the development of chaordic agency. For example, further research should corroborate the impacts that we have observed so far, and determine the extent to which they reflect general trends. Furthermore, the literature suggests that designing games 'about' a specific domain can be a powerful means to foster learning, even outside the scope of game design education (e.g. Kafai, 2006). Hence, it would be important to study the application of EAL-based game design activities outside game design courses (e.g. leveraging game design activities to engage engineering students with wicked problem scenarios). By extension, it would be interesting to explore how our strategy could be applied in other disciplinary domains through other types of project-based activities.

In conclusion, we believe that much work is still needed to align formal education with the needs of our modern, global world. We also believe that such work can be conducted and positive changes achieved even within ineluctable boundaries defining the scope of formal education. We hope that EAL and our game design-based strategy will contribute to promote such changes.

References

Ackoff, R. L. (1978). *The art of problem solving*. New York, NY: Wiley.

Ackoff, R. L. (1994). Systems thinking and thinking systems. *System Dynamics Review, 10*(2–3), 175–188.

Argyris, C. (1977). Double loop learning in organizations. *Harvard Business Review, 55*(5), 115–125.

Bedny, G., & Karwowski, W. (2006). *A systemic-structural theory of activity: Applications to human performance and work design.* Boca Raton, FL: CRC Press & Taylor & Francis.

Beinhocker, E. D. (1997). Strategy at the edge of chaos. *McKinsey Quarterly, 9*(1), 109–118.

Bruner, J. S. (1977). *The process of education.* Cambridge, MA: Harvard University Press.

Buchanan, R. (1992). Wicked problems in design thinking. *Design Issues, 8*(2), 5–21.

Camazine, S., Deneubourg, J. L., Franks, N. R., Sneyd, J., Theraulaz, G., & Bonabeau, E. (2003). *Self-organization in biological systems.* Princeton, NJ: Princeton University Press.

Checkland, P., & Scholes, J. (1990). *Soft systems methodology in action.* Chichester: Wiley.

Davis, B., & Sumara, D. (2006). *Complexity and education: Inquiries into learning, teaching, and research.* Mahwah, NJ: Lawrence Erlbaum.

Dorst, K. (2011). The core of "design thinking" and its application. *Design Studies, 32*(6), 521–532.

Dunne, D., & Martin, R. (2006). Design thinking and how it will change management education: An interview and discussion. *Academy of Management Learning & Education, 5*(4), 512–523.

Engeström, Y. (2014). *Learning by expanding: An activity-theoretical approach to developmental research* (2nd ed.). New York, NY: Cambridge University Press.

Fabricatore, C. (2007). Gameplay and game mechanics design: A key to quality in video games. In *Proceedings of the OECD-CERI Expert Meeting on Videogames and Education* (pp. 1–18). Santiago: OECD. Retrieved from http://www.oecd.org/dataoecd/44/17/39414829.pdf

Fabricatore, C., & López, X. (2012). Sustainability learning through gaming: An exploratory study. *Electronic Journal of E-Learning (EJEL), 10*(2), 209–222.

Fabricatore, C., & López, X. (2014a). A model to identify affordances for game-based sustainability learning. In C. Busch (Ed.), *Proceedings of the 8th European Conference on game based learning* (pp. 99–109). Reading: Academic Conferences and Publishing International Limited.

Fabricatore, C., & López, M. X. (2014b). Complexity-based learning and teaching: A case study in higher education. *Innovations in Education and Teaching International, 51*(6), 618–630.

Fabricatore, C., & López, X. (2015). Higher education in a complex world: Nurturing "chaordic" influencers. In *Proceedings of the Sixth Advanced International Colloquium on Building the Scientific Mind (BtSM2015)* (pp. 1–11). Jupiter, FL: Learning Development Institute.

Fabricatore, C. (2018). Underneath and beyond mechanics: An activity-theoretical perspective on meaning-making in gameplay. In B. Suter, M. Kocher, & R. Bauer (Eds.),

Games and rules: Game mechanics for the "magic circle" (pp. 87–111). New York, NY: Columbia University Press.

Franks, N. R. (1989). Thermoregulation in army ant bivouacs. *Physiological Entomology, 14*(4), 397–404.

Gentner, D., & Smith, L. (2012). Analogical reasoning. In V. Ramachandran (Ed.), *Encyclopedia of human behavior* (2nd ed., pp. 130–136). Burlington, VT: Academic Press.

Johnson, S. (2002). *Emergence: The connected lives of ants, brains, cities and software.* London: Penguin Books.

Jonassen, D. H. (2002). Learning as activity. *Educational Technology, 42*(2), 45–51.

Jones, P. H. (2014). Systemic design principles for complex social systems. In G. S. Metcalf (Ed.), *Social systems and design* (pp. 91–128). Tokyo: Springer.

Kafai, Y. B. (2006). Playing and making games for learning: Instructionist and constructionist perspectives for game studies. *Games and Culture, 1*(1), 36–40.

Kaptelinin, V. (1996). Distribution of cognition between minds and artifacts: Augmentation of mediation? *AI & Society, 10*(1), 15–25.

Kokotovich, V. (2008). Problem analysis and thinking tools: An empirical study of non-hierarchical mind mapping. *Design Studies, 29*(1), 49–69.

Leont'ev, A. N. (1978). *Activity, consciousness, and personality.* Englewood Cliffs, NJ: Prentice-Hall.

McDaniel, R. R., & Driebe, D. J. (2005). Uncertainty and surprise: And introduction. In R. R. McDaniel & D. J. Driebe (Eds.), *Uncertainty and surprise in complex systems* (pp. 3–11). Heidelberg: Springer.

Mitchell, M. (2009). *Complexity: A guided tour.* New York, NY: Oxford University Press.

Nardi, B. A. (1996). Studying context: A comparison of activity theory, situated action models, and distributed cognition. In B. A. Nardi (Ed.), *Context and consciousness: Activity theory and human-computer interaction* (pp. 69–102). Cambridge, MA: MIT Press.

Oxman, R. (2004). Think-maps: Teaching design thinking in design education. *Design Studies, 25*(1), 63–91.

Radomsky, M. V. (2013). Assessing context: Personal, social, cultural. In M. V. Radomski & C. A. Trombly Latham (Eds.), *Occupational therapy for physical dysfunction* (7th ed., pp. 284–309). Philadelphia, PA: Lippincott Williams & Wilkins.

Rittel, H. W. J., & Webber, M. M. (1973). Dilemmas in a general theory of planning. *Policy Sciences, 4*(2), 155–169.

Runco, M. A. (2004). Creativity. *Annual Review of Psychology, 55*(1), 657–687.

Ryan, R. M., & Deci, E. L. (2000). Self-determination theory and the facilitation of intrinsic motivation, social development, and well-being. *American Psychologist, 55*(1), 68–78.

Scheer, A., Noweski, C., & Meinel, C. (2012). Transforming constructivist learning into action: Design thinking in education. *Design and Technology Education, 17*(3), 8–19.

Sipos, Y., Battisti, B., & Grimm, K. (2008). Achieving transformative sustainability learning: Engaging head, hands and heart. *International Journal of Sustainability in Higher Education, 9*(1), 68–86.

Third Nobel Laureate Symposium. (2011). The stockholm memorandum: Tipping the scales towards sustainability. *Ambio, 40*(7), 781–785.

Visser, J. (2001a). Integrity, completeness and comprehensiveness of the learning environment: Meeting the basic learning needs of all throughout life. In D. Aspin, J. Chapman, M. Hatton, & Y. Sawano (Eds.), *International handbook of lifelong learning* (pp. 447–472). Dordrecht: Springer.

Visser, J. (2001b). *Learning communities: Wholeness and partness, autonomy and dependence in the learning ecology.* Proceedings of the International Symposium on Learning Communities, Barcelona, Spain. Retrieved from http://learndev.org/dl/Barcelona2001ext.pdf

Visser, J. (2008). Constructive interaction with change: Implications for learners and the environment in which they learn. In J. Visser & M. Visser (Eds.), *Learners in a changing learning landscape* (pp. 11–35). Dordrecht: Springer.

Visser, J. (2012). Reflections on a definition: Revisiting the meaning of learning. In D. N. Aspin, J. D. Chapman, K. Evans, & R. Bagnall (Eds.), *Second international handbook of lifelong learning* (pp. 163–179). Dordrecht: Springer.

INTERMEZZO 17

Where Science Ends

Jan Visser

Reflections inspired by James Lees: 'HIV, medical science and the call to greater humanness'

I forget who it was among the experienced medical doctors whose lectures I listened to over time, but one of them once mentioned an occasion when he stood at the bedside of a terminally ill patient. There was no hope to keep the patient alive. Yet, he took out his stethoscope and held it against the patient's chest, listening attentively. There was no medical reason to do so. Medical science and technology had reached their end. But where science ends there is still human life that wants to be touched. The doctor's simple gesture elevated the meaning of using a familiar medical instrument to expressing love and care now that it had lost its usefulness for scientific purposes.

Oxford researcher Max Roser[1] provides a remarkable service[2] to those who seek to understand the world while wanting to ground their understanding in solid data. He does so in the spirit—much appreciated by this author—of returning to taxpayers the results of the research they funded. Under the heading of 'Life Expectancy,' Roser notes:

> Since 1900 the global average life expectancy has more than doubled and is now approaching 70 years. No country in the world has a lower life expectancy than the ... countries with the highest life expectancy in 1800.[3]

We grow older thanks to better knowledge about what is good for us and thanks to the spectacular advances that have been made in medical science and technology. I owe it to such progress that I am able to still write these words. But I remember very well how important it was, at moments when I was at risk, that there were loved ones and friends who cared about me, one of them calling me from some 8000 km away, with the simple message: "We don't let you go." Humanness is as simple as that.

We all meet circumstances in which we come face to face with our vulnerability. Both science and humanness are then crucial. Advances in longevity go hand in hand with increased dependence on medical services. Shamefully, the

 | DOI: 10.1163/9789004416802_034

quality of and access to medical care are inequitably spread around the world. In regions of the world where medical and pharmaceutical science and industry advance to unheard of heights, tragically, access to their benefits becomes more and more driven by market forces and less and less by a culture of care, by all and for all. Much must still change.

The call for 'greater humanness' in James Lees's chapter is well made.

Notes

1 See for context: https://www.theguardian.com/media/2015/nov/15/scientists-with-reasons-to-be-cheerful

2 Called *OurWorldInData*, available at https://ourworldindata.org/

3 https://ourworldindata.org/life-expectancy

CHAPTER 18

HIV, Medical Science and the Call to Greater Humanness

James Lees

> Doctors who see medicine as science only, you don't want them by your side when your bleeding won't stop or your child is screaming in pain. The clinicians follow the book. The artists follow their gut. The artists feel your pain and they go to extremes to make it stop. Extreme measures. That's where science ends and art begins.
>
> SHONDA RHIMES (writer and creator of Grey's Anatomy, Meredith Grey voice-over in *Grey's Anatomy*, Season 7, Episode 11)

1 Introduction

In his deeply influential 1970 book, *Pedagogy of the Oppressed*, Brazilian educator and philosopher Paulo Freire began with the premise that *humanization*—the process of becoming more fully human—has always been humankind's central problem (Freire, 1970). While humanization and dehumanization are both real alternatives for Freire, humanization is, he contends, the people's vocation. In the preface to a much later edition, Freire captured what he hoped would endure from his lifetime's work: trust in people, faith in men and women, and the creation of a world in which it will be easier to love. In this chapter, and after a lifetime of work with the HIV epidemic, I express my sadness that medical science, while saving bodies from death, quickly became a dominant discourse of the HIV epidemic that grew increasingly cynical about men and women. In my darkest moments, I worry that medical science's solutions to HIV have kept us from looking honestly at ourselves as a species and allowed us avoid the most difficult but most necessary questions about ourselves and the ways we are willing to dehumanize ourselves and each other. With less and less trust and faith in people, medical science blinded us from the brief moment of grace the HIV epidemic brought us when our species might have learned how to love better.

 | DOI: 10.1163/9789004416802_035

"I wish I did not have HIV, and I do not wish for anyone to have it," says Bonile Peter, former HIV health educator at the University of the Western Cape in South Africa. "But I am a better person because of having it" (personal communication, May 2013). In my 30 years working with the epidemic, it is a story I have heard countless times from people with HIV. For them, HIV has been a very private and personal teacher, not the ugly, evil enemy being fought by legions of soldiers in the cartoon versions of HIV education manuals that still predominate today. For three decades, the world has been systematically taught to fight and kill HIV. Why then should we be surprised if after all those years people with HIV became objects of scorn?

"You people are rude," said Dr. Eduardo Duran, psychologist and traditional Native American healer (personal communication, October 2006). I had been accompanying him on visits to Sangomas (traditional healers) in KwaZulu Natal during the height of the HIV epidemic in South Africa. Questioning who were the 'you people' in Ed's comment, he readily answered, "Western white people." Biting the fat worm of a big hook, I asked Ed to go on. "Western science wants to fight and kill HIV. You have a war on it. You call it the enemy. But in my cosmology, HIV as a virus is a life form, and as a life form, it demands reverence and respect. For us, when illness comes, we treat that illness with the respect we would show to any visitor to our home. We ask it to sit in our best chair, offer it a glass of water, sit quietly with it for a time, listen. And when the moment is right, we respectfully ask, 'So, why have you come?'"

Ed's hook sank deep. I immediately remembered the early 1990's in San Francisco when we knew the epidemic was making demands on the quality of our love, forcing us to recognize how we do and do not care. I remembered the street children I worked with in Mumbai who told me that, "It does not matter how much you know about HIV. If you don't care about yourself, you are going to get it." I remember preparing my 20-year-old street youth client in San Francisco for his impending death. He saw the lump in my throat. "Don't worry," Gary said to me. "Dying is the easy part. It is living that's the hard part." I thought about 18-year-old Jake who intentionally infected himself with HIV so that he could join our support program because at 18, his need for love and belonging trumped his fear of death. I heard the voice of my friend who directed a drug detox program for HIV-infected men in San Francisco as he told me that 40% of the men intentionally infected themselves. "It is a way of killing oneself, a way of belonging, an identity, an expression of grief over those they have lost and an expression of unhappiness with their own lives," Chris explained. And I remembered the many bottles of unopened AZT[1] we collected from our young clients' cheap hotel rooms after they died, the only available treatment for HIV that they politely accepted from our nurse but never took. I thought about Diane Di Prima's (n.d.) poem *Rant* where she writes:

> *The only war that matters is the war against the imagination.*
> *All other wars are subsumed in it.*
> *There is no way out of a spiritual battle.*
> *The war is the war for the human imagination.*
> *Men die every day for the lack of it.*

HIV, why have you come? It is a question that shifted my focus from the virus-centered construction of HIV to placing people at the center of the epidemic. I put the "H" back into HIV. Ed's revelation changed my practice, for how we conceptualize and define a problem says everything about how we will seek to resolve it. It is a question that eventually led me to Father Michael Kelly of Zambia who has written about the HIV epidemic as our 'moment of grace.'

In a blog linked to *Psychology Today*, Samantha Smithstein (2016) writes about moments of grace having two components. One part, she says, has nothing to do with us and is about magic showing up when it shows up. It is, she says, about something bigger. The second part of a moment of grace has everything to do with us personally. For Smithstein, it is simply and powerfully about being willing to show up authentically, with an open mind and heart, to the moment at hand.

Showing up with an open mind and heart to HIV. What would that mean? My experience is that we were never allowed to fully find out. While many of us on the ground around the world knew that the HIV epidemic was about far more than a struggle for physical health, science very quickly medicalised the epidemic and along with it our human existence. When we had a chance for a revolution in how we love each other, we got 'big pharma' instead. When we needed to be open and honest about our brokenness, we got tablets. And we failed to show up authentically, with an open mind and heart, to the moment at hand.

Medical science continues with its myopic fight *against* HIV. Big pharma enjoys greater and greater sales. Neither are they engaged in the question who we are and who we can be as human beings. They are not fighting *for* human beings or for the vision of a greater humanness the epidemic could have inspired us toward. It is, perhaps, a lesson in what happens when science becomes the dominant lens and is allowed to rule. I do not say that medical science is not important. It certainly is. What I do ponder is how medical science can learn to think beyond its margins and engage in a far larger species project than simply fighting germs. Can medical science in any way understand what Eduardo Duran understands as healing? Do we need to look beyond the *mindset* of the scientist and impregnate the *scientific mind* with the—as yet 'unwanted'—language of love and concerns about our own human development? Should the scientific mind be telling cautionary tales to those looking solely toward technology and medicine to take us forward? Maybe so.

Broadening conversations between disciplines requires common language. Common language is what I did not find between my professors of agriculture and cultural anthropology when I put them in the same room at university in 1977 to discuss world hunger. What also limited the conversation was who paid the bills. The agricultural scientists depended on the big agro-chemical companies for funding. Solutions were to be found in miracle seeds that required more nitrogen fertilizers, more insecticide sprays and more water – things poor farmers in poor nations did not have. They did not know how to talk about inequality, unequal distribution of wealth, economic injustice and global capital. They would not see the social, cultural and environmental disruption their new seeds would sow. I found myself in trouble with my plant nutrition professor one day when I suggested that if his new seeds were such a miracle, we should be able to throw them on the floor of the dusty lecture hall on Friday afternoon and come back Monday morning to find a full crop ready to harvest! What distinguished the two groups of academics is that the cultural anthropologists, as part of their discipline, turned their microscopes on themselves. The agricultural scientists had no similar self-reflective imperative built into their discipline. Nor does medical science.

I draw a loose and flawed parallel with the recent approval and global push for the use of Pre-Exposure Prophylaxis (PrEP) medication to prevent HIV infection. Approved by the United States Food and Drug Administration in 2012, PrEP is yet another in a long line of technical interventions into the epidemic that involves the daily consumption of Truveda, an anti-retroviral drug that blocks HIV from incorporating its genetic material into the DNA of the host genome. I do not mind this technical solution that is recently credited with reducing new infections among gay men in London and is actively promoted on university campuses in South Africa. But it feels like an incomplete victory. I need to track backwards in my own life to discover why.

The earliest days of the HIV epidemic in New York and San Francisco, before the virus was identified, were filled with fear, blame and rejection. It took great courage for men with the red blotches of Kaposi's sarcoma on their faces, necks and hands to come out into public, to sit in restaurants, parks and movie theatres while others stared. The onlookers' discomfort betrayed their own fears that the red blotches and thinning, weakened bodies could soon be their own.

As human beings, we wanted to know why friends, family members, lovers and co-workers were dying around us. We made up stories about life styles and co-factors that helped us sleep at night. It was the poppers. The alcohol. The

24-hour parties. Fear trumped empathy more often than the reverse. And Elton John got married—to a woman.

When the culprit was identified as a virus, we were relieved. Finally, we had something to hate. Something to fight against. HIV was the enemy and we its victims. We were good. It was bad. And we started a war. The war on AIDS. But we failed to track the disease further back, failing to sense the bigger and deeper dis-ease far older than the HI virus itself. We settled too soon for the lens of science and became too comfortable with its curative tools to recognize the moment of grace.

There are endless theories, beliefs and religious positions about the origins and causes of HIV and AIDS. Some of these were ascribed to by one of South Africa's former presidents and a former health minister. Others saw conspiracy theories and an attempt by the white CIA to reduce the African black population. They are important stories, their constructions symbolic of other matters and to be carefully understood, not dismissed as they were labeled distractions. But we have settled on one story of HIV, the dominant story of a virus that kills and requires abstinence, faithfulness and condoms to overcome. That dominant story conquered the world faster than any colonizer of the past and just as effectively eliminated both local knowledge and the creation of local responses. The dominant story itself has been an infectious agent that infected the cultural and familial immune systems of localities around the world.

My discomfort at the bio-medical and big pharma solution to the epidemic and the recognition of the complexity of our position as human beings vis-à-vis the virus was piqued by first century B.C. Roman orator Marcus Tullius Cicero (1914) who wrote,

> … we denounce with righteous indignation and dislike men who are so beguiled and demoralized by the charms of pleasure of the moment, so blinded by desire, that they cannot foresee the pain and trouble that are bound to ensue; and equal blame belongs to those who fail in their duty through weakness of will, which is the same as saying shrinking from toil and pain. These cases are perfectly simple and easy to distinguish. In a free hour, when our power of choice is untrammelled and when nothing prevents our being able to do what we like best, every pleasure is to be welcomed and every pain avoided. But in certain circumstances and owing to the claims of duty or the obligations of business it will frequently occur that pleasures have to be repudiated and annoyances accepted. The wise man therefore always holds in these matters to this principle of selection; he rejects pleasures to secure other greater pleasures, or else he endures pain to endure worse pain.

The age of AIDS has not been a 'free hour.' It has not been a time to welcome every pleasure. It is a time of duty and obligation when "pleasures may have to be repudiated and annoyances accepted." A simple condom. Knowing one's HIV status. Honesty. Not mountains to climb or raging rivers to ford. Our unwillingness to adapt to these minor annoyances lead to greater pain. Over 2000 years ago, Cicero alerted us to the complexities of being human and the consequences of our ability to be blinded by desire.

In 1984, the epidemic quietly growing, musical artist Leonard Cohen told a story in the form of a hit song, *Hallelujah.* Cohen told the story of David, second king of the United Kingdom of Israel and Judah. David was well known among his people as a great warrior, poet and musician who composed many of the psalms later contained in the *Book of David.* King David taught us to praise the Lord in song. By telling David's story, Cohen was unknowingly telling the story of the HIV epidemic and presenting the world with a way forward.

Hallelujah begins with a reference to the "baffled" Kind David composing *Hallelujah* or praises to the Lord.[2] This great king, warrior and poet who had many wives and concubines, saw the beautiful Bathsheba bathing on a nearby roof. Moved by his own desire, Bathsheba was soon brought to him and he seduced her. "Her beauty in the moonlight overthrew you," sings Cohen. "She tied you to a kitchen chair. She broke your throne and she cut your hair. And from your lips she drew the hallelujah." Blinded by desire, David took Bathsheba. It is the story of so many men and the story of AIDS. Married, with girlfriends or both, a man sees another beautiful woman and has to have her.

There are consequences in a time that is not free for such behavior. There were consequences for David, a king who already had multiple wives and concubines. Having seduced Bathsheba, the deeply reverent David broke a commandment of the Lord. His power and throne were weakened. Worse, his joyous connection to the Lord was broken.

In 1984 when the song was first released, Cohen himself was stuck in the trap of blaming women for the downfall of men in the way that African women were blamed for the spread of HIV. "She broke your throne and she cut your hair. And from your lips she drew the Hallelujah." *She* did all this? There is a long history of women being blamed for AIDS in Southern Africa, either as witches or whores, and being put to violent deaths. In Cohen himself we see a reflection of blame and male power relating to illness and the death of women.

The song moves on with Bathsheba voicing the next lines:

I've seen your flag on the marble arch
Love is not a victory march
It's a cold and it's a broken Hallelujah

Bathsheba is talking to David about love. Love is not about conquering, she says. There are no victory marches. It is not signified by one's flag planted in the marble arch of the vanquished enemy's town. But is it? Much of male sexuality remains socially constructed around the hunt and capture. Getting to third base and then stealing home. 2,000 years ago, Bathsheba is bringing the feminine to the mighty king and warrior so that he might understand love.

David replies in a following stanza:

Maybe there's a God above
But all I've ever learned from love
Was how to shoot at someone who outdrew you
It's not a cry you can hear at night
It's not somebody who has seen the light
It's a cold and it's a broken Hallelujah

David does begin to think about love. "All I've ever learned from love is how to shoot at someone who outdrew you." But David has already ordered Bathsheba's husband, an officer in his troops, into battle, a move David knew would, and did, result in his death. David has now committed two transgressions against the Lord he so joyfully used to praise. Has this not been the story of HIV in South Africa so far? Man does not really know much about love. Man sees beautiful woman. Man has to have beautiful woman. Someone, as a result, dies.

Cohen sets us up for David's solution and perhaps a way forward to the epidemic with the following lines:

There's a blaze of light in every word
It doesn't matter which you heard
The holy or the broken Hallelujah

Cohen wanted us to understand what the *broken Hallelujah* might be.

In the final refrain, David kneels before the Lord of Song and tells the truth.

I did my best, it wasn't much
I couldn't feel, so I tried to touch
I've told the truth, I didn't come to fool you
And even though it all went wrong
I'll stand before the Lord of Song
With nothing on my tongue but Hallelujah

"I couldn't feel, so I tried to touch." A product of patriarchy and masculinity. Men divorced from their feelings reaching out to touch with their hands and bodies. David gets it. And when he does, he kneels before the Lord telling the truth, praising the Lord not in his joyous way, but telling the truth of his brokenness. David's broken Hallelujah and his joyous Hallelujah are equal, says Cohen.

I do not believe the Lord of Song is our solution to the HIV epidemic. But I do wonder what would happen if each of us finally learned to tell the truth to each other and to ourselves. And if we finally admitted our brokenness as a species. "Yes, I do have HIV. Yes, I suppose I might have HIV because I have never been tested. Yes, to tell you the truth, I am being sexual with other people. Yes, you are not the only one. I have not been using condoms. Truthfully, I don't really care about you but I do want to have sex with you. No, I don't love you. I am broken and I need to stop taking my brokenness out sexually on others." Or perhaps, "Yes, like HIV, we are destroying our host."

Would we have an HIV epidemic if people told the truth? Yes, we would, because the truth is that many people are broken and hold onto the world by the finest thread. But we would have less of an epidemic and we might finally be honest with ourselves about the reasons why people are broken and know that none of us can be whole until all of us are whole.

Elie Wiesel (1986) wrote the following in his Nobel Peace Prize acceptance speech:

> As long as one dissident is in prison, our freedom will not be true. As long as one child is hungry, our life will be filled with anguish and shame. What all these victims need above all is to know that they are not alone; that we are not forgetting them, that when their voices are stifled we shall lend them ours, that while their freedom depends on ours, the quality of our freedom depends on theirs.
>
> This is what I say to the young Jewish boy wondering what I have done with his years. It is in his name that I speak to you and that I express to you my deepest gratitude. No one is as capable of gratitude as one who has emerged from the kingdom of night. We know that every moment is a moment of grace, every hour an offering; not to share them would mean to betray them. Our lives no longer belong to us alone; they belong to all those who need us desperately.

King David found his moment of grace. Bonile Peter and many people with HIV have found their own moments of grace. We need medical science to be a part of ending the HIV epidemic. But we also need to reclaim this moment of grace. Eduardo Duran told me that for traditional healers, the medicine is in the wound. Could it be possible that the wound of HIV is the medicine that brings attention to where we are already wounded? Or is the medicine too strong, reflecting back to us what we are not strong enough to see?

How do we make sense of the global HIV epidemic that medical science has reduced to a disease of our bodies? Anisur Rahman (1993) sees the human species as primarily creativist. Our humanness, believes Rahman, is expressed in the daily creation of solutions to the situations we face. Rahman argues that our creativity and opportunities to express it should be at the center of development efforts. For Rahman, constant use of our creativity is at our human core. It is what makes us human. Throughout his book, Rahman contrasts this to what he coins consumerist development which takes ownership of the problems people face, puts them in the hands of 'experts' to develop 'interventions' and 'solutions,' and sells those fixes back to the people with the problem for financial profit. This is certainly the story of medical science and HIV in Africa.

I live in a world obsessed with 'taking solutions to scale' and of 'evidence-based interventions' that leave little room for local voices. Rahman teaches us that collectively seeking to understand and find solutions for local challenges by the very people facing those challenges is part of the solution. To do otherwise is about power and can dehumanize. I sometimes wonder what would happen if some of the voices I would really like to listen to when it comes to understanding ourselves and this epidemic went through the blind review process for their papers to be accepted by the big international AIDS conferences. Would the words of the Buddha, Confucius, Jesus and Martin Luther King be deemed to be based in the appropriate evidence?

There is a tremendous difference for individuals between being asked what they might do to solve a problem and being told that they should apply a particular solution. As part of my teaching at university and in South African communities, I do not teach *about* HIV but ask people to explain the epidemic in their terms and work together to create solutions to what they discover. I am always struck by how much their understanding aligns with the views of the young street children I worked with in Mumbai years ago. It is, they say, in large part about how much one cares about oneself. It is about building a sense of worthiness and healing, something the apartheid regime in South Africa and the colonizers before them systematically undermined. What good is a condom if the potential users care and respect for themselves and for their partners, so violently undermined by colonial and apartheid rule, has not been healed? Part of the 'solution' to the epidemic is to understand these vulnerabilities and figure out ways to help families and communities create better ways of watching out for each other and never take a chance with the miracle gift of life they have been given.

Our vulnerabilities are a messy business. Being human is messy. Our vulnerabilities and our messiness are not places for medical science to lead. It is a place for artists, not clinicians. But when medical science distributes tablets that divert us from having to deal with our messiness, as it has done within the HIV epidemic, I feel the sadness I spoke of above. Medical science creates a new layer of silence that only deepens the many silences that are already a part of why we have the epidemic in the first place. By wresting control of the epidemic and its solutions, foreign funders have once again colonized and dehumanized African people.

HIV is not just a disease of the body. It is an expression of the dis-ease expressed when our bodies meet. How do we make it through the 21st century intact as a species with a planet that is struggling to host us? We need to let medical science play its appropriate role but not let it take the lead. HIV is not the problem. Global warming is not the problem. We are the problem. And we are the only possible solution. If we can reframe HIV as having come here to teach us, and to inspire us to do better before it is too late. If we can do that, maybe we will also see that getting to the future is a human challenge where science and art play a vital role, jointly inspiring a new scientific mind.

As human beings, we seem to find comfort in having something to fight against. We learn to triangulate on the school yard from age seven, pitting friend against friend to stay popular and in power. In September of 2017, a United States Congressman was interviewed on Fox News after the shooting of civilians in Las Vegas was attributed to a single American individual with no links to terrorism or religious ideology. "The American people," he stated, "are confused by this. They do not know who to blame or who to hate." Confused when we do not know who to blame or hate? That is messy!

Winners and losers. The good and the bad. To love and to hate. Binaries and polarities. Bonile Peter and so many like him do not hate the virus in their bodies. Part of their own healing has been to find a new relationship with what they were taught to despise. In looking at themselves, they have found peace with the virus and in real ways, peace within themselves because of what the virus taught them. It is what Ed Duran says, that the medicine is in the wound.

Can we still claim our moment of grace HIV brought to us? I am not hopeful. I recently saw a poster campaign promoting PrEP (Pre-Exposure Prophylaxis pills) that said, "HIV shouldn't stand in the way of love." Thirty years after we knew that love was the answer to HIV, a pill has side lined the messenger and who we are does not change.

Were Michel Foucault alive—his own death resulting from HIV infection—perhaps he could help us understand medical power in relation to the colonization of Africa by HIV interventionism. The scientific mind itself should carry with it a deep analysis of power, capital and the social construction of the other. I think of the years in South Africa watching my university students die while listening to foreign powers defending the absence of antiretroviral medications in Africa with the words, "These medications have to be taken on a rigorous time schedule and African people do not have watches. Even if they did, they cannot tell time."

Kenneth Kaunda, first elected president of Zambia, wrote on 1966, "Let the West have its technology and Asia its mysticism. Africa's gift to world culture must be in the realm of Human Relationships" (Kaunda, 1966). Recently deceased musician Hugh Masekela spoke of a life-long Indian friend talking about Africa. "The day you [Africa] look inside, it will be the greatest education the whole world can get" (Masekela, 2015).

Medical science and technology will not advance us. They will not advance South Africans. To believe in people is to see them as human. When people are allowed the opportunity to create solutions to the challenges they face, their humanness is recognized and nurtured. In a nation and a continent that has been systematically dehumanized by the colonizing north, HIV and the moment of grace it brings cannot be separated from Freire's question of humanization and dehumanization. Nor can the global tension between capital, humanness and inequality be ignored. African Elder Bev Reeler says we must look at our inner journeys, not simply at policy solutions. "There are different ways to touch the wounds and what is underneath that transcend policies," she says. "You don't jump to solutions. Rather, it is a continuous sensing" (personal communication, November 12, 2017).

When I sense into my 17 years working on the HIV epidemic in Africa and the 13 years before elsewhere, I remain convinced that it is about the quality and character of our love. We think we test for HIV by drawing blood. But HIV has been testing us for over three decades. Marianne Williamson writes that "The way of the miracle-worker is to see all human behavior as one of two things: either love or a call for love."[3] Perhaps this moment of grace HIV has also bought us a chance for colonizers and oppressors to understand love better. Perhaps by taking on love as its own subject, the scientific mind will take us forward to a deeper sense and experiencing of the beauty of our own humanness.

2 Addendum: A Personalized Way forward with HIV

Throughout this chapter, I have intentionally used the first person within the narrative. I do this because I offer my own experience without making claims of universality or speaking for others. Nor do I speak for HIV. HIV, as a living entity, has its own story, different chapters of which are told through the evolution of the global epidemic. Every person on the planet is affected by that story and the more they understand *how* they are affected by it, the less likely they are to become infected *by* it.

The story of HIV in the United States was being silently written in the late 1970's when I was a teenager struggling to learn about sex, relationships and love. As I emerged into the world of dating and love in the early 1980s, so, too, did HIV.

Ed Duran suggested to me on that seminal day in 2006 in KwaZulu Natal that we must respectfully ask HIV, "Why have you come?" I have understood since then that for me, HIV prevention is about understanding how the story of the virus interacted with my own. It is about seeing how HIV has been a part of my own story though I have never been infected by it. What meanings have I given HIV as I searched for meanings in my own life? What part of my own story has HIV forced me to see?

Influential thinker on education and the organization of schooling Will Richardson (2011) wrote

> I do believe that the emphasis will turn back to the learning process, not the knowing process. And while I don't think schools go away in the interaction, the "new normal" will be a focus on personalization not standardization, where we focus more on developing learners, not knowers, and where students will create works of beauty that change the world for the better. At some point, we'll value that more than the SAT.

Medical science and public health created the still-dominant story of HIV and *instructed* people around the world what to do and what not to do. HIV interventionism, as practiced by northern funding agencies in Africa, has remained true to this singular story of HIV and has spent decades and many dollars trying to take instructions to scale. But standardization of the global HIV curriculum has simply not worked. Not completely. Those struggling the most on the economic and emotional margins of life—the people I have spent most of my adult working life with—have remained the most vulnerable to HIV. I believe that we must find a 'new normal' in our response to HIV that, learning from Richardson, "will be a focus on personalization not standardization," where the attention will be more on responding to people's whole lives and perhaps even the love that they seek.

I have known thousands of people living with HIV all over the world. I refuse to count the number of people I have cared about who died of AIDS. But at least among the economically and emotionally marginalized people I have always worked with, there is a common story of the moment of HIV infection: when the need for love, even if just for one hour or night, becomes more important and a stronger impulse than the need to protect life itself. I have known this impulse and need within myself. HIV revealed its depths. It is through learning about myself, not HIV, through *developing* myself in the way Richardson speaks of, and in thankfully doing this in time, that I never became infected with HIV.

Though cautioned against it in 1992 by Susan Sontag, the dominant metaphor of HIV response remains a war. War allows only one story to be told. Medical science speaks of pathogens and their battle against them. Stephan Hausner (2011), quotes psychologist Wolfgang Giegerich as saying, "The more one refuses to hear the message and meaning in pathology, the more pathological it becomes." Sontag and Giegerich's words are caution and prognostication. They are among the many voices that speak to our human journey, which medical science does not know what to do with. Again, the papers of Jesus, the Buddha, Confucius and even Martin Luther King would not be accepted for the biennial AIDS conferences in blind review. Yet, theirs are the voices we need, to understand our human journey—in the same way in which HIV helps us to understand it—and make decisions about where we want it to go.

There is a concept in Judaism called Tikkun olam, translated as *repair of the world*. Tikkun olam captures the notion that the world is full of brokenness and that it is our job to put it back together again. Tikkun olam assumes brokenness. Our brokenness. If we begin with an assumption of our collective brokenness, then perhaps we are able to see dynamics such as HIV, the current wide-spread opiate use in the Unites States, gangsterism, racism, inequalities, religious persecution and even war itself as symptoms of that brokenness and possibly listen to them as messengers trying to get us to think, critically reflect and make decisions on what it means and what is should mean to be human. I was not safe from HIV infection until I recognized my own brokenness and learned how to heal it.

Perhaps HIV is telling us that the way we educate our children in schools is part of the problem. Most young people are instructed and tested rather than experiencing their imaginations being set free. In South Africa, even today, learners are regularly shouted at and told they are stupid. School systems speak of educational goals, but do they speak of human goals? Are the fears, judgments, physical attacks and murder of people with HIV around the world not a reflection of *who* schooling has taught people to be? Are schools part of what breaks many? If, as Benu Gupta (2016) contends, the knowledge of unconditional love is the greatest gift education gives us, has education failed us?

HIV kills its hosts. Ironically, HIV will survive because human beings kill their messengers. As human beings, we are forever in a process of becoming.

It is our human journey. There are, perhaps, resting spots within our greater journey where we can simply be and enjoy where we have come. Today is not a resting spot. It is not a 'free hour' where 'every pleasure is to be welcomed and every pain to be avoided.' Today, our human species teetering on disaster, HIV is but one tiny messenger urging us to greater questions about us. It is my hope that the scientific mind can move exactly to this place and begin to consider the place of love within the human system. To the place where we all, as Richardson (2011) writes, "create works of beauty that change the world for the better."

Notes

1 Azidothymidine.
2 Original Leonard Cohen lyrics found at https://www.azlyrics.com/lyrics/leonardcohen/hallelujah.html
3 Marianne Williamson is author of *A Course in Miracles*. Quote retrieved from https://www.goodreads.com/quotes/459068-the-way-of-the-miracle-worker-is-to-see-all-human

References

Cicero, M. T. (1914). *De finibus bonorum et malorum* [First Century B.C. text] (H. Rackham, Trans.). Retrieved from https://en.wikipedia.org/wiki/Lorem_ipsum

DiPrima, D. (n.d.). *Rant*. Retrieved from https://www.youtube.com/watch?v=bw8lUZCihzA

Freire, P. (1970). *Pedagogy of the oppressed*. London: Penguin.

Gupta, B. (2016). *Olympic value-based education to ensure health lives and promote well-being*. Paper presented at 13 Joint International Session for Presidents and Directors on the National Olympic Academies. International Olympic Academy, Olympia, Greece.

Hausner, S. (2011). *Even if it costs me my life*. New York, NY: Routledge & Taylor & Francis.

Kaunda, K. A. (1966). *Humanist in Africa*. London: Longmans.

Masekela, H. (2015). *Decolonizing the mind*. Lecture to University of South Africa's Annual Spring Law Conference. Retrieved from http://aidc.org.za/decolonising-mind-hugh-masekelas-lecture-unisas-annual-spring-law-conference/

Rahman, M. A. (1993). *People's self-development: Perspectives on participatory action research: A journey through experience*. London: Zed Books.

Richardson, W. (2011, April). *The "new" normal.* Retrieved from https://www.edutopia.org/blog/will-richardson-the-new-normal-future-of-education

Smithstein, S. (2016). Moments of Grace. Do we make them happen or are they simply luck or magic? *Psychology Today Blog*. Retrieved from https://www.psychologytoday.com/us/blog/what-the-wild-things-are/201608/moments-grace

Sontag, S. (1998). *AIDS and its metaphors.* New York, NY: Farrar, Straus & Giroux.

Wiesel, E. (1986). *Acceptance speech.* Retrieved from https://www.nobelprize.org/nobel_prizes/peace/laureates/1986/wiesel-acceptance_en.html

Index

www.ingramcontent.com/pod-product-compliance
Lightning Source LLC
Chambersburg PA
CBHW071413230926
51846CB00001B/69
* 9 7 8 9 0 0 4 4 1 6 7 8 9 *